# Microfluidics in Food Processing

This book serves as a comprehensive introduction to the principles of microfluidization and its diverse applications in the food industry. It explores the use of microfluidics in processing various types of beverages derived from plant products, milk and milk products, cereal-based products, nut-based products, and meat- and egg-based products. Additionally, it delves into the application of microfluidics in formulating nano- and bio-sensors, food micro- and nano-delivery systems, seed protein isolates, and bioactive compounds, among other biomaterials. The initial chapter provides a thorough introduction to the concept of microfluidization, offering readers a comprehensive overview of the underlying principles and techniques involved in this transformative technology. The book highlights the role of microfluidics in the extraction of bioactive ingredients from food sources and explores the use of microfluidic systems for ensuring food safety, including the detection of molecular interactions in food samples. Furthermore, the book explores the application of microfluidics in the fabrication of nanomaterials with tailored properties. With its comprehensive coverage of microfluidization in food processing, this book serves as a valuable resource for researchers, scientists, and professionals in the food industry.

# Sustainable Industrial and Environmental Bioprocesses
*Series Editor: Dr Ashok Pandey*

This book series aims to provide a comprehensive collection of books focusing on bioprocesses in industrial and environmental biotechnology. The multidisciplinary content encompasses the chemical and biochemical engineering, industrial microbiology, and energy biosciences, all with a central theme of sustainable development and circular economy principles. The books align with the sustainable development goals (SDGs) and offer state-of-the-art information and in-depth knowledge on the subject matter. The book in these series also emphasizes on the application of emerging tools, such as machine learning and artificial intelligence, for the advancement of bioprocesses. While primarily targeting academicians and researchers, the series is valuable for policy planners and industry professionals, with carefully tailored contents to cater to their specific needs and interests.

**Professor Ashok Pandey** is currently Distinguished Scientist at the Centre for Innovation and Translational Research, CSIR-Indian Institute of Toxicology Research, Lucknow, India. His major research and technological development interests are industrial and environmental biotechnology and energy biosciences, focusing on biomass to biofuels and chemicals, waste to wealth and energy, etc. He has 16 patents, 120 books, more than 1000 papers and book chapters, etc. with an h index of 131 and more than 74,000 citations. Professor Pandey is the recipient of many national and international awards and honors, which include Fellow, The World Academy of Sciences (TWAS), Highly Cited Researcher (Top 1% in the world), Clarivate Analytics (since 2018–till date), Rank #1 in India in Biology and Biochemistry and #417 in the world, Research.Com (2023); Rank #1 in India in Microbiology under Enabling and Strategic Technologies sector, Elsevier (2021); Rank #1 in India in Biotechnology and #8 in the world, Stanford University Report (2020–2021–2022), etc.

*Pharmaceuticals in Aquatic Environment: Remediation Technologies and Future Challenges*
Vinod Kumar Garg, Ashok Pandey, Navish Kataria, Caterina Faggio

*Biomass Hydrolysing Enzymes: Basics, Advancements, and Applications*
Reeta Rani Singhania, Anil Kumar Patel, Héctor A. Ruiz and Ashok Pandey

*Decentralized Sanitation and Water Treatment: Concept and Technologies*
R.D. Tyagi, Ashok Pandey, Patrick Drogui, Bhoomika Yadav, Sridhar Pilli and Jonathan W.C. Wong

*Decentralized Sanitation and Water Treatment: Treatment in Cold Environments and Techno-Economic Aspects*
R.D. Tyagi, Ashok Pandey, Patrick Drogui, Bhoomika Yadav, Sridhar Pilli and Jonathan W.C. Wong

*Biodegradation of Toxic and Hazardous Chemicals: Detection and Mineralization*
Kashyap K. Dubey, Kamal. K. Pant, Ashok Pandey and Maria Angeles Sanromán

*Biodegradation of Toxic and Hazardous Chemicals: Remediation and Resource Recovery*
Kashyap K. Dubey, Kamal K. Pant, Ashok Pandey and Maria Angeles Sanromán

*Waste Management in Climate Change and Sustainability- Lignocellulosic Waste*
Sunita Varjani, Izharul Haq, Ashok Pandey, Vijai Kumar Gupta, and Xuan-Thanh Bui

*Ashwagandha for Quality of Life: Scientific Evidence*
Sunil C Kaul & Renu Wadhwa

*Microfluidics in Food Processing: Technologies and Applications*
Ayon Tarafdar, Ranjna Sirohi, Barjinder Pal Kaur, Ashok Pandey, and Claude-Gilles Dussap

For more information about this series, please visit https://www.routledge.com/Sustainable-Industrial-and-Environmental-Bioprocesses/book-series/SIEB

# Microfluidics in Food Processing
## Technologies and Applications

Edited by

### Ayon Tarafdar
ICAR-Indian Veterinary Research Institute

### Ranjna Sirohi
Sri Karan Narendra Agriculture University, Jobner, Rajasthan, India

### Barjinder Pal Kaur
NIFTEM

### Ashok Pandey
Centre for Innovation and Translational Research,
CSIR-Indian Institute of Toxicology Research, Lucknow, India

### Claude-Gilles Dussap
Applied Thermodynamics and Biosystems, Institute Pascal
University Clermont Auvergne - CNRS

CRC Press is an imprint of the
Taylor & Francis Group, an **informa** business

Designed cover image: Shutterstock

First edition published 2025
by CRC Press
2385 NW Executive Center Drive, Suite 320, Boca Raton FL 33431

and by CRC Press
4 Park Square, Milton Park, Abingdon, Oxon, OX14 4RN

*CRC Press is an imprint of Taylor & Francis Group, LLC*

© 2025 selection and editorial matter Ayon Tarafdar, Ranjna Sirohi, Barjinder Pal Kaur, Ashok Pandey, and Claude-Gilles Dussap, individual chapters, the contributors

Reasonable efforts have been made to publish reliable data and information, but the author and publisher cannot assume responsibility for the validity of all materials or the consequences of their use. The authors and publishers have attempted to trace the copyright holders of all material reproduced in this publication and apologize to copyright holders if permission to publish in this form has not been obtained. If any copyright material has not been acknowledged please write and let us know so we may rectify in any future reprint.

Except as permitted under U.S. Copyright Law, no part of this book may be reprinted, reproduced, transmitted, or utilized in any form by any electronic, mechanical, or other means, now known or hereafter invented, including photocopying, micro-filming, and recording, or in any information storage or retrieval system, without written permission from the publishers.

For permission to photocopy or use material electronically from this work, access www.copyright.com or contact the Copyright Clearance Center, Inc. (CCC), 222 Rosewood Drive, Danvers, MA 01923, 978-750-8400. For works that are not available on CCC please contact mpkbookspermissions@tandf.co.uk

*Trademark notice*: Product or corporate names may be trademarks or registered trademarks and are used only for identification and explanation without intent to infringe.

*Library of Congress Cataloging-in-Publication Data*
Names: Tarafdar, Ayon, editor. Title: Microfluidics in food processing : technologies and applications /
edited by Ayon Tarafdar, Ranjna Sirohi, Barjinder Pal Kaur, Ashok Pandey, and Claude-Gilles Dussap.
Description: First edition. | Boca Raton : CRC Press, 2025. | Series: Sustainable developmental series |
Includes bibliographical references and index. | Summary: "This book serves as a comprehensive
introduction to the principles of microfluidization and its diverse applications in the food industry.
It explores the use of microfluidics in processing various types of beverages derived from plant products,
milk and milk products, cereal-based products, nut-based products, and meat and egg-based products.
Additionally, it delves into the application of microfluidics in formulating nano- and bio-sensors,
food micro- and nano-delivery systems, seed protein isolates, bioactive compounds among other biomaterials.
The initial chapter provides a thorough introduction to the concept of microfluidization, offering
readers a comprehensive overview of the underlying principles and techniques involved in this
transformative technology. The book highlights the role of microfluidics in the extraction of bioactive ingredients
from food sources and explores the use of microfluidic systems for ensuring food safety, including the detection
of molecular interactions in food samples. Furthermore, the book explores the application of microfluidics
in the fabrication of nanomaterials with tailored properties. With its comprehensive coverage of
microfluidization in food processing, this book serves as a valuable resource for researchers, scientists,
and professionals in the food industry"-- Provided by publisher. Identifiers: LCCN 2024023599 (print) |
LCCN 2024023600 (ebook) | ISBN 9781032609812 (hardback) | ISBN 9781032632575 (paperback) |
ISBN 9781032632599 (ebook) Subjects: LCSH: Food--Water activity. | Food--Composition. |
Microfluidics--Industrial applications. Classification: LCC TX553.W3 M53 2025 (print) | LCC TX553.W3 (ebook) |
DDC 664/.07--dc23/eng/20241030
LC record available at https://lccn.loc.gov/2024023599
LC ebook record available at https://lccn.loc.gov/2024023600

ISBN: 978-1-032-60981-2 (hbk)
ISBN: 978-1-032-63257-5 (pbk)
ISBN: 978-1-032-63259-9 (ebk)

DOI: 10.1201/9781032632599

Typeset in Times
by SPi Technologies India Pvt Ltd (Straive)

# Contents

About the Editors ............................................................................................................ vii
List of Contributors ....................................................................................................... viii

**Chapter 1**     Introduction to Microfluidization ............................................................... 1

                     *Ayon Tarafdar, Ranjna Sirohi, and Ashok Pandey*

**Chapter 2**     Nucleic Acid Detection of Pathogenic Microorganisms on Chip ............ 10

                     *Xinrui Zhang, Xueming Zhu, Jessica Hu, and Dan Gao*

**Chapter 3**     Paper-Based Microfluidic Devices in Food Processing ......................... 55

                     *Soja Saghar Soman, Shafeek Abdul Samad, Priyamvada Venugopalan, Nityanand Kumawat, and Sunil Kumar*

**Chapter 4**     Microfluidic Nanodetection Systems for Molecular Interactions ........... 82

                     *Manman Du, Xin Wang, Xin Meng, Yaohua Du, and Xinwu Xie*

**Chapter 5**     Microfluidic Devices for Synthesizing Nanomaterials ......................... 106

                     *Khairunnisa Amreen, Ramya K., and Sanket Goel*

**Chapter 6**     Microfluidization of Juice Derived from Plant Products........................ 128

                     *Divya Arora, Sanjana Kumari, and Barjinder Pal Kaur*

**Chapter 7**     Microfluidization of Milk and Milk Products ....................................... 142

                     *Anit Kumar, Kumar Sandeep, Kanchan Kumari, Prem Prakash, M. A. Aftab, and Rachna Sehrawat*

**Chapter 8**     Microfluidics-Based Food Micro- and Nanodelivery Systems ............. 155

                     *Monika Chand, Pratima Raypa, Deepak Joshi, Nitu Rani, and Narashans Alok Sagar*

**Chapter 9**     Microfluidization of Cereals-Based Products......................................... 177

                     *Jithender Bhukya, R. Nisha, Sophia Chanu Warepam, Harsh Dadhaneeya, Raj Singh, and C. Nickhil*

**Chapter 10**    Microfluidization of Meat- and Egg-Based Products.............................. 199

                     *Rajat Suhag and Atul Dhiman*

**Chapter 11** Microfluidization-Assisted Extraction of Bioactive Compounds from Biological Resources ............................................................................................212

*Meemansha Sharma, Rakesh Karwa, S. Ilavarasan, Mamta Meena, Manju Gari, Ranjna Sirohi, Anshuk Sharma, and Thakur Uttam Singh*

**Chapter 12** Microfluidization of Nut-Based Proteins: Modulation of Structural, Physicochemical, and Functional Properties ............................................................223

*Geetarani Loushigam, S. Prithya, T. P. Sari, and Prarabdh C. Badgujar*

**Chapter 13** Microfluidization of Seed Storage Proteins ............................................................241

*Neeraj Ghanghas, Yogesh Kumar, and Rajat Suhag*

**Index** ............................................................................................256

# About the Editors

**Dr. Ayon Tarafdar** is currently working as an ARS Scientist at ICAR-Indian Veterinary Research Institute under the Ministry of Agriculture and Farmers Welfare. He has also worked as a fully funded researcher in Washington University in St. Louis, Missouri, USA. His main research focus has been on non-thermal food processing operations, microfluidization, bioprocess modeling and optimization, machine learning, food waste valorization, advanced food characterization tools and precision agriculture.

**Dr. Ranjna Sirohi** is currently working as an Assistant Professor at Sri Karan Narendra Agriculture University, Jobner, Rajasthan, India. She has worked as Research Professor at Korea University, Seoul, Republic of Korea, and at École Polytechnique Fédérale de Lausanne, Switzerland. Her major research interests are in bioprocess technology, food & food waste valorisation, waste to wealth, biopolymers and biofuels.

**Dr. Barjinder Pal Kaur** is currently working as an Assistant Professor in the Department of Food Engineering NIFTEM since 2013. She is associated with various journals of international repute as an editorial board member and reviewer. Her area of expertise is novel food processing technologies such as high-pressure processing, microfluidization, drying technology, application of hydrogels and liposomes in food.

**Prof. Ashok Pandey** is currently Distinguished Scientist at the Centre for Innovation and Translational Research, CSIR-Indian Institute of Toxicology Research, Lucknow, India. He is HSBS National Innovation Chair (Biotechnology) and Visiting/Distinguished Professor in UPES, Dehradun; IIT, Roorkee; KHU, South Korea; UCA France, etc. His major research and technological development interests are industrial and environmental biotechnology and energy biosciences, focusing on biomass to biofuels and chemicals, waste to wealth and energy, etc.

**Prof. Claude-Gilles Dussap** is presently Emeritus Professor at Institute Pascal (Engineering sciences lab of University Clermont Auvergne – CNRS). His research interests are on the analysis of the relationships, which exist between the physiological responses of microorganisms and bioreactors, including a thorough analysis of bioreactor performances regarding the mass, heat, light-energy transfer and mixing properties of bioreactors. He has strong experience in mathematical modeling of biological kinetics, thermodynamic equilibrium properties of aqueous biological solutions and of reactor characteristics.

# Contributors

**M. A. Aftab**
Department of Food Science and Postharvest
  Technology, Bihar Agricultural College,
  Bihar Agricultural University
Sabour, Bhagalpur, Bihar, India

**Khairunnisa Amreen**
MEMS, Microfluidics and Nanoelectronics
  (MMNE) Lab, Birla Institute of Technology
  and Science (BITS)
Pilani, Hyderabad Campus, Hyderabad, India
Department of Electrical and Electronics
  Engineering, Birla Institute of Technology
  and Science (BITS)
Pilani, Hyderabad Campus, Hyderabad, India

**Divya Arora**
Department of Food Engineering, National
  Institute of Food Technology
  Entrepreneurship and Management
Kundli, Haryana, India

**Prarabdh C. Badgujar**
Department of Food Science and Technology,
  National Institute of Food Technology
  Entrepreneurship and Management
Kundli, Sonipat, Haryana, India

**Jithender Bhukya**
Department of Food Science and Technology,
  School of Science, GITAM University
Hyderabad, India

**Monika Chand**
Department of Food Science and Technology,
  National Institute of Food Technology
  Entrepreneurship and Management
Sonipat, Haryana, India

**Harsh Dadhaneeya**
National Institute of Food Technology
  Entrepreneurship and Mangement
Kundli, Haryana, India

**Atul Dhiman**
Department of Food Science and Technology,
  Dr. Y S, Parmar University of Horticulture
  and Forestry
Solan, Himachal Pradesh, India

**Manman Du**
Medical Support Technology Research
  Department, Systems Engineering Institute,
  Academy of Military Sciences, People's
  Liberation Army
Tianjin, PR China
School of Environmental Science and
  Engineering, Tianjin University
Tianjin, PR China

**Yaohua Du**
Medical Support Technology Research
  Department, Systems Engineering Institute,
  Academy of Military Sciences, People's
  Liberation Army
Tianjin, PR China
National Bio-Protection Engineering Center
Tianjin, PR China

**Dan Gao**
The State Key Laboratory of Chemical
  Oncogenomics, Shenzhen International
  Graduate School and Open FIESTA,
  Shenzhen International Graduate School,
  Tsinghua University
Shenzhen, PR China

**Manju Gari**
Division of Pharmacology and Toxicology,
  ICAR-Indian Veterinary Research Institute
Izatnagar, Bareilly, Uttar Pradesh, India

**Neeraj Ghanghas**
Department of Food and Human Nutritional
  Sciences, University of Manitoba
Winnipeg, Manitoba, Canada

# Contributors

**Sanket Goel**
MEMS, Microfluidics and Nanoelectronics (MMNE) Lab, Birla Institute of Technology and Science (BITS)
Pilani, Hyderabad Campus, Hyderabad, India
Department of Electrical and Electronics Engineering, Birla Institute of Technology and Science (BITS)
Pilani, Hyderabad Campus, Hyderabad, India

**Jessica Hu**
The State Key Laboratory of Chemical Oncogenomics, Shenzhen International Graduate School and Open FIESTA, Shenzhen International Graduate School, Tsinghua University
Shenzhen, PR China

**S. Ilavarasan**
Division of Pharmacology and Toxicology, ICAR-Indian Veterinary Research Institute
Izatnagar, Bareilly, Uttar Pradesh, India

**Deepak Joshi**
Department of Food Science and Technology, National Institute of Food Technology Entrepreneurship and Management
Sonipat, Haryana, India

**Rakesh Karwa**
Division of Pharmacology and Toxicology, ICAR-Indian Veterinary Research Institute
Izatnagar, Bareilly, Uttar Pradesh, India

**Barjinder Pal Kaur**
Department of Food Engineering, National Institute of Food Technology Entrepreneurship and Management
Kundli, Haryana, India

**Anit Kumar**
Department of Food Science and Postharvest Technology, Bihar Agricultural College, Bihar Agricultural University
Sabour, Bhagalpur, Bihar, India

**Sunil Kumar**
Division of Engineering, New York University
Abu Dhabi

**Yogesh Kumar**
Department of Agricultural and Food Sciences, University of Bologna
Cesena, FC, Italy

**Kanchan Kumari**
Department of Food Science and Postharvest Technology, Bihar Agricultural College, Bihar Agricultural University
Sabour, Bhagalpur, Bihar, India

**Sanjana Kumari**
Department of Food Engineering, National Institute of Food Technology Entrepreneurship and Management
Kundli, Haryana, India

**Nityanand Kumawat**
Division of Engineering, New York University
Abu Dhabi

**Geetarani Loushigam**
Department of Food Science and Technology, National Institute of Food Technology Entrepreneurship and Management
Kundli, Sonipat, Haryana, India

**Mamta Meena**
Division of Pharmacology and Toxicology, ICAR-Indian Veterinary Research Institute
Izatnagar, Bareilly, Uttar Pradesh, India

**Xin Meng**
College of Biotechnology, Tianjin University of Science and Technology
Tianjin, PR China

**C. Nickhil**
Department of Food Engineering and Technology, Tezpur University
Assam, India

**R. Nisha**
Department of Agricultural Engineering, Nehru Institute of Technology
Coimbatore, Tamil Nadu, India

**Ashok Pandey**
Centre for Innovation and Translational Research, CSIR-Indian Institute of Toxicology Research
Lucknow, Uttar Pradesh, India

Centre for Energy and Environmental
  Sustainability, Lucknow, Uttar Pradesh, India
Sustainability Cluster, School of Engineering,
  University of Petroleum and Energy Studies
Dehradun, Uttarakhand, India

**Prem Prakash**
Department of Food Science and Postharvest
  Technology, Bihar Agricultural College,
  Bihar Agricultural University
Sabour, Bhagalpur, Bihar, India

**S. Prithya**
Department of Food Science and Technology,
  National Institute of Food Technology
  Entrepreneurship and Management
Kundli, Sonipat, Haryana, India

**Ramya K.**
MEMS, Microfluidics and Nanoelectronics
  (MMNE) Lab, Birla Institute of Technology
  and Science (BITS)
Pilani, Hyderabad Campus, Hyderabad, India
Department of Electrical and Electronics
  Engineering, Birla Institute of Technology
  and Science (BITS)
Pilani, Hyderabad Campus, Hyderabad, India

**Nitu Rani**
Department of Biotechnology, University
  Institute of Biotechnology, Chandigarh
  University
Mohali, Punjab, India

**Pratima Raypa**
Department of Biochemistry, GBPUAT
Pantnagar, Uttarakhand, India

**Narashans Alok Sagar**
Department of Biotechnology, University
  Institute of Biotechnology, Chandigarh
  University
Mohali, Punjab, India
University Centre for Research and
  Development, Chandigarh University
Mohali, Punjab, India

**Shafeek Abdul Samad**
Division of Engineering, New York University
Abu Dhabi

**Kumar Sandeep**
Department of Food Science and Postharvest
  Technology, Bihar Agricultural College,
  Bihar Agricultural University
Sabour, Bhagalpur, Bihar, India

**T. P. Sari**
Department of Food Science and Technology,
  National Institute of Food Technology
  Entrepreneurship and Management
Kundli, Sonipat, Haryana, India

**Rachna Sehrawat**
Department of Food Process Engineering,
  National Institute of Technology
Rourkela, Odisha, India

**Anshuk Sharma**
Division of Pharmacology and Toxicology,
  ICAR-Indian Veterinary Research Institute
Izatnagar, Bareilly, Uttar Pradesh, India

**Meemansha Sharma**
Division of Pharmacology and Toxicology,
  ICAR-Indian Veterinary Research Institute
Izatnagar, Bareilly, Uttar Pradesh, India

**Raj Singh**
Department of Food Engineering and Technology,
  Tezpur University (A Central University)
Tezpur, Assam, India

**Thakur Uttam Singh**
Division of Pharmacology and Toxicology,
  ICAR-Indian Veterinary Research Institute
Izatnagar, Bareilly, Uttar Pradesh, India

**Ranjna Sirohi**
Sri Karan Narendra Agriculture University
Jobner, Rajasthan, India

**Soja Saghar Soman**
Division of Engineering, New York University
Abu Dhabi

**Rajat Suhag**
Faculty of Agricultural, Environmental and
  Food Sciences, Free University of Bolzano,
  Piazza Università
Bolzano, Italy

**Ayon Tarafdar**
Livestock Production and Management Section, ICAR-Indian Veterinary Research Institute
Izatnagar, Bareilly, Uttar Pradesh, India

**Priyamvada Venugopalan**
Division of Engineering, New York University
Abu Dhabi

**Xin Wang**
College of Biotechnology, Tianjin University of Science and Technology
Tianjin, PR China

**Sophia Chanu Warepam**
National Institute of Food Technology Entrepreneurship and Mangement
Kundli, Haryana, India

**Xinwu Xie**
Medical Support Technology Research Department, Systems Engineering Institute, Academy of Military Sciences, People's Liberation Army
Tianjin, PR China
National Bio-Protection Engineering Center
Tianjin, PR China

**Xinrui Zhang**
The State Key Laboratory of Chemical Oncogenomics, Shenzhen International Graduate School and Open FIESTA, Shenzhen International Graduate School, Tsinghua University
Shenzhen, PR China

**Xueming Zhu**
The State Key Laboratory of Chemical Oncogenomics, Shenzhen International Graduate School and Open FIESTA, Shenzhen International Graduate School, Tsinghua University
Shenzhen, PR China

# 1 Introduction to Microfluidization

*Ayon Tarafdar, Ranjna Sirohi, and Ashok Pandey*

## 1.1 BACKGROUND AND HISTORY

Microfluidization is a rapidly advancing field in science and engineering that deals with the manipulation and control of fluids at microscale levels. This technology finds applications across various disciplines including biology, chemistry, physics, and engineering. At its core, microfluidization focuses on the behavior of fluids confined to micrometer-scale channels, where unique phenomena arise due to the dominance of surface forces over inertial forces. This chapter serves as an introduction to the principles, techniques, applications, and challenges associated with microfluidization and intends to provide the readers with basic understanding of the microfluidics process before delving into the diverse applications of this technology.

The origin of microfluidization can be traced back to the early 20th century when the study of fluid dynamics began to encompass phenomena at smaller length scales. Werner Jacobi, a German engineer regarded as the "father" of microfluidics, created the initial IC prototypes in 1949 (Choi & Mody, 2009). According to Whitesides (2006), microfluidics broke out from microelectronics and semiconductor technologies to become its own field. However, it was not until the 1990s that microfluidics gained significant attention with the advent of microfabrication techniques such as photolithography and soft lithography. These techniques allowed for the precise fabrication of microscale channels and structures, enabling the manipulation of fluids at the micron level. Therefore, before we delve into the understanding of microfluidics, it is important to look into the physics behind the process. A timeline of the origin and advancement of microfluidic research is shown in Figure 1.1.

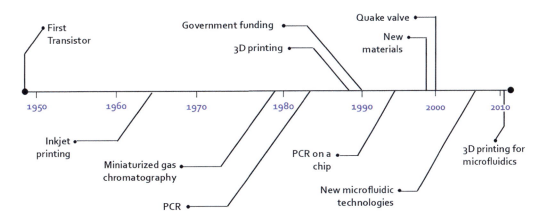

**FIGURE 1.1** The history of advances in microfluidics research. (Reproduced with permission, Convery & Gadegaard, 2019.)

## 1.1.1 Principles and Physics of Microfluidization

Microfluidics operate on principles distinct from those governing macroscale fluid dynamics. At the microscale, fluid behavior is dominated by surface tension, viscosity, and interfacial effects rather than inertial forces. As a result, phenomena such as laminar flow, capillary action, and electrokinetic effects play significant roles in microfluidic systems. Understanding these principles is crucial for designing and controlling microfluidic devices.

The dominant forces on a fluid confined at the microscale are different from those which are observed at macroscale (Convery & Gadegaard, 2019). However, assuming an incompressible fluid, Navier Stoke's equation defining the space–time evolution of the velocity field of the fluid can be defined as:

$$\begin{cases} \dfrac{\partial u}{\partial t} + \left(u.\nabla\right)u = -\dfrac{1}{\rho}\nabla P + v\nabla^2 u + f \\ \nabla.u = 0 \end{cases} \tag{1.1}$$

where (m/s) is the fluid velocity field, (kg/m$^3$) is the density, (Pa) is the pressure field, (m$^2$/s) is the kinematic viscosity, and (m/s$^2$) is an external acceleration field.

For a turbulent regime, the left half of equation (1.1) dominates, for which the governing equation can be expressed as:

$$\frac{\partial u}{\partial t} + \left(u.\nabla\right)u = f \tag{1.2}$$

Whereas, for a laminar regime, the right half of equation (1.1) is dominant for which the governing equation can be expressed as:

$$v\nabla^2 u = -\frac{1}{\rho}\nabla P - f \tag{1.3}$$

This is the typical regime followed in the microchannel of the microfluidizer, where viscous forces dominate the inertial forces. Considering the cylindrical geometry with diameter, $d$ (radius, $r$) and length, $l \gg d$ of the microfluidic channel (Figure 1.2), the flow can be expressed mathematically in cylindrical coordinates $(r, z)$ as:

$$\frac{1}{r}\frac{\partial}{\partial r}\left(r\frac{\partial u_z}{\partial r}\right) = -\frac{1}{v\rho}\frac{\partial P}{\partial z} \tag{1.4}$$

The solution of equation (1.4) is a parabolic velocity profile as follows:

$$\begin{cases} u_z = u_{max}\left(1 - \dfrac{r^2}{a^2}\right) \\ u_{max} = \dfrac{a^2}{4v\rho}\dfrac{\Delta P}{l} \end{cases} \tag{1.5}$$

# Introduction to Microfluidization

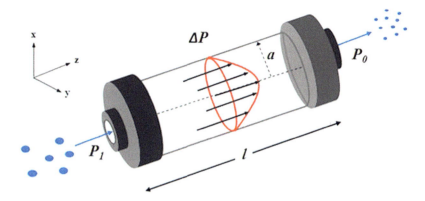

**FIGURE 1.2** Representative flow of fluid through cylindrical microchannel of a microfluidizer.

The volumetric flow rate ($Q$) is directly proportional to the pressure drop ($\Delta P$) in the microfluidizer channel. Introducing proportionality constant ($R_h$), the following equation can be derived from the definition:

$$Q = R_h \Delta P \qquad (1.6)$$

Here, $R_h$ represents the hydrodynamic resistance to flow in the microchannel of the microfluidizer. Therefore, given the geometry of the microchannel and the fluid viscosity, the microfluidic resistance can be determined.

## 1.2 MICROFLUIDIC DEVICES AND COMPONENTS

Microfluidic devices consist of a network of microchannels, chambers, valves, and pumps fabricated on a chip-scale platform. These devices can be made from various materials including glass, silicon, and polymers. Key components of microfluidic systems include microchannels, valves, and pumps. Microchannels are the primary conduits through which fluids flow. Microchannels can be straight, curved, or branched, depending on the desired application. Valves control the flow of fluids within the system. They can be passive (e.g., capillary valves) or active (e.g., pneumatic valves). Pumps generate the necessary pressure or flow to drive fluid through the system. Common types include syringe pumps, peristaltic pumps, and electrokinetic pumps.

The materials used in early microfluidic devices were usually silicon, quartz, or glass, which could be easily fabricated using well-established photolithography, deposition, and etching techniques (Harrison, 2018). Some of the most commonly used techniques in microfluidics include droplet and continuous flow microfluidics, electrophoresis, and microfluidic mixing. Droplet microfluidics involves the generation, manipulation, and analysis of discrete liquid droplets in microchannels. It finds applications in high-throughput screening, drug delivery, and single-cell analysis. In continuous flow microfluidics, fluids flow continuously through microchannels, enabling precise control over reaction conditions and residence times. This technique is widely used in chemical synthesis, bioassays, and sample processing. Electrophoresis is a technique used to separate charged particles or molecules in a microfluidic channel under the influence of an electric field. It is commonly employed in DNA sequencing, protein analysis, and particle sorting. In microfluidic mixing, different fluids can be mixed at the microscale and pose unique challenges due to the dominance of laminar flow. Various techniques such as passive mixing structures, chaotic advection, and electro-kinetic mixing are employed to achieve efficient mixing in microfluidic systems.

### 1.2.1 Fabrication of Microfluidic Devices

One of the most important steps in microfluidic applications is selecting the ideal material for the device's fabrication. The platform material is probably going to have an impact on the characteristics of produced nanomaterials, because the qualities are considerably more magnified on a microscale surface. Metals have several benefits that make them appropriate for the production of microchips. They can tolerate high pressure, high heat loads, and hazardous substances. These are inexpensive, widely available, and simple to operate. The production of size-tunable methacrylic nanoparticles in a stainless steel multi-lamination micromixer provides evidence that metal microfluidic devices are useful for synthesizing nanomaterials.

Silicon is a material of choice for fabricating microfluidic systems because of its easy accessibility, chemical compatibility, and thermostability (Singh et al., 2010). For many years, silicon was the primary material used for microfluidic platforms due to its ease of manufacture, flexibility in design, semiconducting qualities, and suitability for surface changes.

Glass is easily surface functionalized, electrically insulating, stiff, biologically compatible, chemically inert, thermostable, and electrically insulating. Because of these characteristics, glass-based microreactors can be used to conduct chemical reactions that call for harsh circumstances, such as high pressure, high temperatures, and aggressive solvents.

Microfluidic systems can be created using aluminum oxide-based low-temperature cofired ceramic. Ceramics are favored for microchips due to their distinct surface chemistry, strong resistance to corrosive conditions, and exceptional stability at elevated temperatures.

The most popular polymers used in the creation of microfluidic devices include polydimethylsiloxane (PDMS). PDMS, an elastomer with superior qualities for microchip production, is among the most representative materials in this class. PDMS exhibits biocompatibility, optical transparency, gas permeability, natural hydrophobicity, low autofluorescence, and high elasticity and is inexpensive, useful for prototyping, and easy to mold (Nielsen et al., 2019). As compared to PDMS, polymethylmethacrylate (PMMA) is an amorphous thermoplastic with no small-molecule absorption. PMMA has good mechanical qualities and is optically transparent.

### 1.2.2 Chip Fabrication Methods

The unique properties of the material involved and the limited specifications of the product require the adaptation of fabrication techniques. The cost is a crucial consideration when selecting the fabrication method. This is crucial for microfluidic platforms, because they are typically disposable and hard to clean. The technique that is selected for one-time chips needs to be financially viable. Microfluidic fabrication techniques include chemical methods, mechanical methods, and laser-based methods.

#### 1.2.2.1 Chemical Methods

The two modes of operation for microfluidic manufacturing processes are material removal and material depositing. While silicon surface micromachining, inkjet three-dimensional (3D) printing, direct writing, powder 3D printing, lithography, and two-dimensional virtual hydrophilic channels are material depositing techniques, electrochemical discharge machining, wet etching, and dry etching (physical chemical and reactive ion dry etching) are engaged in material removal procedures (Waldbaur et al., 2011). In order to remove a specific depth of material from a substrate, etching involves attacking one side while protecting the other. Their concept was to replace the traditional lateral flow paper strips with a single platform that could detect various tests. Furthermore, photolithography was selected due to its ease of use (Lim et al., 2019).

## 1.2.2.2 Mechanical Methods

In mechanical methods, parts move physically to allow the channel surface to be in close contact or to be connected or disconnected. Air-jet machining with microabrasive, microgrinding, micromilling injection molding, and hot embossing are examples of material deposition techniques; micro-abrasive water jet machining, ultrasonic machining, and xurography are examples of material removal techniques.

## 1.2.2.3 Laser-Based Methods

This technique includes the following processes: stereolithography, laser direct machining, selective laser sintering, absorbent material processing, ultra-short pulse processing, laser direct machining, and two-photon polymerization. A hydrophobic paper surface treated with a laser made it possible to generate microfluidic patterns as small as 62 μm (Lim et al., 2019).

Microfluidic devices have a wide range of uses that aim to get beyond the problems or obstacles that come with conventional tests. There is a lot of promise in customized medicine, chemical screening, DNA sequencing, drug delivery, cell treatment, cell culture, cell separation, and culture (Mancera-Andrade et al., 2018).

## 1.3 MICROFLUIDIZATION OF FOOD

Microfluidization (MF) has emerged as a non-thermal food processing technology that employs high-pressure homogenization principles. This process involves impact, shear, hydrodynamic cavitation, and rapid pressure drop forces within a specially designed interaction chamber (Li et al., 2022). MF induces significant and beneficial changes in food matrices, altering physicochemical properties, enhancing rheological and sensory attributes, causing structural modifications, reducing particle size, and increasing surface area, thereby improving bioavailability and bioaccessibility without compromising original flavor (Kumar et al., 2022). It yields stable micro-/nanoemulsions with high encapsulation efficiency, enhanced antioxidant properties, and improved product stability. Moreover, MF can effectively deactivate microbes and enzymes, making it a superior method for encapsulating bioactive components and advancing preservation and production techniques.

### 1.3.1 COMPONENTS OF A MICROFLUIDIZER USED FOR FOOD PROCESSING

The microfluidizer comprises two crucial components: interaction chamber (Y or Z type) and intensifier pump, which directly impact processing efficiency. In operation, the feed enters via a hopper and is accelerated by the high-pressure intensifier pump to speeds of up to 400 m/s before entering the interaction chamber. This pump, capable of pressurizing up to approximately 2500 bar or 35000 psi, oscillates between suction and compression modes. In the Y-type chamber, the pressurized feed divides into microstreams and collides within the chamber, primarily used for liquid–liquid dispersions. Conversely, the Z-type chamber directs the high-speed liquid through a zigzag microchannel for dispersal of solid materials and cell structure breakdown (Mert, 2020).

The efficiency of MF depends on factors like flow channel design, treatment parameters (pressure, cycle pass, and temperature), and material type. The interaction chamber, designed as a continuous microchannel, facilitates energy dissipation, turbulent mixing, and particle size homogenization, leading to a consistent pressure profile. High-pressure emulsification in the microfluidizer involves cavitation, impact, turbulence, and shear, with shear and impact forces playing crucial roles. Shear force, occurring between channel walls and product streams, reduces particle size and disperses agglomerates, while impact force results from high-velocity stream collisions. The chamber is equipped with a cooling coil to maintain product temperature and an optional processing module for further refinement (Kumar et al., 2022). Microfluidizer operation involves a combination of inertial forces in turbulent flow and hydrodynamic cavitation, impacting droplet disruption and particle dispersion within the chamber.

### 1.3.2 ADVANTAGES OF MICROFLUIDIZATION

- MF is highly regarded in the food industry as an innovative green processing technique with diverse applications such as encapsulation, preservation, emulsification, homogenization, and structural modification (Kavinila et al., 2023).
- Microfluidizer technology is capable of producing exceptionally stable emulsions with droplet sizes below 0.1 μm, ensuring a homogeneous particle size distribution and consistent product quality (Kavinila et al., 2023).
- MF processes can be conducted in both batch and continuous modes, offering flexibility and efficiency in food production (Kavinila et al., 2023).
- A number of studies have emphasized the advantages of MF on product attributes like texture, appearance, rheology, nutritional content, and techno-functional properties, showcasing its superiority over other methods (Kavinila et al., 2023).
- Unlike high-pressure homogenization, microfluidizers operate using a fixed geometry aperture and intensifier pump, maintaining a constant pressure that facilitates uniform shear force and the production of extremely small units with narrow size distributions (Tobin et al., 2015).
- The shearing rate efficiency of microfluidizers surpasses traditional grinders, ensuring greater repeatability and control over particle size, while the presence of a cooling jacket aids in temperature control during processing (Kumar et al., 2022).

### 1.3.3 LIMITATIONS OF MICROFLUIDIZATION

- MF often requires a pre-emulsification step to make a coarse emulsion for achieving uniform droplet size and enhanced stability of the emulsion. Dual-channel microfluidizers can address this by allowing separate feeding of two phases, as demonstrated by Bai et al. (2016).
- MF is susceptible to over-processing, leading to recoalescence and increased droplet size once a critical point is reached. This challenge can be mitigated by using suitable surfactants, optimizing pressure, and passing cycles to ensure efficient emulsification, as highlighted by Ozturk and Turasan (2021).
- Microfluidizers demand high-energy and feature small-sized microchannels (<300 μm), making them prone to blockages and challenging for continuous production. Designing flow channels meticulously is crucial to managing these limitations effectively (Li et al., 2022).
- The non-linear dimensions of microfluidizer channels contribute to sample residue settling, making cleaning processes difficult and increasing the risk of microbial contamination and product spoilage with new batch processing.
- Cost-effectiveness is a concern, as microfluidizers are significantly more expensive (20–30 times) than high-pressure homogenizers with similar processing efficiency. This factor requires careful consideration during MF scale-up processes due to high manufacturing costs.
- Temperature variations during high-pressure processing can lead to sample degradation, necessitating a consistent cooling system throughout the process to maintain temperature uniformity from the inlet to the outlet.
- Sample loss due to settling inside microfluidizer channels and potential pressure blockages and fluctuations from equipment noise are significant limitations that require attention and appropriate management in MF applications.

# 1.4 APPLICATIONS OF MICROFLUIDIZATION

The versatility of microfluidic technology has led to its widespread adoption in diverse fields. There are many notable applications of microfluidization. This section provides a general overview of these applications, the details of which have been covered in other chapters of this book.

## 1.4.1 BIOMEDICAL DIAGNOSTICS

Microfluidic devices are used for point-of-care diagnostics, biomarker detection, and disease monitoring. Microfluidics is revolutionizing diagnostics by enabling the miniaturization and integration of complex analyses on a single chip. These microfluidic chips manipulate tiny volumes of samples precisely within microchannels, allowing for rapid tests with minimal sample requirements. This facilitates point-of-care diagnostics, enabling healthcare workers to diagnose diseases like infections or perform blood tests in resource-limited settings or near the patient, drastically improving healthcare accessibility and speeding up treatment decisions.

## 1.4.2 DRUG DELIVERY

Microfluidic systems enable precise control over drug formulation, release kinetics, and targeting, leading to improved therapeutic outcomes. Microfluidic devices precisely control the formation of micro- and nanocarriers for drugs, enabling uniform size and properties. This allows for controlled drug release kinetics and targeted delivery to specific sites within the body, enhancing therapeutic efficacy while minimizing side effects. This technology holds promise for improved treatment of various diseases by optimizing drug delivery.

## 1.4.3 ENVIRONMENTAL MONITORING

In environmental monitoring, microfluidics empowers researchers with miniaturized lab-on-a-chip devices. These chips handle minuscule fluid volumes precisely within microchannels, facilitating on-site analysis of environmental samples. This enables rapid detection of pollutants like heavy metals, bacteria, or microplastics in water and soil. The compact size, low reagent consumption, and potential for automation make microfluidic devices ideal for real-time environmental monitoring in remote or resource-limited settings. Microfluidic sensors are employed for detecting pollutants, monitoring water quality, and studying environmental contaminants.

## 1.4.4 FOOD AND BEVERAGE INDUSTRY

Microfluidics is making waves in the food and beverage industry by offering precise control over food structures and analysis. Microfluidic devices can generate monodisperse emulsions and encapsulates, leading to novel textures, improved stability of functional ingredients, and controlled release of flavors or nutraceuticals. Additionally, these chips enable rapid on-site analysis of food and beverages for quality control, detection of contaminants, and monitoring of vital parameters like sugar content or spoilage markers, ensuring food safety and optimizing production processes.

## 1.4.5 PARTICLE SIZE REDUCTION

Microfluidics provides a novel method for particle size reduction. Microfluidic devices employ precisely controlled high-pressure and well-defined microchannels to create intense shear forces. As the particle suspension flows through these channels, the shear forces break down the particles into smaller and more uniform sizes. This technique offers advantages such as a narrow size distribution, minimized heating (protecting heat-sensitive materials), and continuous processing, making it a valuable tool for producing nanoparticles for pharmaceuticals, cosmetics, and other fields.

### 1.4.6 NANOTECHNOLOGY

Microfluidic platforms facilitate the synthesis, manipulation, and characterization of nanoparticles for various applications including drug delivery, catalysis, and imaging. Nanotechnology and microfluidics form a powerful synergy. Microfluidic devices act as miniaturized reactors, offering exquisite control over the synthesis of nanoparticles. The precise manipulation of fluids within microchannels allows for uniform mixing of precursors, tight control over reaction parameters such as temperature and flow rates, and continuous processing. This facilitates the production of nanoparticles with well-defined size, shape, and surface properties, crucial for applications in targeted drug delivery, diagnostics, and catalysis. By enabling scalable and precise nanoparticle synthesis, microfluidics is propelling the frontiers of nanotechnology.

## 1.5 CONCLUSIONS AND PERSPECTIVES

Despite significant advancements, microfluidization faces several challenges including scaling up production, integration with existing technologies, and standardization of fabrication techniques. Future directions in microfluidization include the development of multifunctional microfluidic platforms, integration with artificial intelligence and machine-learning algorithms, and exploration of novel materials and fabrication methods to overcome existing limitations. Microfluidization has emerged as a powerful tool for manipulating fluids at the microscale, with applications spanning diverse fields ranging from biomedicine to nanotechnology. By harnessing principles of fluid dynamics at small length scales, microfluidic devices offer unprecedented control over fluid behavior, enabling novel applications and discoveries. Continued research and innovation in microfluidization hold the promise of revolutionizing various industries and advancing our understanding of complex biological and chemical systems.

## REFERENCES

Bai, L., Huan, S., Gu, J., & McClements, D. J. (2016). Fabrication of oil-in-water nanoemulsions by dual-channel microfluidization using natural emulsifiers: Saponins, phospholipids, proteins, and polysaccharides. *Food Hydrocolloids*, 61, 703–711.

Choi, H., & Mody, C. C. (2009). The long history of molecular electronics: Microelectronics origins of nanotechnology. *Social Studies of Science*, 39(1), 11–50.

Convery, N., & Gadegaard, N. (2019). 30 years of microfluidics. *Micro and Nano Engineering*, 2, 76–91.

Harrison, D. J. (2018). The development of microfluidic systems within the Harrison research team. arXiv preprint arXiv:1802.05598.

Kavinila, S., Nimbkar, S., Moses, J. A., & Anandharamakrishnan, C. (2023). Emerging applications of microfluidization in the food industry. *Journal of Agriculture and Food Research*, 12, 100537.

Kumar, A., Dhiman, A., Suhag, R., Sehrawat, R., Upadhyay, A., & McClements, D. J. (2022). Comprehensive review on potential applications of microfluidization in food processing. *Food Science and Biotechnology*, 31, 17–36.

Li, Y., Deng, L., Dai, T., Li, Y., Chen, J., Liu, W., & Liu, C. (2022). Microfluidization: A promising food processing technology and its challenges in industrial application. *Food Control*, 137, 108794.

Lim, H., Jafry, A. T., & Lee, J. (2019). Fabrication, flow control, and applications of microfluidic paper-based analytical devices. *Molecules*, 24(16), 2869.

Mancera-Andrade, E. I., Parsaeimehr, A., Arevalo-Gallegos, A., Ascencio-Favela, G., & Parra-Saldivar, R. (2018). Microfluidics technology for drug delivery: A review. *Frontiers in Bioscience (Elite Ed.)*, 10, 74–91.

Mert, I. D. (2020). The applications of microfluidization in cereals and cereal-based products: An overview. *Critical Reviews in Food Science and Nutrition*, 60(6), 1007–1024.

Nielsen, J. B., Hanson, R. L., Almughamsi, H. M., Pang, C., Fish, T. R., & Woolley, A. T. (2019). Microfluidics: Innovations in materials and their fabrication and functionalization. *Analytical Chemistry*, 92(1), 150–168.

Ozturk, O. K., & Turasan, H. (2021). Applications of microfluidization in emulsion-based systems, nanoparticle formation, and beverages. *Trends in Food Science & Technology*, 116, 609–625.

Singh, A., Malek, C. K., & Kulkarni, S. K. (2010). Development in microreactor technology for nanoparticle synthesis. *International Journal of Nanoscience*, 9(01n02), 93–112.

Tobin, J., Heffernan, S. P., Mulvihill, D. M., Huppertz, T., & Kelly, A. L. (2015). Applications of high-pressure homogenization and microfluidization for milk and dairy products. *Emerging dairy processing technologies: Opportunities for the dairy industry*, 93–114. Wiley.

Waldbaur, A., Rapp, H., Länge, K., & Rapp, B. E. (2011). Let there be chip – Towards rapid prototyping of microfluidic devices: One-step manufacturing processes. *Analytical Methods*, 3(12), 2681–2716.

Whitesides, G. M. (2006). The origins and the future of microfluidics. *Nature*, 442(7101), 368–373.

# 2 Nucleic Acid Detection of Pathogenic Microorganisms on Chip

*Xinrui Zhang\*, Xueming Zhu\*, Jessica Hu\*, and Dan Gao*

## 2.1 INTRODUCTION

At present, foodborne pathogens pose an enormous threat to human life and health, causing huge economic losses to the entire society. According to a report by the World Health Organization (WHO), 600 million people (almost 1 in 10 people) worldwide fall ill each year due to consuming contaminated food, and 420,000 people die each year, resulting in a loss of 33 million healthy life years (DALYs). Moreover, children under the age of 5 bear 40% of the burden of foodborne diseases,[1] with 125,000 deaths annually, making it necessary to detect foodborne pathogens. Foodborne pathogens are of different types, such as viruses, bacteria, fungi, and parasites.[2] The most common foodborne pathogens are *Listeria*, *Vibrio cholerae*, *Salmonella*, and *Escherichia coli*. Diseases caused by foodborne pathogens are often accompanied by uncomfortable reactions such as diarrhea, abdominal pain, hemolytic uremia, fever, vomiting, and, in severe cases, may lead to death.[3] It is worth our attention that many pathogens have very low infection doses.[4] For example, the infection doses of *Escherichia coli* O157: H7 and *Salmonella* are as low as about 50 cells[5] and $10^3$ CFU/mL,[6] respectively, which makes people easily infected. Therefore, it is necessary to develop fast, accurate, and sensitive methods to detect food pollutants to reduce the impact of foodborne pathogens on human health.

There are several methods for foodborne pathogen detection, such as culture-based methods, biochemical methods, flow cytometry, and PCR. These methods are known as the gold standard methods.[7] Although their results are accurate, they still face the drawbacks of long analysis time and low throughput, requiring instruments and well-trained technical persons. Therefore, it is not suitable for on-site rapid detection. In order to address these issues, many new nucleic acid detection technologies have emerged, such as reverse transcription polymerase chain reaction (RT-PCR), loop-mediated isothermal amplification (LAMP),[8] and enzyme-linked immunosorbent assay (ELISA). ELISA based on the immune response between antigens and antibodies is often used for food analysis due to its specificity, speed, and high-throughput ability. However, it is often affected by limited antibody sources, insufficient sensitivity, and cross-contamination. In addition, the number of pathogens commonly present in food is relatively small, and sample pretreatment is required to reduce interference with the food substrate and improve detection sensitivity.[9] Due to the complexity of food matrices, many methods require trained operators and expensive equipment to process samples, which is labor-intensive and not suitable for on-site detection.

---

\* These authors contributed equally to this work.

DOI: 10.1201/9781032632599-2

Gene editing technology like CRISPR/Cas12a can effectively characterize foodborne pathogens with rapid response and high specificity compared to traditional detection methods, making it less prone to false positives. However, its signal amplification is not high and often needs to be combined with nucleic acid amplification methods.[10] The nucleic acid amplification methods mainly include those based on enzyme-mediated nucleic acid amplification and those based on non-enzyme-mediated nucleic acid amplification. Enzyme-mediated nucleic acid amplification methods mainly include recombinase polymerase amplification (RPA), branch rolling ring amplification (RCA), and recombinase-assisted amplification (RAA) methods. This type of method has certain limitations, such as the need for recombinant enzymes, polymerase, and other proteins, which may potentially lead to contamination. Moreover, the operation is cumbersome, and the storage conditions for enzyme proteins are harsh, making them difficult to preserve. The methods based on enzyme-free mediated nucleic acid amplification mainly include catalytic hairpin self-assembly (CHA) method, hybrid chain reaction (HCR) method, etc.[11] They have advantages of good signal amplification effect, easy to operate, and good selectivity. Therefore, using enzyme-free nucleic acid amplification combined with CRISPR/Cas12a detection technology to detect foodborne pathogens is a new and effective approach.

In recent years, microfluidic technology has shown a promising and powerful biological detection tool due to its advantages such as miniaturization, automation, portability, parallel analysis ability, and small reagent consumption.[12] Most importantly, all operational steps, including sample pretreatment, nucleic acid purification, and amplification, as well as real-time quantification, can be integrated into a microfluidic platform for on-site testing applications.[13] The miniaturization, integration, and automation of these devices involving multiple processes have made microfluidics popular in various fields. So far, a large number of microfluidic platforms have been combined with different types of technologies for detecting foodborne pathogens.[14] Based on the requirements of practical applications, many integrated microfluidic systems have been developed to detect multiplex pathogens, reduce costs, and improve efficiency.

In this chapter, we summarized the latest important papers on biosensors for microfluidic-based detection of foodborne pathogens over the past five years, including methods for sample preparation and signal conversion of foodborne pathogens, detection of foodborne pathogens using enzyme-mediated nucleic acid amplification methods, detection of foodborne pathogens using enzyme-free nucleic acid amplification methods, and their respective combination with gene editing techniques (CRISPR/Cas12a). First, in the microfluidic sample preparation section, we introduced several methods for capturing and enriching foodborne pathogens, as well as methods for nucleic acid sample preparation. Second, the current status and prospects of different nucleic acid amplification techniques for detecting foodborne pathogens were introduced in detail, and the opportunities and challenges brought by the emergence of gene editing techniques were also discussed. We hope that these methods can provide insights for future scientific research and practical applications.

## 2.2 MICROFLUIDIC SAMPLE PREPARATION

The role of sample preparation in detecting pathogenic microorganisms is crucial, as it significantly influences the effectiveness and sensitivity of subsequent detection procedures, particularly in complex sample analysis. This procedure can be broadly categorized into two primary steps: pathogen enrichment and nucleic acid sample preparation. Microfluidics, serving as a potent tool for pathogen detection, facilitates the seamless integration of lab-on-a-chip sample preparation units for pathogen enrichment and pretreatment. This section discusses the most crucial sample preparations for pathogen detection that can be integrated into microfluidic devices.

### 2.2.1 PATHOGENIC MICROORGANISM ENRICHMENT

Pathogen enrichment involves isolating and capturing microorganisms from diverse matrices to concentrate the samples for detection purposes. Many approaches to attain this objective can be categorized into physical and biochemical strategies, which are determined by the inherent chemical and physical properties of microorganisms. The following section presents an overview of the latest developments in enrichment techniques.

### 2.2.1.1 Physical Methods

Membrane filtration is a physical process wherein a permeable membrane segregates smaller particles from larger particles.[15] In contrast to traditional enrichment methods like centrifugation, which are time-consuming and challenging to integrate with microfluidic modules,[16] filter membranes offer a more straightforward structure and the ability to rapidly and effectively concentrate pathogens. Thus, it may be effortlessly integrated into microfluidic devices to simplify the enrichment process. For example, Jo et al.[17] integrated a membrane filter and an embedded absorbent pad into a finger-powered microfluidic device to concentrate bacteria–magnetic nanoparticle (MNP) complex from unbound MNPs. As shown in Figure 2.1, the device allowed filtered bacteria–MNP complexes to generate colorimetric signals on the membrane filter, reducing analytical errors when unbound free MNPs coexist in bacterial separation. The suggested device exhibited the capability to detect *Escherichia coli* (*E. coli*) at a concentration of 102 colony-forming units per milliliter (CFU/mL). However, membrane filtration is hindered by the challenges of limited fluid throughput and membrane

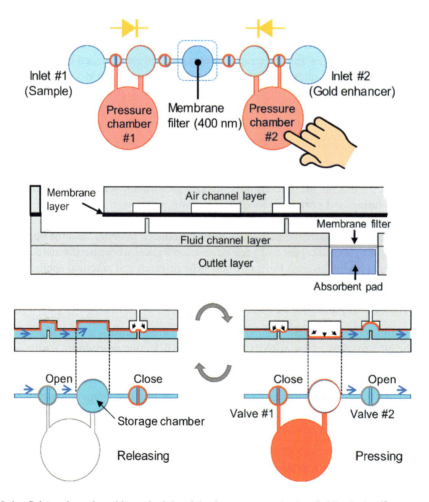

**FIGURE 2.1** Schematic and working principle of the finger-powered microfluidic device.[17]

# Nucleic Acid Detection of Pathogenic Microorganisms on Chip

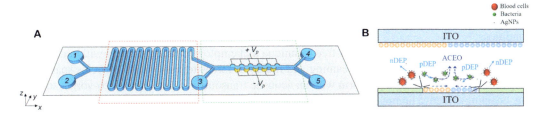

**FIGURE 2.2** Different DEP-based microfluidic chips. (a) The schematic diagram of live and dead bacteria sorting.[21] (b) The schematic of the operational process of ACEO chip.[23]

obstruction. Fang et al.[18] proposed an integrated microfluidic device with a membrane-based filtering module and a micro-mixer to isolate Gram-positive and Gram-negative bacteria from blood to address this issue. This innovative design allowed for the continual agitation of cells, hence preventing the occurrence of filter clogging. The device can remove all white blood cells and 99.5% of red blood cells from bacteria, achieving capture rates of approximately 85% in 90 min. Although membrane filtration is widely used for pathogen enrichment due to its simplicity, it has limitations such as filter clogging. Besides, its reliance on fluid temperature and viscosity can also affect separation efficiency and particle capture.[16]

Dielectric electrophoresis (DEP), a force acting on a polarized substance in a non-uniform electric field, is now widely used for the isolation of pathogens from microfluidic chips due to its label-free, effective, and particle-manipulable nature.[19,20] When a particle moves in a non-uniform electric field, it can be either affected by positive DEP (pDEP) or negative DEP (nDEP) based on the properties of fluids and particles. pDEP causes particles to move toward the stronger side of the electric field gradient, and nDEP causes particles to move toward the weaker electric field, thus separating pathogens from the sample matrix.[20] For example, di Toma et al.[21] developed a DEP-based on-chip platform to enrich live from dead *E. coli*. Through the skillful design of the microfluidic channels, live bacteria affected by p-DEP flowed to the top outlet, while dead bacteria affected by n-DEP flowed through the bottom outlet, thus enriching live cells (Figure 2.2a). This study exhibited 98% sorting efficiency in 5 s on a chip footprint of approximately 86 mm$^2$ and was able to sort bacteria with similar membrane properties as *E. coli*. However, since the DEP force commonly decreases exponentially above the electrode. It is only efficient at the surface of the electrode or within tens of micrometers, making it ineffective for low-concentration sample enrichment (103 CFU/mL).[22,23] Chen et al.[22] applied three-dimensional alternative current electrokinetic/surface-enhanced Raman scattering, integrating alternating current electroosmosis flow (ACEO) and DEP onto a microfluidic platform. As shown in Figure 2.2b, ACEO generated the flow motion directed toward the electrodes, while the pDEP force attracted the bacteria to the collecting wells, and nDEP repelled blood cells from the collector, thereby filtering cells at the electrode edge. Based on this platform, bacteria can be concentrated thousands of times in about 2 min with a limit of detection (LOD) of 3 CFU/mL, breaking the traditional DEP methods. Although pDEP generally works on Gram-positive and Gram-negative bacteria, cell recovery is often reduced due to the adhesion of cells and electrodes, as the electrodes are directly in contact with the fluid during enrichment. Yoon et al.[24] applied hydrodynamics to a DEP-based microfluidic platform for rare bacteria enrichment. By periodically controlling the force between hydrodynamic pDEP, bacteria were prevented from attaching to the electrodes and isolated continuously from the blood cells. The result found that the recovery rate of bacteria was increased to 91.3%, and the limit concentration was 100 CFU/mL, significantly improving the capability of bacteria enrichment. However, the above methods cannot evade the loss caused by adhesion during cell enrichment. Besides, the direct contact between the electrode and the fluid will also lead to bubbles caused by fluid electrolysis.[25] Overall, the DEP-based microfluidic platform described above reduced sample loss, increased the detection sensitivity, and achieved high sample sorting efficiency. However, pathogen enrichment using the DEP method is susceptible to voltage-induced Joule heating, especially in smaller viral particles enriched under solid electric field conditions.[26,27]

On-chip acoustophoresis is a label-free technique that employs an acoustic field within microfluidic channels to induce acoustic streaming and radiation forces to manipulate particles spatially.[28] Nilghaz et al.[29] introduced a platform with an acoustic transducer to generate micro-vortex streaming, enabling effective homogenization and purification of food samples. The system did not require external tubing and pumping, making it more desirable for point-of-care testing (POCT). Although the combination of acoustophoresis and microfluidic platforms offers new approaches to sample preparation, conventional methods still face many challenges. Significant advancements have been made in acoustophoresis-based sample separation recently. Ning et al.[30] proposed a high-throughput method for bacterial isolation from blood samples. The proposed device integrated serpentine geometry microchannels and standing surface acoustic waves (SSAWs), which multiplied the operating distance of SSAW in a limited chip size. This experiment enabled high-throughput sorting while maintaining cell separation sensitivity and specificity (92.7% purity and 97.5% recovery rate for *E. coli*), providing a new solution to improve sample processing capacity. In another study, Van Assche et al.[31] utilized bulk acoustic waves to reduce acoustic streaming by introducing a gradient in an acoustic impedance to enrich submicron particles. With this device, 500 nm bacteria can be isolated from 200 nm particles with fivefold relative enrichment, which significantly facilitates the enrichment of micron- and submicron-sized particles in acoustic fluid platforms. Overall, the acoustophoresis-based on-chip platform shows a remarkable capacity in point-of-care system development.

### 2.2.1.2 Biochemical Methods

In addition to employing physical properties for capturing and enriching pathogens, certain biochemical affinity agents such as antibodies or aptamers have also been widely utilized in pathogen enrichment procedures. The immunomagnetic separation (IMS) technique employs immunomagnetic beads (IMBs) immobilized with specific antibodies to capture target bacteria from the sample matrix.[32] Hussain et al.[32] utilized IMB to isolate target pathogens from the sample matrix with the limit of concentrations of 102 CFU/mL within 10 min. Based on the specificity of IMB-coated antibodies, IMBs usually bind and capture only one pathogen exclusively. Fang et al.[18] constructed "flexible neck" regions of mannose-binding lectin proteins-coated IMBs (FcMBL-coated IMBs) to capture five kinds of blood pathogen simultaneously, thereby extending the coverage of the same disease-causing pathogens. Based on the specific binding properties of IMBs to target pathogens, IMBs are often coupled with other assays to engage in the pathogen capture enrichment and detection processes. Freitas et al.[33] demonstrated a rapid magneto-immunoassay methodology to detect the capsid protein from the virus. As shown in Figure 2.3, IMBs can be further modified by horseradish peroxidase enzyme (HRP) to form a sandwich immunoassay in microfluidic channels with antibody-modified electrodes. HRP later acted as an electrochemical marker to obtain the electrochemical analytical signal. Since this method used IMS for pathogen capture and isolation, the matrix effect encountered in the DEP method can be entirely avoided.

Except for antibodies, aptamers are also affinity agents that target specific molecules. Aptamers are synthetic single-stranded DNA or RNA molecules smaller than antibodies (approximately 12–30 kDa).[34] Compared to antibodies, aptamers are easier to synthesize, cheaper, have higher affinities, better selectivity, and controllable orientation specificity.[35] Thus, aptamer assays are widely integrated into microfluidic devices to enable POCT applications. Recently, Su et al.[35] proposed a dual aptamer assay automatically performed on the chip without external equipment. The assay concentrated bacteria with aptamer-coated magnetic beads and an aptamer conjugated with quantum dots to quantify the bacterial amount. *E. coli* was detected with concentrations as low as LOD of 105 CFU/reaction, exhibiting higher sensitivity than other POCT devices (103 CFU/reaction). In another work, by forming enzyme-linked aptamers on IMBs, *E. coli* was detected with concentrations as low

# Nucleic Acid Detection of Pathogenic Microorganisms on Chip

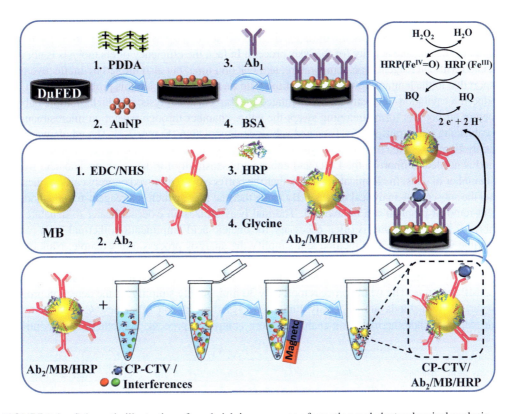

**FIGURE 2.3** Schematic illustration of sandwich immunoassay formation and electrochemical analysis.

as $1.2 \times 10^1$ CFU/mL in 2 h, lowering the detection limit.[12] The above literature suggests that aptamers are biochemical compounds that have attractive receptors in samples due to their high binding affinity and selectivity.

### 2.2.2 Target Preprocessing

#### 2.2.2.1 Nucleic Acid Sample Preparation

The enriched samples need to be pretreated before performing nucleic acid amplification. The general sample processing methods include cell lysis and nucleic acid extraction for subsequent on-chip nucleic acid amplification.[36] Traditional extraction approaches generally require large amounts of samples and reagents and are highly labor-intensive and time-consuming. Therefore, improvements in pretreatment methods are currently directed toward improving the yield of isolated nucleic acids, reducing impurities, shortening sample preparation time and sample volume, and integrating into a chip for "sample-in to answer-out" analysis.[37] The following subsection will describe recent research in the aspect of cell lysis and nucleic acid extraction.

The essence of cell lysis lies in obtaining sufficient content for subsequent detection processes and ensuring that downstream stages, such as extraction and amplification, are not inhibited.[38] Lysis methods can be broadly categorized as chemical, thermal, mechanical, and electrical, according to their nature.[36,37] Among them, chemical and thermal lysis approaches are easy and require little instrumentation, making them ideally integrated into the chip.[36] Chemical lysis utilizes lysing agents, such as organic solvents, enzymes, or detergents, to disrupt cell membranes, digest proteins, and create hydrophobic environments that facilitate adhesion to silica surfaces.[37,39] Dignan et al.[40] integrated direct-from-swab cellular lysis into a centrifugal nucleic acid extraction platform. Recovery of lysate from swabs was maximized by rotating the microdevice, and proteinase K with guanidine hydrochloride was used to denature the protein during lysis. The whole device was incubated at 56 °C for 10 min to stimulate the enzyme activity for the lysis process. In fact, the efficiency of chemolysis is not only associated with the reagent type, temperature, and reaction time, but also limited by ineffective mass transport due to the laminar flow character of the fluid on chip.[41] To tackle this problem, Kaba et al.[39] introduced a cavitation-microstreaming-facilitated on-chip chemical lysis, in which multiple bubbles trapped in air pockets form streaming swept the entire chamber through cavitation microstreaming, thus enhancing mass transport for lysis and subsequent extraction (Figure 2.4). The whole workflow took only 20 min, and the final sample lysis efficiency can be up to 76.9%. Further optimization of the method is needed to improve the LOD and enhance its sensitivity for detecting nucleic acid from a small number of cells. In chemical lysis, whether the buffer can be removed entirely can significantly impact the subsequent analytical results.[36] By using thermal lysis, which is a reagent-free method, the reagent removal problem can be avoided. Thermal lysis utilizes an external heater to denature cell membranes.[36] Since heating is often necessary for nucleic acid amplification, thermal lysis can be compatible with the detection method and simplify the analysis process. For example, Naik et al.[42] combined the sample preparation process of thermal lysis with LAMP on a paper-based microfluidic chip, where the samples were directly heated to 65 °C without other reagents and subsequently used for LAMP detection. The whole process took only 30 min, greatly simplifying the detection process. However, these methods are all easily limited by lysis speed and efficiency. Compared to the above methods, photothermal lysis enables a more controlled process. Photothermal lysis utilizes

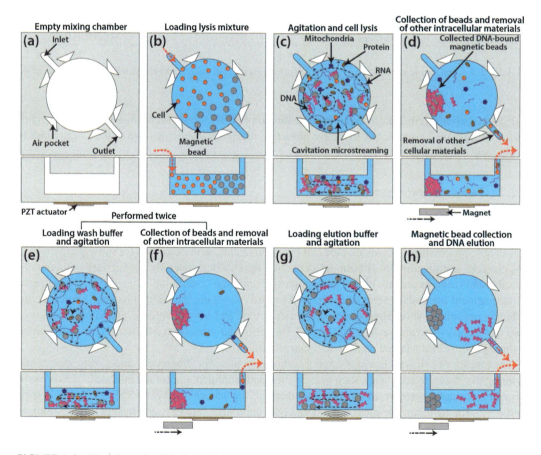

**FIGURE 2.4** Workflow of cell lysis and DNA extraction in the multilayer microfluidic cell-lysis and DNA-extraction chip.[39]

Nucleic Acid Detection of Pathogenic Microorganisms on Chip

photothermal heating assorted light-absorbing materials to enable ultrafast heating and cooling, on-site heat control, and specific target constituent damage.[43] Yu et al.[44] recently proposed a photothermal lysis platform based on strongly absorbing plasmonic Au nanoislands (SAP-AuNIs). The platform consists of a PDMS-based microfluidic chamber and SAP-AuNIs on a borosilicate glass substrate. During lysis, plasma membranes were exposed to photothermal heat and localized plasmonic heat of SAP-AuNIs, which induced lipid bilayer lysis. This device exhibited more than 90% of lysis efficiency without any thermal degradation. Mechanical lysis and electrical lysis are other effective chemical-free lysis methods. The mechanical lysis approach utilizes shear stress to tear or puncture membrane structure, which is usually integrated into a microfluidic device via various nanostructures or microbubbles.[45] Liu et al.[46] designed a microbubble array chip which lysed cells by applying shear stress to the cell on the bubble surface, causing membrane lysis. The microbubble array tool can achieve cell lysis within 1 min, with a lysis efficiency of 97.62%, providing an efficient and rapid cell lysis strategy. The microchannel design of electrical lysis is simpler than that of mechanical lysis, which utilizes a high voltage to create irreversible punctures on the cell membrane.[47] An interlocking spiral electrode and an AC voltage facilitate the electrical lysis method employed by Banovetz et al.[48] After pDEP and ionic liquid isolation capture cells were used, an AC voltage of 148 Vpp at 70 kHz was applied in three 20 s pulses to lyse the cells. The interlocking spiral design of the wireless electrodes ensured the lysis process regardless of the cell's position, thus enhancing lysis efficiency. Although electrical lysis has a simple channel configuration and membrane selectivity, like mechanical lysis, this method also has the risk of joule heating, which may cause damage to the sample.[49] Sample lysis can be made further improvements by combining the methods mentioned above. Therefore, the ideal lysis method could depend on the sample type and the required quality.

Nucleic acid extraction is another crucial part of sample preparation, and its effectiveness will impact the purity and quality of the tested nucleic acids, thus directly affecting the follow-up detection procedures.[50] Generally, extraction methods can be categorized into liquid–liquid extraction (LLE) and solid-phase extraction (SPE). Applying the LLE method on a microfluidic platform requires continuous flow through microchannels to facilitate the contact between phases. However, such devices must be operated at a specific flow rate, which makes it difficult to control the flow of a single phase in such devices, thus reducing the extraction efficiency.[51] In this regard, Paul et al.[51] proposed an electrowetting-on-dielectric digital microfluidic platform, which adapts LLE in drop-to-drop (DTD) format by dropwise flow control. This solution used organic dyes as solutes, enabling parallel and serial extraction and thus extracting DNA selectively. This device eliminated the need for external magnets and simplified DNA isolation. However, the most significant limitation of this method is its limited 17%–18% extract efficiency. In fact, the time consumption, presence of artificial manipulation steps, usage of toxic organic solvents, and need for expensive glassware of LLD make it challenging to integrate into microfluidic platforms. Therefore, SPE has been popularized for on-chip nucleic acid extraction. SPE separates solutes from unwanted components by adsorbent filtration via the physical and chemical properties of the solutes.[52] Commonly used on-chip SPE methods include magnetic beads[40] and silica and glass beads.[51] For example, Kaba et al.[39] used cavitation-microstreaming to facilitate chemical lysis. DNA released after lysis is attached to the silica-coated magnetic beads, and a magnet was used to spiral above the chamber to collect the nucleic acids. After removing the solution from the chamber, the DNA was washed and collected by controlling the perturbation of the magnets. The whole extraction efficiency was up to 76.9%, significantly improving the ineffective mass transport on-chip. However, the above system often requires complicated external operating systems for fluid control and has many operation limits. To solve this problem, Fan et al.[53] further proposed a gas-liquid immiscible phase extraction system based on the mentioned magnetic beads extraction method. By manipulating the magnetic beads passing through the gas–liquid interface, the system enhances the recovery rate of the magnetic beads to more than 90% and the extraction effect of about 80% in a low-surfactant system, therefore, simplifies the need for complex cassettes and shortens the time from approximately 30 to 7 min (Figure 2.5). The above nucleic acid extraction methods are generally integrated into the nucleic

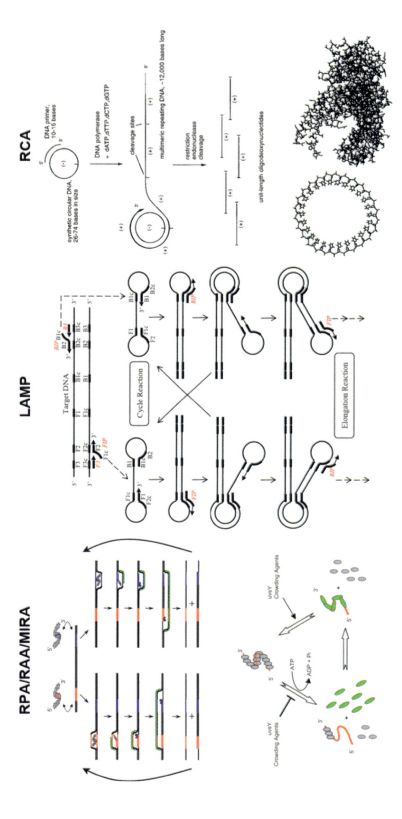

**FIGURE 2.5** Working principles of different isothermal nucleic acid amplification methods. RPA method, LAMP method, and RCA method (from left to right).[55]

acid amplification analysis afterward. In order to simplify the process, Zai et al.[54] proposed an extraction-free on-chip nucleic acid amplification method. This microfluidic platform contains merely a lysis chamber and an amplification chamber, which can directly amplify SARS-CoV-2 RNA without purification. The whole process took only 30 min, greatly simplifying the detection process. The existing sample preparation process, regardless of cell lysis or NA extraction, tends to be integrated into a microfluidic device to develop a "sample-to-answer" platform, which is more portable and efficient.

### 2.2.2.2 Nucleic Acid Signal Conversion using Aptamer

A nucleic acid aptamer is an oligonucleotide sequence (DNA or RNA). Usually, oligonucleotide fragments are obtained from nucleic acid molecular libraries using in vitro screening techniques—systematic evolution of ligands by exponential enrichment (SELEX). Often, nucleic acid adaptors are designed to tightly bind to proteins, targeting the target protein for subsequent research.[56] With the development of science and technology, in the field of pathogen detection, aptamers have been converted into nucleic acid signals to improve the detection limit.[57] Taking CHA reaction as an example, CHA reaction is an enzyme-free nucleic acid amplification reaction, mainly consisting of S chain, H1 chain, and H2 chain. First, the S chain will combine with the H1 chain to form a S-H1 complex, which will form a CHA product with H2. How to perform signal conversion? Taking CHA-CRISPR detection as an example, aptamer and S chain will first complement with each other.[58] In the presence of target bacteria, they will compete to bind to aptamer and release the S chain, which is completed in the serpentine channel. Further to the right is the microcolumn array, on which H1 chains are fixed. After all the raw materials are mixed, S will open the H1 chain to form the S-H1 complex, which will further open the H2 chain to form the CHA product and release S. The CHA product is then fixed on the microcolumn array. At this moment, rinse the entire flow channel with a buffer solution to clean the unreacted raw materials and eliminate any potential impact of the raw materials on the detection.[59] By adding the CRISPR system, when CHA products are present, crRNA will bind to them, activate the trans cleavage activity of Cas protein, and non-specifically cleave single-stranded DNA molecules to generate fluorescence. This design concept has gradually become popular recently, and Kang's team has proposed an integrated design strategy, establishing an innovative approach $Fe_3O_4$@AuNRs adaptor sensor, used for efficient detection of *Staphylococcus aureus*, can enrich and eliminate bacteria under near-infrared light radiation.[60] Detection analysis relies on fitness and $Fe_3O_4$@AuNRs. The binding of [$Fe_3O_4$] AuNRs aptamer induces the release of c-aptamer (c-Apt) in the presence of *Staphylococcus aureus*, thereby triggering the CHA reaction and restoring the fluorescence of the reporter gene. The aptamer sensor can be used to detect *Staphylococcus aureus* in milk, with a recovery rate of 90.3–95.6% at concentrations between $1.0 \times 10^2$ and $1.0 \times 10^4$ CFU/mL. This provides a new approach for detecting foodborne pathogens.[61] Yang's team has developed a uniform naked eye and surface-enhanced Raman scattering (SERS) dual-mode sensor for aflatoxin B1 (AFB1) in food samples, triggered by aptamer recognition and assisted signal amplification without enzyme-catalyzed hairpin assembly (CHA). This sensor is highly sensitive and portable. The recognition of AFB1 by aptamers induces the production of the HP1–AFB1 complex, which hybridizes with Ag+-labeled hairpin DNA (HP2) and releases Ag+. Subsequently, an enzyme-free CHA reaction is initiated through a designed helper DNA (HP3) to form double-stranded DNA (HP2–HP3), accompanied by the release of the HP1–AFB (1) complex.[62] The released HP1–AFB (1) complex is once again recognized by HP2 and HP3 to trigger cascade cyclic amplification, leading to the production of a large amount of free Ag+ and dsDNA. Then, methylene blue is inserted as a Raman tag into dsDNA and $Fe_3O_4$@Au. At the same time, free Ag+ induces the aggregation of AuNPs and causes a visible color transition from red to black-blue.

## 2.3 DETECTION OF PATHOGENIC MICROORGANISMS USING NUCLEIC ACID AMPLIFICATION METHOD ON CHIP

### 2.3.1 Enzyme-Based Nucleic Acid Amplification Method

#### 2.3.1.1 RPA

RPA is an isothermal nucleic acid amplification method that relies on a set of enzymes to facilitate amplification in a constant temperature range of 38–42 °C. These enzymes include a recombinase, which is capable of pairing primers with homologous sequences in double-stranded DNA to form a recombinase–primer complex, a single-stranded DNA-binding protein (SSB) binding to displaced strands of DNA, stabilizing the complex, and strand-displacing polymerase initiates DNA synthesis[63] (Figure 2.5). Compared to other isothermal amplification platforms, RPA has a faster reaction rate (<20 min) and a lower reaction temperature, making it an ideal tool to integrate into the microfluidic platform for rapid on-site nucleic acid detection.[63] Rani et al.[64] integrated RPA into lateral flow test strips (LFS) to detect *E. coli* O157: H7 in milk. The method was sensitive and rapid with LOD down to 4–5 CFU/mL. However, when applying RPA to detect specific pathogens in multi-pathogen samples, the detection ability of RPA is very limited due to the homology-directed repair properties of the enzymes.[65,66] Therefore, RPA is generally used for multi-bacterial detection. Li et al.[67] performed parallel analysis of six genetic markers for five common pathogens from a single sample via on-chip RPA, enabling sensitive and specific multimicrobial analysis (Figure 2.6a). In another study, Shang et al.[68] developed a fully integrated micro-platform system to eliminate the need for bulk equipment and skilled personnel based on 100% specificity of detecting eight foodborne bacteria.

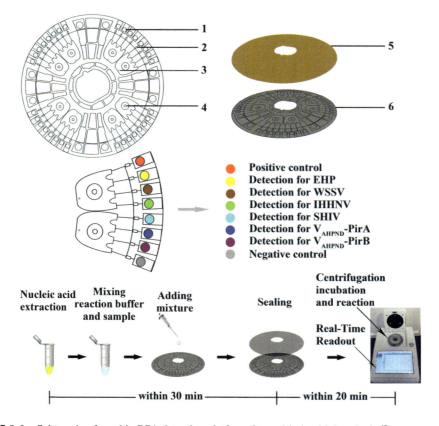

**FIGURE 2.6** Schematic of on-chip RPA detection platform for multimicrobial analysis.[67]

Traditionally, RPA products are analyzed by gel electrophoresis or DNA-binding dye, which requires a bulky detector and is difficult to detect weak positive results.[69] Therefore, with the advent of the POCT model, most RPA-based devices, such as the above platform proposed by Li et al.,[68] integrated real-time detection methods, such as colorimetric and fluorometric methods, for product analysis.[55] Fluorometric tools are one of the most commonly used methods that utilize fluorescent reporters to label the amplicons and thus monitor the amplification process. The platform proposed above by Shang et al.[69] even visualized the fluorescence signal curve in a smartphone, further enhancing the device's portability. In addition, since RPA is widely integrated into lateral flow immunochromatographic assay (LFIA) strips, biosensors such as colorimetric are also commonly employed as the detection method for RPA assays. In LFIA, the detection process generates labeled amplification products by tagging primers or probes. These labeled amplification products form amplicons–bioreceptor–colorimetric indicator complexes with colorimetric indicators (e.g., AuNPs) to achieve pathogen detection.[70,71] The above examples enabled amplified signal output using various biosensors. However, these methods are not sufficient for accurate quantification of samples. Due to the reliance on a linear standard curve for nucleic acid quantification in the RPA reaction, the need for complex primer screening and probe optimization in single-molecule assays, and the susceptibility to nonspecific amplification of background signals, it is not easy to perform accurate quantification for target molecules.[72]

To overcome these limitations, digital RPA (dRPA) has been proposed, which allows absolute quantification of nucleic acid concentrations in just one test. Compared with the traditional RPA method, dRPA can achieve accurate quantification without complicated standard curves, enhance sensitivity to the single-molecule level by dividing nucleic acid into separate compartments, and remove potential inhibitors interference to amplification.[72–74] These benefits have led to its widespread applications in various microarrays to improve the sensitivity and specificity of detection. For example, Yin et al.[75] proposed a "sample-in-multiplex-digital-answer-out" system based on a dRPA chip, which divided 12,800 chambers into four dRPA regions and utilized vacuum-based self-priming introduction to inject sample reagents into the reaction regions passively (Figure 2.6b). The chip can lower the detection limit to 10 cells per type of pathogen by multiplex dRPA and fluorescence detection.

For viral RNA amplification, an additional reverse transcription (RT) is needed. Recently, RT-RPA has been widely used for POCT of SARS-CoV-2. Li et al.[76] proposed a centrifugal microfluidics-based multiplex RT-RPA assay for parallel multigene testing of conserved regions of the SARS-CoV-2 genome. With sensitivity and specificity among different genes higher than 90%, false-negative results were maximally excluded. In addition, RT-RPA assay was also integrated into microfluidic platforms along with different on-chip techniques. For example, Liu et al.[77] and Bender et al.[78] have proposed microfluidic-integrated lateral flow RT-RPA assay and RT-RPA assay with paper-based isotachophoresis for RNA extraction, respectively, aiming to enhance the sensitivity and reduce the detection time. Based on the needs of dRPA for absolute quantification of target molecules and RT-RPA for viral detection, Seder et al.[79] have innovatively proposed conjugating dRNA with RPA to accurately quantify viral RNA recently. The novel RT-dRPA system utilizes a vacuum system to drive the inlet of an oil-phase compartmentalized reagent. Following the RT-RPA reaction and fluorescence signal detection of the sample, the absolute quantification of viral RNA can be achieved in 37 min with the sensitivity as low as 10 RNA copies/µL.

Briefly, the on-chip RPA methods mentioned above leverage the advantages of low reaction temperature and rapid amplification of RPA, thus achieving remarkable progress by enabling multipathogen detection, optimizing the RPA performance, expanding the application on the biosensor, and integrating the one-chip RPA, which significantly expands its application in the prospect of POCT.

### 2.3.1.2 Loop-Mediated Isothermal Amplification

The LAMP reaction is catalyzed by a Bst polymerase with displacement activity and two to three sets of specific primers. The two sets of primers include outer primers and inner primers, which are generally sufficient to amplify the target DNA. In contrast, the loop forward primer and loop backward primer are optional to enhance the specificity and efficiency of the reaction (Figure 2.5). Amplification of nucleic acids can be accomplished at a constant temperature (60–66 °C) within 60 min and allows multiple detection methods, including colorimetric, fluorometric, gel electrophoresis, and turbidity methods.[80]

LAMP has been widely applied in microfluidic technology, especially for the POCT, owing to its high sensitivity, rapid reaction time, without the need for complicated thermal cyclers, and accurate detection of concentration samples.[81] Liu et al.[82] introduced a centrifugal chip that integrated the entire foodborne microbial detection process, including cell lysis, LAMP, and colorimetric detection. The amplification process was heated by an aluminum panel underneath the chip, and centrifugal force was used to achieve precise liquid manipulations, including liquid mixing, dosing, and transferring to the reaction chamber. The results were detected by a color sensor. The system could detect five foodborne microorganisms in 70 min, with a LOD lower than 100 copies/μL. LAMP amplicon analysis can be categorized into two types: endpoint detection and real-time monitoring of the LAMP reaction. The example mentioned above utilized the endpoint detection method.[83] Fluorescence real-time detection, as another popular method, is more sensitive than the former method and can be applied to real-time monitoring.[84] Jin et al.[85] further proposed using a dual-sample on-chip LAMP assay based on the aforementioned single-sample centrifugal microfluidic chip to simultaneously detect multiple waterborne bacteria in multiple samples through fluorescence real-time detection. The device achieved 93.1% sensitivity and 98.0% specificity within 35 min using only a minimal volume of 22 mL per each half of a single chip. The LOD was reduced from $7.92 \times 10^3$ to $9.54 \times 10^1$ pg, significantly improving the throughput, detection limit, detection time, detection sensitivity, and detection accuracy. Although real-time fluorescence LAMP detection can achieve high sensitivity, it is not applicable in resource-limited places. In this regard, Xiao et al.[86] proposed a hand-held sample-in-answer-out system to detect six foodborne microorganisms in meat products. By incorporating colored LAMP into microfluidics, the device enabled LAMP reaction at 65 °C, which could detect meat with an adulteration level as low as 1% within 1 min, and the positive samples would transform from pink to yellow that could be directly observed with the naked eye. This device significantly improved the portability of the LAMP assay. However, evaporation phenomena and air bubbles may occur in the chambers during the heating process, resulting in sample reduction, making the device unsuitable for low-sample quantity tests.

To fulfill the need for RNA virus detection, researchers added a reverse transcriptase enzyme to LAMP for reverse transcriptase LAMP (RT-LAMP). Wang et al.[87] designed an RT-LAMP system incorporating six parallel microfluidic chambers to detect Zika Virus in the whole blood. The device consisted of a two-stage microfluidic chamber for sample processing and nucleic acid amplification, a compartment that heats from below, and LEDs that illuminate from above. The smartphone camera can dynamically monitor the fluorescence signal. This dynamic image analysis could reduce the detection time from the conventional 30–45 min to 22 min, and the LOD of the virus could be as low as $2.70 \times 10^2$ copies/μL, enabling rapid and sensitive detection of viruses. However, RT-LAMP assays usually require time-intensive and laboratory-based sample pretreatment before detection.[88] As a result, assays without nucleic acid extraction are being developed. In this regard, Lalli et al.[89] proposed a colorimetric RT-LAMP assay for the detection of extraction-free SARS-CoV-2 RNA in saliva. In this study, heat treatment of undiluted saliva was used to reduce the inhibitors, and the LOD was raised to 10 particles/reaction. Nevertheless, another study also pointed out that the extraction-free test is only suitable for clinical samples with fewer inhibitors and cannot be applied in complex samples.[90]

In order to achieve quantitative analysis of samples, a digital LAMP CD-like chip has been proposed to detect and differentiate gene-deleted type and wild-type viruses simultaneously.[90] This system can efficiently and consistently differentiate swine pathogens, achieving a LOD of $10^2$ to $10^1$ copies/μL. The emphasis of dLAMP is on generating numerous individual droplets, as the number of droplets can directly impact the accuracy of dLAMP results.[91] Hence, Wu et al.[91] proposed a

# Nucleic Acid Detection of Pathogenic Microorganisms on Chip

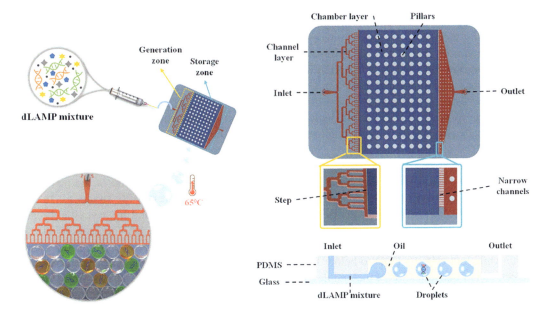

**FIGURE 2.7** Schematic illustration of droplet generation via a pump-free duplex droplet microfluidic device.[91]

pump-free duplex droplet microfluidic device designed based on stepwise emulsification (Figure 2.7). The device consists of a droplet generation zone and a droplet storage zone, which contain 64 parallel nozzles that can be manually injected to generate 48,100 uniform droplets at a total volume of 25 µL due to the difference in Laplace pressure. Subsequently, the droplets entered the storage area for LAMP, which was previously filled with 10% 008-fluorosurfactant oil to prevent melting and evaporation of the droplets during amplification. The amplification signal was detected using a fluorescent probe to achieve absolute quantification of the *E. coli* DNA down to 19.8 copies/µL, removing the need for standard curves.

The above study presented the application of on-chip LAMP in bacteria and viruses in recent years. Overall, there have been breakthroughs in the detection speed, detection accuracy, and detection limit of LAMP in the past few years, and the development of portable and integrated POCT devices can be seen as a development trend.

### 2.3.1.3 RCA

RCA is another isothermal amplification method that amplifies circular DNA by strand displacement polymerase. The amplification process is performed when a singer primer binds with the template strand and thus produces ssDNA. This newly synthesized strand then replaces the original template for the primer to bind. The newly synthesized DNA contains multiple copies of the target fragment, which can be specifically bound by ligated padlock probes for sensitive detection of RCA products (Figure 2.5).[55,92,93] Like any other isothermal amplification method, RCA reacts exponentially at a constant temperature of about 37°C, quickly producing a large amount of DNA, thus eliminating the need for an external heat source and improving sensitivity and specificity.[55] Recently, many studies have proposed integrating RCA with padlock probes on microfluidic chips for nucleic acid detection. However, since RCA methods have disadvantages, such as being labor-intensive, involving expensive probe solutions, and target binding is restricted to a single interaction, its application on microfluidics is considerably less than that of LAMP and RPA.[92]

To address these shortcomings, Ishigaki et al.[94] integrated RCA with fluorescence in situ hybridization (FISH)-based padlock probe into a chip. The reaction was performed under stopped-flow conditions, and a circular-shaped microchamber was used to reduce sample residue and increase the consumption of expensive probe solution from 10 μL to approximately 3.5 μL. Whereas an ultrasonic mixing method was used to reduce the sample residue, it shortened the fluorescent probe hybridization time by 5 min (fourfold superior to the standard protocol), thus significantly optimizing the efficiency of the padlock and RCA reactions. However, this method involves extensive loading procedures and manual intervention. Garbarino et al.[95] proposed multi-step RCA for synthetic influenza virus on-chip analysis to simplify the operation. The device employed magnetic microbeads to transport the target molecule between three linked reaction chambers. The product amplified by phi29 polymerase was immobilized by magnetic beads and read through optomagnetic detection. The proposed automated on-chip assay enabled sample detection with LOD of 20 pM within 45 min.

In the above studies, PLP targets specific sites in a single conserved region of the pathogen, resulting in a low specificity of detection for hypervariable viruses in real-life situations. In this regard, Ciftci et al.[96] skillfully designed nine padlock probes for Class I and II Newcastle disease virus (NDV) based on the alignment of 335 sequences. These probes cocktail was amplified and then subjected to two rounds of RCA to detect highly mutating viral strains, preserving 100% specificity while lowering the LOD to below 10 copies of viral RNA, meeting the need for multi-virus detection. For quantification of the samples, the above device further leveraged a paper-based microfluidic system to enrich and digitally quantify fluorophore-tagged RCA products. The advantage of digital measurement of RCA signal is that the amplicons remain undiluted or unaffected by one another and are capable of counting the number of amplified molecules, thereby minimizing amplification bias. With the benefit of these techniques, the detection limit as low as ten copies of the virus RNA was achieved in this study. This technique was later applied to detect both EBOV RNA types and the multiplexing of tropical virus panels, expanding the target pathogen of this assay.[97]

Unlike the RCA methods described above, capture RCA (cRCA) is also widely used in microfluidic platforms, where the technology enhances target capture with long single-stranded DNA products with repeat tandem aptamer sequences.[98] In Li et al.'s work,[99] cRCA products featuring repeat tandem aptamers specific for *E. coli O157:H7* were used to modify the inner surface of microfluidic channels. By varying *in situ* RCA product modifications, the capturing efficiency and specificity of *E. coli O157:H7* were greatly increased compared to the unit-aptamer approach. In another study, cRNA was coupled with signaling RCA (sRCA) for sensitive detection of *E. coli O157:H7* in an aptamer-based microfluidic chip.[100] Dual-RCA detection was performed by *in situ* cRNA to capture target *E. coli O157:H7* cells and by sRCA to amplify and enhance the detection signals. The experimental results show that the dual-RCA approach can improve the detection sensitivity about 250 times, with a LOD of 80 cells/mL, making it favorable for high-throughput and high-sensitivity pathogen detection.

Conclusively, although the RCA method has some limitations, such as the complexity of primer and PLP and the cumbersome design of the chip, it is still another powerful tool for in situ nucleic acid detection in microfluidic platforms based on the specificity of the assay and a wide range of applications.

## 2.3.2 ENZYME-FREE NUCLEIC ACID AMPLIFICATION METHOD

### 2.3.2.1 Catalytic Hairpin Self-Assembly (CHA)

It is undeniable that in recent years there has been a surge of interest in utilizing enzyme-free amplification strategies for detection of pathogens due to their high sensitivity, high selectivity, and rapid result feedback. CHA is one of the widely used methods in the detection of *Escherichia coli* O157:H7, which caused many foodborne diseases such as hemorrhagic colitis and hemolytic uremic syndrome, and even deaths of animals and humans.[101–105] CHA amplification technology, a nucleic acid-based toehold-mediated strand displacement reaction driven by the free energy of base pair

# Nucleic Acid Detection of Pathogenic Microorganisms on Chip

formation can help to ensure food safety and public health with the advantage of being isothermal, having high catalytic efficiency, and having great amplification ability[106,107] in comparison with the standard plate colony counting, which is currently a more accurate and reliable detection method for pathogen detection. Its advantages of rapidity, sensitivity, and cost-effective abilities made it a promising tool for application as a biosensor in the future.[108,109]

In recent years, scientific research by Luo et al. and Sun et al. clearly demonstrated the rapid and sensitive detection of *E. coli* with CHA amplification, however with the incorporation of different types of analytical tools like electrophoresis and use of chemiluminescence (CL) biosensors. In the case of the study by Sun et al., sulfhydryl nucleic acid and HRP-modified gold nanoparticles were used as the bridging material for CHA cycling and CL reaction.[110,111] In the presence of *E. coli*, they will competitively bind to a specifically designed aptamer for *E. coli* (Apt-E), and this reaction will then trigger the release of another specially designed complementary nucleic acid DNA chain (S). This release of S chain will subsequently initiate the CHA reaction assembling two hairpin probes (H1 and H2) into an H1/H2 complex. This complex together with the modified gold nanoparticles will catalyze the CL reaction between luminal and $H_2O_2$ and thus ultimately form a high-intensity fluorescent signal in the presence of the bacteria (Figure 2.8).[107] In case when there are no bacteria,

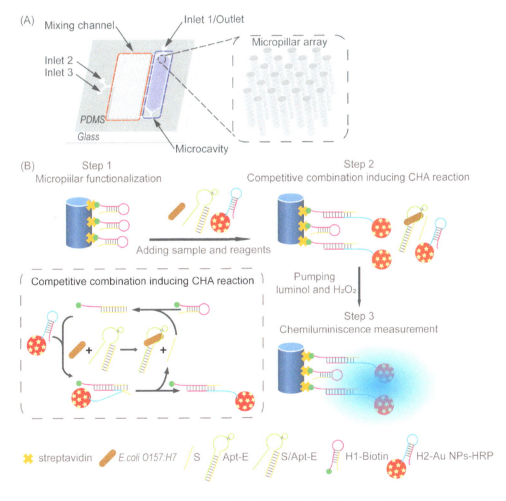

**FIGURE 2.8** An overall demonstration of the microfluidic chemiluminescence (MCL) biosensor using the multiple signal amplification method to detect target bacteria. (a) A graphical abstract of the microfluidic chip. (b) A diagram illustrating the principal mechanism behind the multiple signal amplification strategy.

the S chain will tightly bind to the Apt-E and thus there will be no S chain to induce the formation of the H1/H2 duplex. In essence, this method can quickly and sensitively detect nucleic acid which ultimately turn into a rapid detection for bacteria (Figure 2.9a)[107].

A new analytical method using CHA in combination with microchip electrophoresis (ME) was presented by Luo et al. to detect *E. coli O157:H7* in defatted milk. To maintain the original properties of the probes, SYBR gold solution was used to label ssDNA or dsDNA of the ME system.[112] Like the study performed by Sun et al., a specific bacteria recognition aptamer (apt-E) was used.

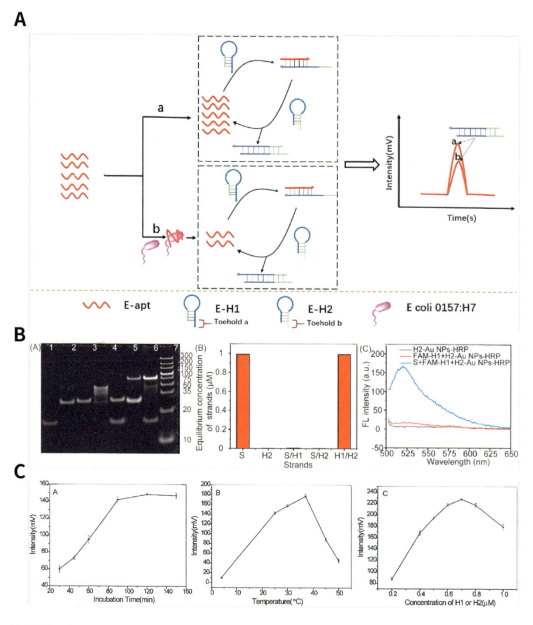

**FIGURE 2.9** (a) A graphical demonstration for the combination of the CHA-based signal amplification method with MCE for sensitive detection of target bacteria. (b) The overall performance of the designed hairpin probes and the CHA amplification method. (c) Optimization conditions were tested for the sensitive detection of bacteria.

Nucleic Acid Detection of Pathogenic Microorganisms on Chip

The Apt-E can trigger the catalyzed reaction of the hairpin assembly of H1 and H2 for signal amplification to form an H1/H2 complex.[112] A strong H1/H2 complex signal peak can be detected using electrophoretograms. In the presence of *E. coli*, the bacteria will specifically bind to apt-E causing a reduction in the amount of free apt-E and thus causing a reduction of H1/H2 complex peak height in comparison with the absence of bacteria.

Luo et al. also used the technique of electrophoresis to verify the specificity between bacteria and apt-E and apt-E's ability to induce CHA amplification. If bacteria were put together with apt-E before apt-E reaction with H1 and H2, the band of H1/H2 complex was darker compared with the H1/H2 complex band without the incubation of bacteria. There was no H1/H2 complex band when bacteria were with apt-E/H1, apt-E/H2, or solely H1 and H2 and thus concluding that it was apt-E that was responsible for inducing the assembly of H1 and H2.[112] The method also showed good selectivity in that only *E. coli* O157:H7 illustrated a strong signal, while others showed a signal close to the blank sample ensuring the selectivity of apt-E.[113] In summary, the combination of CHA and ME techniques demonstrated by Luo et al. serves to have an ultrasensitive ability to quantify the amount of *E. coli* in defatted milk. This research may be enhanced and modified in the future and provide a more diverse approach to detect more types of bacteria.

With all pros being illustrated, there are still some cons that may be modified to enhance the overall microchip-CHA method in the future. In the research by Sun et al., three different kinds of hairpin oligo-structure (H1-1, H1-2, and H1-3) were previously predicted and designed by a NUPACK software package to obtain the best hairpin structure (H1-3) with the highest conjugation efficiency with the S-DNA chain. The mechanism behind this was based on the maximum number of complementary pairing bases between S and different H1 structures and the number of Gibbs free energy.[107] In a similar context, three different H2 structures (H2-1, H2-2, and H2-3) were also designed to select the optimal H2 (H2-3) structure for spontaneous interaction between H1 and H2 to form an H1-H2 complex in the absence of the S-DNA chain (Figure 2.9b). The overall preparation process can be very time-consuming, and it requires the assistance of NUPACK software which can be problematic if its performance is impaired by different technical issues. In this research, the author designed three different types of hairpin structures for CHA cycling circuit. But assumptions can be made if the author goes further with its investigation and designs more types of hairpin structure; the effectiveness of the CHA cycling circuit may increase, but the time consumption can be enormous. Another possible flaw for CHA application is that it requires a step of labeling the nucleic acid, which may hinder the performance of aptamers or hairpin probes.[112] Moreover, the CHA method requires the determination of an optimal incubation temperature, time of CHA reaction, and concentration of the hairpin structure to obtain the best sensitivity (Figure 2.9c). This may require a lot of time and tedious work for the researcher.

### 2.3.2.2 Hybrid Chain Reaction (HCR)

Like the CHA method, the HCR method is also an enzyme-free, isothermal, highly efficient signaling amplification strategy with low complexity and low cost.[114,115] This target-trigger amplification system was first demonstrated by Dirks and Pierce in 2004.[116] The working mechanism behind HCR is that the target will initiate a toehold-mediated strand displacement reaction causing the cross-opening of two meta-stable DNA hairpins (H1-domain a and H2-domain c-b*) and form a nicked double helix structure.[114,117] Usually, the hairpin structures are modified with certain signaling molecules to have a higher selectivity of the target.[118] The main difference between HCR and PCR is that HCR is a probe-amplification technique, while PCR is a target amplification technique.[119] Unlike PCR, HCR can reduce the occurrence of false-positive results and cross-contamination from amplicons.[114] However, the design for the hairpin structures is a crucial factor and can be hideous since a small variation in hairpin may lead to failure of the whole HCR process.[114] Over the past 5 years, HCR had been integrated with different analytical techniques for the detection of pathogenic bacteria.

To name a few recent analytical methods in combination with HCR, detection for *Staphylococcus aureus* using a homogeneous electro-chemiluminescence aptasensor with the assistance of HCR and magnetic separation was proposed by You et al. in 2023. The presence of bacteria will bind to the specifically designed aptamer, which allows the opening of the probe DNA. The exposed probe DNA will subsequently trigger the HCR amplification of the DNA duplexes which can be combined with the luminescent molecule Ru (phen) 32+. This method demonstrated high sensitivity and specificity with its luminescent intensity being linearly correlated with bacteria's logarithmic concentration ranging from 10 CFU/mL to $10 \times 10^7$ CFU/mL with a detection limit of 3 CFU/mL. Another pathogenic bacteria detection method based on HCR amplification was performed by Ying et al. in 2021. *Vibrio parahaemolyticus* was the target bacteria and it was specifically detected by a combination method of using synthetic modified magnetic nanoparticles aptamer, HCR, and a lateral flow nucleotide biosensor (LFNB).[120] The principle behind this method is very similar: the initiator probe composed of a trigger sequence of HCR and aptamer was first added to the hairpin mixture to activate the chain between H1-bio and H2-bio to form a long double helix polymer-labeled biotic. Then the target bacteria will be recognized by the aptamer and all the unbounded aptamer and non-target residues will be removed by centrifugation and magnetic beads. The final complex of aptamer–biotin–SA-MBs will be mixed with HCR products and subjected to a sample pad where the reaction between the complex and the AuNP surface occurs for the visual detection of target bacteria which is characterized by the formation of a red band (Figure 2.11b).[120] The detection limit was $2.6 \times 10^3$ cells with the whole process taking 67 min. A DNA nanotechnology method using the nanoprobe supported by HCR to detect bacteria was demonstrated by Tang et al. The special design nanoprobe DTAAT can recognize the bacteria by a way of binding both the antibody and aptamer, which enhanced its sensitivity and specificity. DTAAT consists of an antibody-aptamer conjugate (AA) coupled with DNA tetrahedron (DT) and a trigger strand (A). The design of DT improved the local concentration of HCR products and fluorescent substances. In the presence of *Staphylococcus aureus*, there will be a dual recognition of bacteria's aptamer and antibody protein, and the release of trigger stand will activate the amplification of HCR and form a fluorescent signal at three vertices simultaneously.[121] The detection limit of this method was as low as 26 CFU/mL (Figure 2.12a). In summary, there are multiple advanced analytical methods combined with HCR for the detection and analysis of bacteria. All these research and data showed a substantial development in HCR-based methods and there is great potential for a sensitive and rapid practical application in the future.

In terms of the application of microfluidic chip with the HCR method for the detection of bacteria, Li et al. and Shang et al. demonstrated successful examples that are worth deeper investigation in the future. Li et al. introduced an eye-based microfluidic aptasensor (EA-sensor) for rapid detection of bacteria *E. coli* without instrument.[122] This EA-sensor adopts the concept of volumetric-bar-char chip that has been developed in previous studies to quantify disease-related biomarkers in an equipment-free circumstance.[123–125] The mechanism behind is relatively simpler than previous methods. The EA-sensor used a specially modified aptamer to recognize bacteria *E. coli 0157:H7*, while the HCR method was used to increase the signal amplification (~ by about 100 folds).[122] The initiator of HCR is first conjugated with the substrate and the excess aptamer will bind to the initiator through complementary base pairing.[122] In the presence of bacteria, the bacteria will competitively bind to the aptamer, and this causes the aptamer to dissociate with the initiator and subsequently the initiator triggers the cascade hybridization which formed a nicked double helix of H1 and H2. The H1 and H2 have been labeled with a probe of platinum nanoparticles, which were used to catalyze $H_2O_2$ to produce oxygen, and this oxygen will be converted proportionally compared with the bacteria concentration in the form of visible displacement of ink within a microfluidic channel (Figure 2.10). This method had an LOD of 400 CFU/mL for the detection of spiked *E. coli* in milk sample using the naked eye (Figure 2.11a). Overall, this EA-sensor proved to be fast (75 min), convenient, cost-effective, sensitive, and specific to quantify the presence of bacteria, and thus it possesses great potential to become a POCT tool for detection in resource-scarce environment.

# Nucleic Acid Detection of Pathogenic Microorganisms on Chip

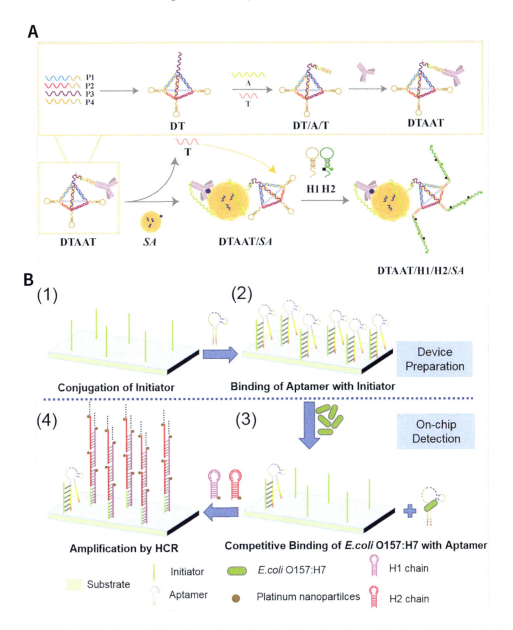

**FIGURE 2.10** (a) A schematic illustration of bacteria analysis based on the combination of DNA tetrahedron with HCR method.[121] (b) The principal mechanism of HCR-based method. The top panel shows the preparation procedures of initiator conjugation and aptamer binding.[122]

As for the research demonstrated by Shang et al., the author used a cloth-based CL biosensor in combination with HCR to produce hemin/G-quadruplex DNAzyme molecules which in turn can enhance the CL signaling for the DNA of *Listeria monocytogenes*. Among the methods of colorimetric, fluorescence, electrochemical, electrochemiluminescence, and CL, the CL method proved to be well-suited to detect DNA sequences with the capability of low detection limit, low background, and simple instrumentation.[126] To further improve the detectability of bacteria, the HCR amplification method was integrated with oxidizing enzyme, and hemin/G-quadruplex DNAzyme was used due to its high stability and cost-effective properties. In addition, it has excellent catalysis ability to oxidize the luminol by $H_2O_2$ and produce CL signal.[127] Pure milk samples spiked with different

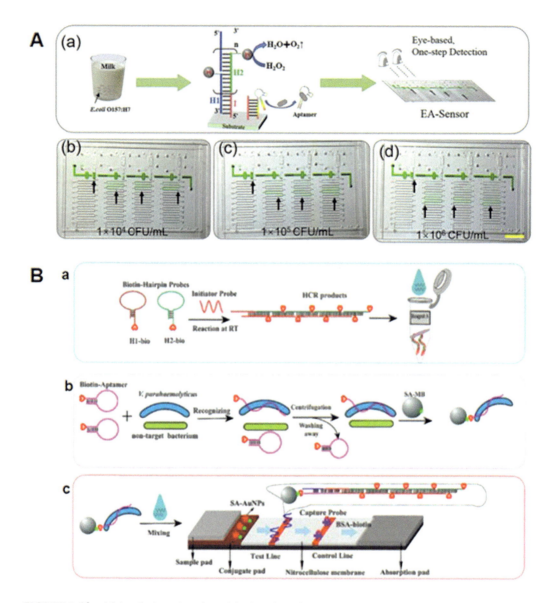

**FIGURE 2.11** (a) A naked eye-based on-chip detection of target bacteria (*E. coli* O157:H7) in infected milk samples. (b) Schematic illustration of aptamer–HCR–LFNB.

bacterial T-DNA concentrations were used to test the sensibility of this cloth-based biosensor. In the presence of bacterial gene hlyA, the biosensor can trigger the HCR-integrated hemin/G-quadruplex DNAzyme for signal enhancement and CL detection (Figure 2.12b). The result demonstrated that the CL intensities changed linearly as the logarithms of bacteria concentrations in the range of $1 \times 10^2$ to $1 \times 10^7$ CFU/mL and a LOD of 50 CFU/mL.

There are still some limitations, such as the possibility of self-assembling or forming other higher-order structures such as hairpin, which may affect its binding capacity with the pathogen and cause error during detection.[128] In addition, the real-life application can also be affected by different sample's properties. Different samples are made with different complex matrixes, the matrix may affect the specificity of aptamer, because it can have a reaction between the sample's protein or other molecules. The problem with POCT is that its resolution and detectability are limited and thus may

# Nucleic Acid Detection of Pathogenic Microorganisms on Chip

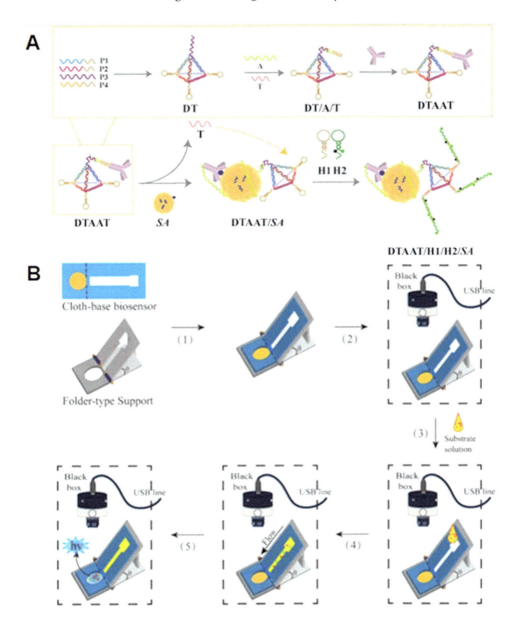

**FIGURE 2.12** (a) A schematic illustration of bacterial analysis based on the combination of DNA tetrahedron with the HCR method. (b) The schematic illustration of the assay procedure of the proposed DNA biosensor.

limit its widespread application for bacterial analysis. Like the limitations of the CHA method, the reparation and design of signal probe can be time-consuming and tedious. Besides, the CL measuring requires an analyzing system and instrument such as a CCD camera and VGIF software to tailor images and thus cannot be portable and used in resource-limited environment. Furthermore, the intensity of CL may be affected by steric hindrance of DNA where DNA molecules are tightly bound to each other in the detection zone, causing areas for HCR amplification to be blocked.[129] CL intensity may also be affected by the concentration of chemical reagents ($H_2O_2$ and luminol), and this phenomenon might be caused by the effect of self-inhibition at higher concentrations and thus $H_2O_2$ was modified to a concentration of more than 0.035 M and luminol at 2 mM. The stability of the biosensor is also a key factor for maintaining its specificity for detecting bacteria. Usually, it is stored within a sealed plastic bag at an environmental temperature of 4 °C. The accuracy of the biosensor could be a problem if the application is in remote regions with long intervals of investigation time. Most importantly, if this method were to be used in real-life applications for food security, the efficacy of the result may be hindered due to the reason that the experiment is very likely performed not under optimal experimental conditions.

### 2.3.2.3 Other Methods

As described in the above sections, the current commonly used signal amplification technology based on the nucleic acid chain reaction that has been applied to the detection of foodborne pathogens mainly includes polymerase chain reaction (PCR), rolling circle amplification reaction (RCA), LAMP, and HCR. However, there are still other methods developed by researchers for signal amplification, such as biotin-SA, click chemistry, cascade reaction, nanomaterials, self-circulating of molecular beacons, etc.

Biotin is a water-soluble vitamin which is known by people as either vitamin H or co-enzyme R, while Avidin is a glycoprotein composed of four subunits extracted from egg white.[130] The bacteria *Streptomyces* produces a protein called "Streptavidin" (SA) which resembles the biological characteristics of avidin protein. The combination of SA and biotin gives its ability to carry out signal amplification with the advantage of different protein or nucleic acid molecules that can bind to biotin or SA. The amplification mechanism behind this is that one SA can bind to four biotics while one protein or nucleic acid molecule can bind to multiple biotics; thus, a multi-stage amplification process is established.[130] The other advantages of the biotin–SA method include its very strong binding ability with 105 to 106 times antigen–antibody affinity and rapid formation speed. In addition, various times of proteins, enzymes, DNA, and antibodies can also bind to either biotic or SA. Different biomolecules can be modified with biotin or SA to achieve specific detection and corresponding signal amplification.[131] To give a research example, Guo et al. combined the methods of HCR, ELISA technology, and biotin-SA for the detection of *E. coli* O157:H7.[132] The bacteria were used to form a double antibody sandwich structure using two recognition antibodies. One side was modified on the surface of colloidal gold particles which were also labeled with hairpin probes. Each probe is connected with horseradish peroxidase polymer (HRP) and when HCR reaction is triggered, a large amount of HRP will be loaded on the double antibody sandwich which catalyzes the reaction of color development caused by tetramethylbenzidine (TMB). The absorbance test can then be performed to detect the amount of bacteria (Figure 2.13a).

On the other hand, a chemical synthesis reaction method called "Click chemistry" proposed by Kolb (2001) showed promising ability for signal amplification.[133] In click chemistry, different molecules can be formed rapidly through splicing of small units. This reaction can take place quickly in water at room temperature, being highly specific and without by-products. The small ligands produced in click chemistry showed no interference with the active-labeled molecules and thus can be coupled with biological molecules such as protein, antibodies, enzymes, and nanoparticles to achieve signal amplification. With these advantages, click chemistry can be used in different biosensors for different purposes and the current four main types of click chemistry include: (1) cycloaddition (1,3-dipolar cycloaddition and copper-free cycloaddition catalyzed by copper (I) of azides and alkynes), (2) nucleophilic ring-opening reaction, (3) carbonyl reaction of nonaldehydes, and (4) addition reaction of carbon–carbon multiple bonds.[134] To give some examples of using click chemistry in the fields of food safety and detection of pathogens, Mou et al. first used bacteria *E. coli* to capture and reduce exogenous $Cu^{2+}$ to $Cu^+$ through metabolic process.[135] The $Cu^+$ can trigger click chemical reaction between modified azides and alkyne functional molecules. This subsequently will lead to the aggregation of monodisperse colloidal gold particles causing a color change. The method was able to detect 40 CFU/mL of bacteria within 1 h. Another example was performed by Liong et al., who used an antibody of *S. aureus* modified with trans-cyclooctene and tetrazine-modified magnetic nanoprobes using the cycloaddition reaction method.[136] A large number of magnetic

# Nucleic Acid Detection of Pathogenic Microorganisms on Chip

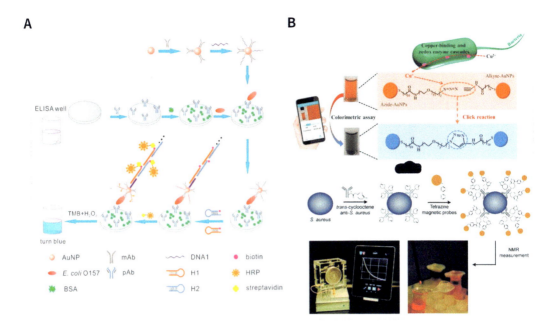

**FIGURE 2.13** (a) A schematic illustration for the application of biotin-SA-based signal amplification strategy for the detection of foodborne pathogens. Demonstration of *E. coli* O157:H7 detection. (b) The application of click chemistry based on signal amplification technology for detection of foodborne pathogens: the principal mechanism for detecting *E. coli*; detecting *S. aureus*.

nanoparticles were gathered on the bacterial surface. The use of click chemistry to amplify the magnetic signals using nuclear magnetic resonance could detect the bacteria as low as $2 \times 10^2$ CFU/mL and displayed a faster rate of assay time (Figure 2.13b).

The cascade reaction can be referred to as multiple reaction occurring simultaneously due to certain intermediate medium within the same reaction system. This single reaction will trigger a cascade of events which ultimately will enhance and amplify the signal that is generated.[137] Currently, the combination of the multi-enzyme catalytic reaction system and signal output conversion system is often used together for an overall signal amplification. This generally will combine with different types of biosensors to achieve detection and analysis in the fields of clinical medicine, food safety, and environmental monitoring.[138] A large number of cascade reaction were demonstrated by Zhang et al. where the multi-enzyme catalytic system was combined with a chemiluminescence biosensor for signal amplification and detection of *E. coli* O157:H7.[139] The mechanism behind this is the use of antibody and recognition of antibody to form a double antibody sandwich with glucose oxidase (GOx), target bacteria, and magnetic nanoparticles. The Gox complex will catalyze the reaction of converting glucose to produce $H_2O_2$. $H_2O_2$ was further hydrolyzed by lactase to induce luminol luminescence to detect target bacteria. The detection limit for this multi-enzyme catalytic system was $1.2 \times 10^3$ CFU/mL. Another experiment conducted by Gao et al. used the ELISA method to compare the efficiency of detecting *S. typhimurium* bacteria between the HRP-based catalytic method and urease cascade reaction-based biosensors.[140] In the presence of the target, HRP catalyzed the color reaction of TMB to detect the concentration of bacteria. The ammonia produced by urease could generate silver on the surface of gold nanorods in the presence of silver nitrate and glucose. The different amounts of silver could cause different colors of the gold nanoparticles, and the concentration of bacteria can be detected by the absorbance test. The cascade reaction method showed two- to threefold increase in sensitivity than the HRP-based catalytic method (Figure 2.14a).

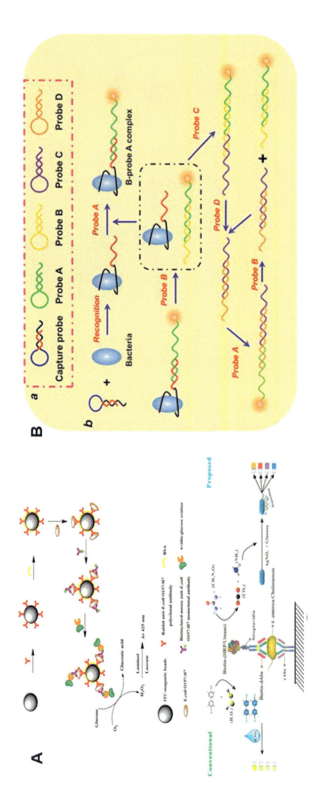

**FIGURE 2.14** (a) Schematic application of cascade reaction-based signal amplification method for detecting foodborne pathogens: the principal mechanism for detecting *E. coli*; detecting *S. typhimurium*. (b) A schematic illustration of the principal mechanism of detecting target bacteria through self-circulation of molecular beacon.

A novel, sensitive, and enzyme-free pathogenic *Staphylococcus aureus* detection method was developed by Dong et al. using self-circulation of molecular beacons. The hairpin structure containing protein A aptamer was used as an identifying agent, and an allosteric capturing probe which was previously designed to recognize the surface protein of target bacteria binds on the surface bacteria through the interaction between the target aptamer and protein A.[141,142] In the presence of bacteria, recognizing protein A by the aptamer will lead to allosterism of capture probe which exposes its initiator section triggering the multiple self-circulation-based signal amplification. For the working mechanism of self-circulation process, the initiator section in allosteric probe can first unfold probe A with two terminals labeled with fluorescence moiety and corresponding quenching moiety.[143] The unfold probe A then activates the assembly of probe A and probe B to form a complex which causes the release of the capture probe. The capture probe then again unfolds probe A to form a signal recycle. On the other hand, the ssDNA section of the probe A and probe B complex can trigger hybridization of probe C and probe D. The probe C and probe D's ssDNA section can catalyze the formation of the probe A and probe B duplex, causing the amplification of the fluorescence signal (Figure 2.14b). Overall, the whole process converted the signal of bacteria into nucleic acid signal, and a linear relationship was obtained between the amount of target bacteria and signals. The signal was then detected by a sensitive fluorescence sensing system with a detection limit of $10^3$ CFU/mL.[143]

Signal amplification technologies based on nanomaterials also showed very promising potential for the detection of pathogenic bacteria. Nanomaterials usually means materials with at least one dimension in the nanometer scale (0.1–100 nm) or in three-dimensional space or composed of these materials as basic units.[130] There are many advantages of using nanomaterials in biosensors for signal amplification, for example, their large specific surface area and surface energy aid in loading a large number of signal molecules which provide an amplification effect. In addition, nanomaterials have great conductivity and biocompatibility which will enhance the function of certain electrochemical biosensors and not affect the stability of biosensors. Like enzymes, nanomaterials also have the ability to catalyze signal conversion, but they do not require certain optimal experimental conditions and thus are more stable than biological enzymes. All these factors make nanomaterials very favorable for application in areas such as food safety. There are four main types of nanomaterials that were used for signal amplification. There are nanomaterials such as colloidal gold particles with a large surface area that can directly be used as signal markers in biosensors. There are nanomaterials that can uphold and release a large number of signal molecules such as $MnO_2$ nanosheets and mesoporous materials. There are nanomaterials that can directly enhance the strength of signal molecules such as nanoflowers, which are nanomaterials with a flower shape and can be formed by self-assembly of inorganic salt ions and organic ligands.[144] Last but not the least, nanomaterials such as metal–organic frameworks can mimic the action of catalytic enzymes and release many signal molecules.[130] Ye et al. demonstrated a great example using protein–protein nanoflowers with specifically designed *Concanavalin A* to recognize the target bacteria, sucrose invertase, and inorganic calcium ions for the detection of *E. coli* O157:H7 in milk.[145] To briefly explain the mechanism behind this, the antibody that captures the bacteria was immobilized on the ELISA plate, and then the nanoflowers were added to the plate to detect target bacteria. The prepared ConA-invertase-CaHPO$_4$ hybrid nanoflowers simultaneously loaded sufficient invertase and enhanced the activity of immobilized invertase. The concentration of bacteria was in a linear relationship with the concentration of glucose converted by invertase using sucrose, and the amount of bacteria can be quantitated using a blood glucose meter (Figure 2.15). The detection limit for this method can go as low as 101 CFU/mL and can be used as a reliable point-of-care detection method for food pathogens in the future.

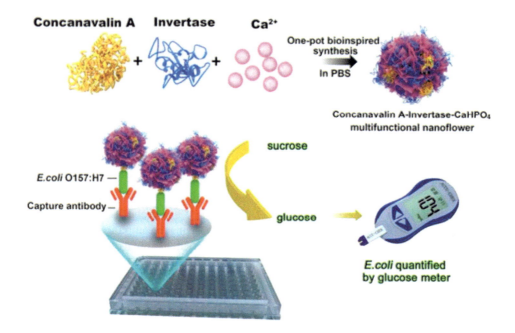

**FIGURE 2.15** Application of nanoflower-based signal amplification technology for detecting food pathogens: detection of *E. coli* O157:H7.

## 2.4 DETECTION OF PATHOGENIC MICROORGANISMS USING NUCLEIC ACID AMPLIFICATION METHOD COMBINED WITH CRISPR METHOD ON CHIP

### 2.4.1 ENZYME-BASED NUCLEIC ACID AMPLIFICATION WITH CRISPR METHOD

The CRISPR-Cas system is a prokaryotic adaptive immune system used to resist interference from foreign genetic components in living organisms. Class 1 CRISPR systems (including types I, III, and IV) contain multiprotein effector complexes, while Class 2 CRISPR systems (including types II, V, and VI) have single effector proteins. CRISPR-Cas12a is another type 2 RNA-guided endonuclease used for gene editing. Cas12a nuclease recognizes DNA target sequences on the opposite chain of 5'-TTTV PAM that complement the crRNA spacer region, without the use of tracrRNA. After recognition, overlapping DNA double-stranded breaks were generated at the RuvC and Nuc domains. Compared to Cas9-producing flat ends, Cas12a-producing staggered ends may be more advantageous for applications such as integrating DNA sequences in precise directions. The CRISPR-Cas12a gene editing system can serve as a great supplement to the Cas9 system and even has advantages over the Cas9 system in certain editing fields. In the detection of foodborne pathogens, the side chain cleavage activity of Cas12a protein is often used in combination with fluorescence detection to detect foodborne pathogens. However, due to the low signal amplification level of CRISPR technology itself, it is usually necessary to combine it with other nucleic acid amplification methods to achieve better real-time detection results.

#### 2.4.1.1 RPA with CRISPR

RPA is an isothermal nucleic acid amplification technique that has emerged in recent years. The design requirements for RPA primer probes are simple and capable of amplification at temperatures ranging from 37 °C to 42 °C. Recombinant enzymes are used to bind to the primers and search for

Nucleic Acid Detection of Pathogenic Microorganisms on Chip

homologous sequences complementary to the primers, unlocking local double-stranded DNA without the need for high-temperature denaturation of the template. A large number of amplification products can be obtained in 20–30 min. The reaction time depends on the number of copies of the starting template and the size of the amplicon. RPA generally tolerates 3–5 mismatched bases, which does not affect the RPA reaction performance. In addition, this technology has the potential to reduce the impact of matrix-related inhibitors compared to other DNA amplification methods. Compared with traditional methods, this method has the advantages of low reaction temperature, short reaction time, no expensive equipment, etc. The reagent is in the form of freeze-dried powder, does not need cold chain storage, and is simpler to operate. The results can be analyzed by gel electrophoresis or real-time monitoring of fluorescence signals. The combination of RPA and CRISPR has many advantages, such as being able to distinguish background signals and experimental values well, as the products produced by the RPA reaction differ greatly from the raw materials. The limitations of complex operation and harsh storage conditions for polymerase are also worth considering. A common binding method is to use isothermal amplification of nucleic acids and the trans cleavage activity of the CRISPR/Cas12a system, using a single-stranded DNA fluorescence quenching agent (ssDNA FQ) reporter and naked eye detection of side flow analysis (LFA) to generate fluorescence signals. Liu's group used RPA-CRISPR technology to detect *Staphylococcus aureus*, which can shorten the detection time to 35 min and increase the high-throughput detection threshold of pathogen DNA larger than five copies. In recent years, PCR chips have emerged, which is a real-time quantitative PCR-based, high-throughput, reliable, and sensitive multiple real-time quantitative PCR technology. It can prepare primers for a series of related genes on a 96-well plate.[146] Through an optimized real-time quantitative PCR system and quality control system, researchers can conduct precise quantitative research on dozens to hundreds of genes simultaneously. Therefore, PCR chips are a powerful and attractive nucleic acid detection tool, but due to the protein adsorption and two-step detection mode of RPA-CRISPR/Cas12, there are still significant challenges in the application of RPA-RISPR/Cas12 systems in self-priming chips. Xia's group developed a nonadsorption self-priming digital chip and established a direct digital double crRNA (3D) detection method for ultrasensitive detection of pathogens on chip. This 3D assay combines the advantages of rapid amplification of RPA, specific cleavage of Cas12a, and accurate quantification of digital PCR,[147] enabling accurate and reliable digital absolute quantification of *Salmonella* in POCT. In the era of evolution and natural selection, there are more and more foodborne pathogens that plague humanity, and brucellosis is one of them. Brucellosis is a long-term disease caused by Brucella that harms humans and livestock. Brucella is an important and dangerous pathogen. Therefore, rapid and accurate detection of Brucella is crucial for reducing infection rates. Since the discovery of the trans cleavage ability of Cas effector proteins, the CRISPR/Cas system has shown great potential in developing next-generation biosensors for biomolecular detection. Xu's group combined the trans cleavage activity of the CRISPR/Cas12a system with recombinant polymerase amplification (RPA) and developed a universal sensing platform for *brucella* detection using two analytical methods.[148] The obtained fluorescent biosensors (F-CRISPR) and electrochemical biosensors (E-CRISPR) based on RPA-CRISPR/Cas12a can detect up to two positive reference plasmids per reaction, which can be applied to milk (food) samples. In the development process of RPA-CRISPR detection research, it provides diversity and reference value for detection methods, but there are still some obstacles in commercial use, such as pollution risk and insufficient sensitivity. Hu's group proposed a robust solution involving photochemical control of CRISPR RNA (crRNA) activation in CRISPR detection. Based on this strategy, RPA and CRISPR-Cas12a detection systems can be integrated into completely enclosed test tubes. CrRNA can be designed to temporarily inactivate, making RPA unaffected by Cas12a cleavage. After the RPA reaction is completed,[149] the CRISPR-Cas12a detection system is activated under rapid light irradiation. This light-controlled, fully enclosed CRISPR diagnostic system avoids the risk of pollution and improves sensitivity by more than two orders of magnitude compared to the traditional one-pot method.

### 2.4.1.2 LAMP with CRISPR

LAMP uses Bst DNA polymerase with chain displacement characteristics to efficiently, quickly, and specifically amplify target sequences under isothermal conditions, achieving $10^9$–$10^{10}$-fold amplification within 15–60 min. Isothermal amplification has been applied in the detection of pathogens, parasites, viruses, diseases, and genetically modified products due to its advantages of fast response, high specificity, simple equipment, and easy identification of results.[150] It has been widely used in clinical infectious disease diagnosis, environmental monitoring, food safety, and other fields, and has become an ideal method suitable for molecular POCT platforms. LAMP detection methods can be divided into various methods, including turbidimetry, pH colorimetry, fluorescent dye method (such as NHB, calcein, SYBR Green, and Syto), and fluorescent probe method. Like the RPA mentioned above, the LAMP nucleic acid amplification reaction also involves the involvement of polymerase, utilizing the trans cleavage activity of Cas12a protein, usually paired with fluorescence detection methods.[93] Xia's group developed an unlabeled and highly sensitive method for *Salmonella enteritidis* (*S. enterica*) detection based on the G-quadruplex probe CRISPR-Cas12 system (referred to as G-CRISPR-Cas). Due to the amplification process induced by LAMP, the G-CRISPR-Cas detection method can detect *Salmonella enteritidis* as low as 20 CFU/mL. The dual recognition process of RNA binding guided by LAMP primers and Cas12a ensured the specificity of pathogenic gene detection. More researchers often combine the isothermal characteristics and high sensitivity of LAMP with CRISPR/Cas systems to detect different biological targets.[151] For *Vibrio parahaemolyticus*, it often comes from seafoods such as fish, shrimp, crabs, shellfish, and seaweed. In clinical practice, acute onset, abdominal pain, vomiting, diarrhea, and watery stools are the main symptoms and are a common foodborne pathogen. Hu's group collaborated with LAMP response and CRISPR/Cas12a to detect *Vibrio parahaemolyticus* with good specificity and high sensitivity. The detection limits of pure culture medium and DNA reached 2.5 CFU/mL and 5 fg/μL, respectively. After enrichment and cultivation for 2 h, it is possible to detect artificially contaminated samples with an initial inoculum of 5 CFU/mL of *Vibrio parahaemolyticus*.[152] After LAMP amplification combined with CRISPR, the entire reaction time was less than 30 min, and the secondary amplification increased the detection sensitivity. LAMP and CRISPR have high detection efficiency, but it is worth studying which substances can enhance their cleavage activity. Regarding this, Li's group optimized several key parameters of reaction chemistry and developed a chemically enhanced CRISPR detection system (called CECRID) for nucleic acids detection. For the signal detection phase based on Cas12a/Cas13a, we determined the buffer conditions and substrate range for optimal detection performance and revealed the key role of bovine serum albumin in enhancing the trans cleavage activity of Cas12a/Cas13a effectors.[153] By comparing several chemical additives, we found that adding L-proline can ensure or improve the detection ability of Cas12a/Cas13a. This has also been validated in clinical pathogenic microorganisms. Ma et al. developed a universal biosensing platform for ultrasensitive detection of pathogenic bacteria called SCENT Cas (nucleic acid testing using silver nanoclusters powered by CRISPR/Cas12a). Simply put, the species-specific invA gene of *Salmonella typhimurium* (*S. typhi*) was amplified using LAMP isothermal amplification, followed by triggering the trans cleavage of CRISPR/Cas12a. Transcleavage nonspecifically degrades any single-stranded DNA (ssDNA).

### 2.4.2 ENZYME-FREE NUCLEIC ACID AMPLIFICATION WITH CRISPR METHOD

### 2.4.2.1 CHA with CRISPR

Unlike the HCR combined with CRISPR method, recent research on studying catalytic hairpin assembly with CRISPR method mainly focused on the area of detecting and analyzing small biomarkers such as microRNA. Traditional methods for miRNA detection such as oligonucleotide microarrays, northern blotting, and quantitative RT-PCR have some drawbacks such as they are time-consuming and expensive due to the use of precise equipment.[154] The sensitive, rapid, and cost-effective properties from CHA clustered with the CRISPR method ignited great interest to expand its application to various fields, especially disease prevention. There were studies done on analyzing the presence of certain viruses such as new coronavirus (SARS-CoV-2); however, studies related to utilizing CHA combined with CRISPR for the detection of pathogenic microorganisms were rather limited.

MicroRNAs (miRNAs) are a kind of noncoding small RNA molecules that can be treated as sensitive biomarkers for different types of diseases due to their association with many biological processes.[155] An abnormal expression of miRNAs can be related to occurrence of cancer disease and thus, developing a rapid and specific method for miRNA detection has great potential for applications in early clinical diagnosis and therapy treatment.[156] In the work demonstrated by Cui et al., CRISPR-Cas13a system was combined with CHA for the detection of miRNA-21 using an electrochemical assay platform. The hairpin DNA1 (H1) probes were assembled on the surface of gold electrode through bonding between gold and sulfur material. This bonding together with 6-mecrapto-1-hexanol (MCH) will block the formation of signals. In the presence of target molecules, the molecule hybridizes with the space region of the Cas13a/crRNA complex to trigger CRISPR-Cas13a trans-cleavage action leading to the cleavage of DNA 0 (H) to release secondary target DNA fragments (ST). The ST stand is the initiator of CHA and would subsequently enhance the amplification process by hybridizing with H1 and DNA 2 (H2) which was previously labeled with methylene blue. This would then release more ST stands and the cycle repeats causing a cascade of signal amplification. There was a good linear relationship between the concentration of miRNA-21 and peak current of methyl blue. With optimal experimental conditions, the detection limit by this electrochemical biosensor can be quantified as low as 2.6 fM (Figure 2.16a).[157] This method also showed an excellent versatility as it can be used to detect other miRNA by just changing the spacer sequence of crRNA.

Another example integrating CRISP-Cas12a with CHA for the detection of microRNA was conducted by Peng et al. The use of CRISP-Cas12a was due to its high efficiency in cleavage action, with approximately 1250 turnovers per second.[158] The CHA circuit is designed to convert and amplify each target molecule into multiple programmable DNA duplexes. The duplexes' protospacer adjacent motif (PAM) and protospacer sequence can be recognized by Cas12a/gRNA complex and subsequently initiate the trans-cleavage action of CRISPR-Cas12 on fluorescent ssDNA, and thus generating fluorescence signals (Figure 2.16b). This method had two important breakthroughs. The first one was that the collateral cleavage-mediated signal amplification of CRISPR-12a was enhanced due to the introduction of unpaired sites into the DNA duplex output. The second one was that it can achieve sensitive detection for different miRNA biomarkers through modification of variable regions in CHA modules.[159] This was proven by other studies, with the use of CRISPR-Cas12 with CHA, isothermal detection was achieved with different miRNAs such as miR-21, miR-141, and miR-144, and the results were significantly better than using Cas13a-based methods.[160,161]

A new perspective based on integrating RNA-based CHA circuit with CRISPR/Cas12a method was presented by Chen et al. for one-step detection of microRNAs. First of all, a spacer-blocking crRNA was designed to bridge between the CHA circuit and CRISPR-Cas12. Without target molecules, the crRNA sequence cannot interact with Cas12a. In the presence of target miRNA, it would specifically trigger RNA-based CHA and cause a configurational change of the blocked crRNAs into precursor crRNAs, which subsequently can be converted into mature crRNAs by the action of Cas12a for DNA targeting. The Cas12a–crRNA complexes will activate their DNase activity to trans-cleave single-stranded oligonucleotide-containing dye and generate fluorescence signal output (Figure 2.16c).[162] The fluorescence signal exhibited a linear relationship with the logarithm of target molecule concentration in the range from 100 to 105 fM. The detection limit for this method was 81.96 fM and it proved to be rapid and cost-effective due to its one-pot reaction system property.

Other than detecting miRNA, a new assay by Yang et al. based on CRISPR/Cas13a and CHA was developed to detect the coronavirus (SARS-CoV-2) that caused the COVID-19 pandemic. This type

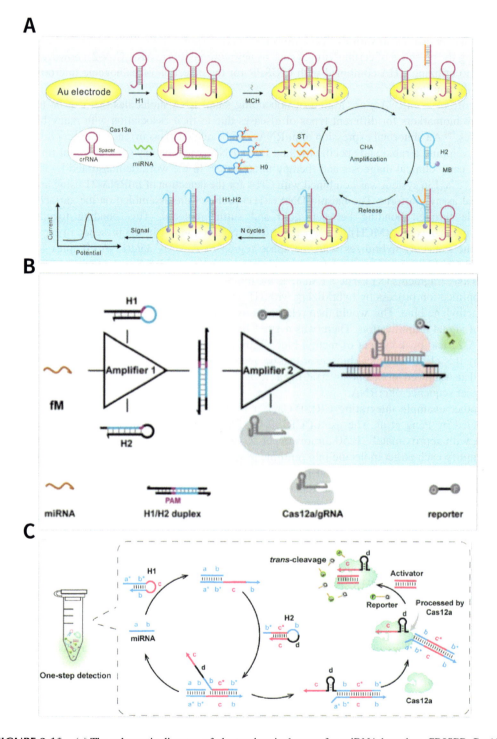

**FIGURE 2.16** (a) The schematic diagram of electrochemical assay for miRNA based on CRISPR-Cas13a and CHA amplification method. (b) Demonstration of the working principle for amplified detection of miRNA by CRISPR-CHA method. (c) A schematic illustration of the proposed circuit for one-step detection of miRNA.

# Nucleic Acid Detection of Pathogenic Microorganisms on Chip 41

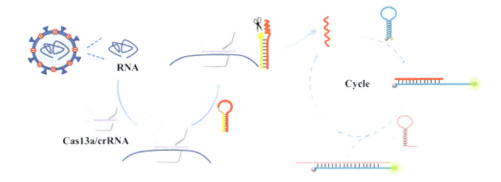

**FIGURE 2.17** A schematic illustration of detecting SARS-CoV-2 RNA based on Cas-CHA assay.

of virus can cause severe respiratory disease and is highly contagious. A rapid and ultrasensitive detection for the coronavirus is needed to prevent the spread of infection and reduce the huge health burden that impacts the society[163] Currently, the gold standard for detecting the virus is quantitative real-time PCR with RT-qPCR.[164] However, this method requires expensive equipment and well-trained experts which can be very time-consuming.

The method developed by Yang et al. proved to be rapid and simple enough to be used for on-site analysis. In the presence of target gene, the spacer region of the Cas13a/crRNA duplex can bind to the target and activate the collateral cleavage action of CRISPR/Cas13a. This will cause the cleavage of uracil ribonucleotide (rU) in the loop of hairpin reporter (reporter is composed of five rU) which releases the primer and triggers the CHA amplification reaction and generation of fluorescent signals (Figure 2.17). The method can specifically distinguish SARS-CoV-2 RNA from other common human coronavirus and can be flexibly modified to detect other viruses. The method is very advantageous with minimum processing time of 35 min, without the use of special equipment, high recovery rate, and achieve a relatively low detection limit of 84 aM with pretreated saliva samples.

Although the integration of CRISPR/Cas and CHA amplification delivered favorable results in different fields of research, there are still potential factors that can hinder its efficacy. The storing environment of biosensing electrodes is very important. Cui et al. developed the CRISPR/Cas combined with CHA for miRNA-21 detection with the LOD of 10 pM. The biosensing electrodes can be stored at 4 °C for 2 weeks. However, a temperature fluctuation in the storage room may occur in real-life application and this can greatly affect the performance of biosensing electrodes and the result. The selection of Cas protein is also very important and that depends on the specific needs of the researcher. Studies suggested that different subtypes of Cas protein have various collateral cleavage abilities; thus, researchers must take the time to investigate the performance of diverse Cas protein toward its own circuit design, and this can be very tedious and delicate.[165] The possibility for false positive result by the CRISPR/Cas and CHA method is not negligible. For instance, for miRNA detection, there are highly homologous miRNA families with only one or few bases difference. Thus, a fully accurate detection in molecular diagnosis remains a problem not to be overlooked in the future. As for the design of the hairpin reporter for the target analyte, Yang et al. demonstrated that the fluorescence signal can decrease with the extension of complementary base number within the reporter. This phenomenon may be explained by the instability of the stem-loop structure which may activate the CHA amplification reaction independently. However, a too stable stem may also lead to the problem of not releasing the primer which may also decrease the fluorescence signal. Despite the potential concerns mentioned above, the combination of CRISPR/Cas and CHA still shows great promise in molecular diagnosis of various fields due to its rapid, sensitive, and cost-effective detection properties.

## 2.4.2.2 HCR with CRISPR

The novel technology of clustered regularly interspaced short palindromic repeats (CRISPR)/CRISPR-associated (Cas) system has brought many new opportunities for a rapid and high sensitivity detection method in the fields of human health and food safety. The CRISPR-Cas system is a complex immune system within prokaryotes and exhibits nonspecific collateral cleavage ability when activated by target nucleic acid under guide-RNA (g-RNA).[158,166,167] Due to its ultrahigh sensitive trans-cleavage ability on genetic material, CRISPR-based diagnostic methods can be regarded as the "next-generation molecular diagnostic technology."[168,169] However, the CRISPR methods are restricted by nucleic acid application and thus the combination of CRISPR detection and enzyme-free nucleic acid amplification can provide unlimited possibilities. In recent years, studies related to utilizing CRISPR in combination with HCR methods to detect pathogenic bacteria had been well-defined by Song et al., Cai et al., and Liu et al.[170,171]

A CRISPR/Cas12a or CHANCE (Cas 12a-HCR evANescent wave fluorescenCE) system for rapid and accurate detection of *E. coli* O157:H7 without target nucleic acid amplification was developed by Song et al.[170] The system was designed to cope with a portable device enabling on-site quantitative detection. In this research, an evanescent wave optical fiber fluorescence sensor was used due to its small size and low cost. The evanescent wave with limited penetration depth was utilized to activate fluorescence. This method improved the sensitivity, providing a strong anti-interference capability and minimizing the requirements of equipment to achieve POCT. The crRNA sequence used in this method was the sequence of *E. coli* O157:H7 rfbE gene from the National Center for Biotechnology Information. The detection mechanisms can be simply explained as when the target bacteria are not present, the Cas12a cleavage protein will not be activated and thus the ssDNA initiator (H0) will remain as an intact structure causing HCR reaction (formation of H1/H2 duplex) to produce fluorescent signal. The hairpin structures were previously labeled with fluorophore (Cy3) and quencher (BHQ2) on the complementary base to reduce interference from background fluorescence signal. Real water samples spiked with different genomic DNA concentrations of bacterial cells were used. In the presence of the target, the Cas12 will cleave the HCR initiation chain leading to a decrease of the fluoresce signal. The final fluorescence signal is inversely proportional to the concentration of the target, and this can be quantitatively measured (Figure 2.18a).[171] In combination with a nucleic acid extraction method, the whole analysis time was within 50 min and its detection limit was 17.4 CFU/mL with a recovery rate between 69.3% and 106%.

Another advanced pathogenic bacteria detection method was demonstrated by Cai et al., synergistically utilizing CRISPR-Cas12a and tetrahedral DNA nanostructure-mediated hyperbranched HCR (TDN-hHCR). A TDN-hHCR was used because it could increase the collision probability and enhance the local concentration of hairpins; in addition, TDN-hHCR has been verified to have about 70-fold faster reaction rate than traditional HCR.[172–174] This method targets a whole *Salmonella* cell by using IMS to separate and recognize bacteria with antibody-modified gold nanoparticles (AUNPs). Subsequently, the labeled bacteria cell can be magnified and converted into bio-barcode DNA which can trigger the trans-cleavage actions of CRISPR-Cas12a. This will lead to the inhibition of TDH-hHCR as Cas12a cleaves its initiator stand. No Forster resonance energy transfer will occur between the fluorophores of hairpins and thus causing a decrease in the fluorescence readout (Figure 2.18b).[175] The detection limit for this method was as low as 8 CFU/mL. For the bacteria spiked in milk and egg white, the linear range of the method was from 17 to 25 CFU/mL. To further verify the selectivity of this detection system, different non-target bacteria such as *Listeria monocytogenes, E. coli* O157:H7, *Vibrio parahaemolyticus, Staphylococcus aureus*, and a bacterial mixture (*Listeria monocytogenes, E. coli* O157:H7, *Vibrio parahaemolyticus, Staphylococcus aureus* with ratio of 1:1:1:1:1) were employed, and the results showed a low fluorescence readout on *Salmonella* while all the other bacteria were detected with high values which were almost equal to the value of the negative control (Figure 2.19a).

# Nucleic Acid Detection of Pathogenic Microorganisms on Chip

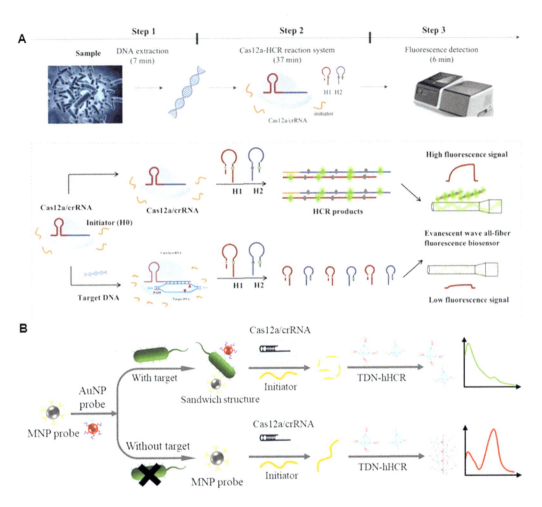

**FIGURE 2.18** (a) A schematic illustration of the proposed CHANCE system for *E. coli* O157:H7 detection. (b) A schematic illustration of the Cas12a-alone detection system for *Salmonella*.

An interesting study using the CRISPR system (Cpf1) coming from a bacterium *Lachnospiraceae* was coupled with HCR methods to detect *Salmonella* based on an electrochemical biosensor. The electrochemical biosensors have been well recognized for their rapid signal readouts, simple operation, cost-effectiveness, and high sensitivity for detecting bacteria and thus have great potential to be employed in the POCT system.[176] To set up the detection process, the nonspecific ssDNA reporter tag was drafted with a methylene blue electrochemical tag for signal transduction, and a thiol moiety to tether was on the sensor to receive electrical signals.[110] The electron will be transferred between the Au electrode and the redox-active species on ssDNA.[160] The polymer double-stranded DNA of the HCR was immobilized on dynabeads (DBs) through the linker probe which was locked through the *Salmonella* aptamer. In the presence of target bacteria, the polymer double-stranded DNAs of the HCR will stay connected through the linkers from DBs, thereby activating the cross-opening of the functional DNA hairpin structures of the HCR to generate a DNA double helix with multiple ssDNA. These ssDNA will initiate the formation of a Cas12a–crRNA–target DNA ternary complex which in turn initiates the CRISPR-Cas12a activity to remove the methylene blue reporter gene from the surface of the gold electrode. This will cause a variation in the electron transfer of electrochemical tag and reduce the transduction signal of the methylene blue probe on the electrode (Figure 2.19b).

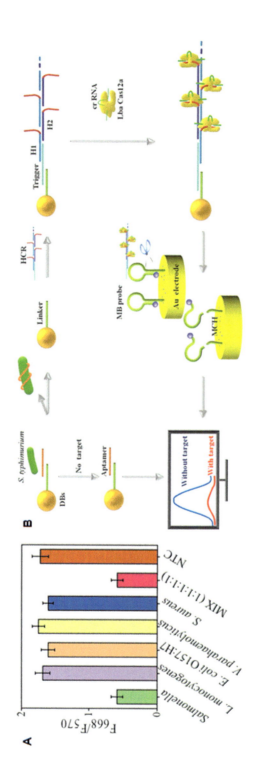

**FIGURE 2.19** (a) Demonstration of the selectivity of the Cas12a-hHCR detection system. The concentrations of the bacteria *Salmonella* and other non-target pathogens were set at $1 \times 10^5$ CFU/mL with NTC as the negative control. (b) The schematic illustration of the electrochemical biosensor for detecting *S. typhimurium*.

This method further verified its feasibility using DPV signals and the results showed that in the presence of the HCR product and target bacteria, there will be a decrease in the current signal. When elements such as crRNA, target bacteria, or Cas12a were missing, there was no significant change in the DPV current signal. Actual milk samples spiked with diluted densities of bacteria were tested for recovery measurements. The results showed that the detection limit was down to 20 CFU/mL.

There are still areas of improvement for the combination of CRISPR with the HCR method. Based on the three specific studies mentioned above, the experimental conditions (incubation time, concentration of substrate, and others) must be carefully controlled to have a functional component such as an aptamer and the Cas12a protein. A temperature deviation may cause denaturation or change in the structure for these protein molecules, thus hindering the specificity for the target sequence and even may cause extensive degradation of hairpins. All these different factors may be difficult to control in actual environmental on-site detection for bacteria or pathogens. In addition, when using Cas12a with real samples, the results can be affected due to the complex matrix system containing proteins, fats, and carbohydrates. Overall, the combination of CRISPR and HCR has great potential to be extended to different application scenarios in the future, including but not restricted to human health diagnosis, food safety assurance, and environmental pollution monitoring.

## 2.5 CONCLUSION AND PERSPECTIVES

Foodborne diseases caused by pathogens pose a significant threat to human health. Early, rapid, and accurate detection of foodborne pathogens is important for food safety. Compared with traditional analytical methods, microfluidic technology has inherent advantages such as integration, automation, low reagent consumption, improved sensitivity, and efficiency, and has become a research hotspot in food analysis. This chapter introduces the preparation of foodborne pathogen samples, nucleic acid amplification methods for detecting foodborne pathogens, and the combination of CRISPR technology. All programs such as sample preparation, signal amplification strategies, and detection techniques are considered to be integrated into a microfluidic system to enhance the analytical performance of pathogen detection. However, the innovation of enzyme amplification in detecting foodborne pathogens, the sensitivity of detection, and whether it meets on-site detection require more research. Enzymatic amplification methods do not involve any addition of bases, only permutation, and combination under existing base conditions. Therefore, the combination of enzyme-free amplification and CRISPR technology still brings high background signals that are worth considering. At present, foodborne pathogens are not all single forms of infection, and many clinical symptoms of diseases show joint infection of multiple pathogens. Based on actual needs, developing low-cost, user-friendly, portable, highly sensitive, and accurate multi-channel detection methods will be one of the future research hotspots.

## REFERENCES

(1) Guldimann, C.; Johler, S. An introduction to current trends in foodborne pathogens and diseases. *Current Clinical Microbiology Reports* 2018, 5, 83–87.

(2) Law, J. W. F.; Ab Mutalib, N. S.; Chan, K. G.; Lee, L. H. Rapid methods for the detection of foodborne bacterial pathogens: Principles, applications, advantages and limitations. *Frontiers in Microbiology* 2015, 5, 1–19.

(3) Bolton, D. J.; Robertson, L. J. Mental health disorders associated with foodborne pathogens. *Journal of Food Protection* 2016, 79, 2005–2017.

(4) Rotariu, L.; Lagarde, F.; Jaffrezic-Renault, N.; Bala, C. Electrochemical biosensors for fast detection of food contaminants – Trends and perspective. *TrAC Trends in Analytical Chemistry* 2016, 79, 80–87.

(5) Vaisocherová-Lísalová, H.; Víšová, I.; Ermini, M. L.; Špringer, T.; Song, X. C.; Mrázek, J.; Lamačová, J.; Scott Lynn, N.; Šedivák, P.; Homola, J. Low-fouling surface plasmon resonance biosensor for multi-step detection of foodborne bacterial pathogens in complex food samples. *Biosensors and Bioelectronics* 2016, 80, 84–90.

(6) Wen, C. Y.; Hu, J.; Zhang, Z.; Tian, Z. Q.; Ou, G.; Liao, Y.; Li, Y.; Xie, M.; Sun, Z.; Pang, D. W. One-step sensitive detection of Salmonella typhimurium by coupling magnetic capture and fluorescence identification with functional nanospheres. *Analytical Chemistry* 2013, 85, 1223–1230.

(7) Rohde, A.; Hammerl, J. A.; Boone, I.; Jansen, W.; Fohler, S.; Klein, G.; Dieckmann, R.; Al Dahouk, S. Overview of validated alternative methods for the detection of foodborne bacterial pathogens. *Trends in Food Science & Technology* 2017, 62, 113–118.

(8) Roy, S.; Mohd-Naim, N. F.; Safavieh, M.; Ahmed, M. U. Colorimetric nucleic acid detection on paper microchip using loop mediated isothermal amplification and crystal violet dye. *ACS Sensors* 2017, 2, 1713–1720.

(9) Xing, G. W.; Zhang, W. F.; Li, N.; Pu, Q. S.; Lin, J. M. Recent progress on microfluidic biosensors for rapid detection of pathogenic bacteria. *Chinese Chemical Letters* 2022, 33, 1743–1751.

(10) Chakraborty, J.; Chaudhary, A. A.; Khan, S.-U.-D.; Rudayni, H. A.; Rahaman, S. M.; Sarkar, H. CRISPR/Cas-based biosensor as a new age detection method for pathogenic bacteria. *ACS Omega* 2022, 7, 39562–39573.

(11) Wang, Y.; Yang, T.; Liu, G.; Xie, L.; Guo, J.; Xiong, W. Application of CRISPR/Cas12a in the rapid detection of pathogens. *Clinica Chimica Acta* 2023, 548, 117520–117525.

(12) Abbas, N.; Song, S.; Chang, M.-S.; Chun, M.-S. Point-of-care diagnostic devices for detection of *Escherichia coli* O157:H7 using microfluidic systems: A focused review. *Biosensors* 2023, 13, 741–760.

(13) Bahavarnia, F.; Hasanzadeh, M.; Sadighbayan, D.; Seidi, F. Recent progress and challenges on the microfluidic assay of pathogenic bacteria using biosensor technology. *Biomimetics* 2022, 7, 175–196.

(14) Zhang, Y.; Hu, X.; Wang, Q.; Zhang, Y. Recent advances in microchip-based methods for the detection of pathogenic bacteria. *Chinese Chemical Letters* 2022, 33, 2817–2831.

(15) Zeid, A. M.; Abdussalam, A.; Hanif, S.; Anjum, S.; Lou, B.; Xu, G. Recent advances in microchip electrophoresis for analysis of pathogenic bacteria and viruses. *Electrophoresis* 2023, 44, 15–34.

(16) Tian, Y.; Zhang, T.; Guo, J.; Lu, H.; Yao, Y.; Chen, X.; Zhang, X.; Sui, G.; Guan, M. A LAMP-based microfluidic module for rapid detection of pathogen in cryptococcal meningitis. *Talanta* 2022, 236, 122827–122835.

(17) Jo, Y.; Park, J.; Park, J. K. Colorimetric detection of *Escherichia coli* O157:H7 with signal enhancement using size-based filtration on a finger-powered microfluidic device. *Sensors* 2020, 20, 2267–2276.

(18) Fang, Y. L.; Wang, C. H.; Chen, Y. S.; Chien, C. C.; Kuo, F. C.; You, H. L.; Lee, M. S.; Lee, G. B. An integrated microfluidic system for early detection of sepsis-inducing bacteria. *Lab on a Chip* 2021, 21, 113–121.

(19) Morani, M.; Mai, T. D.; Krupova, Z.; van Niel, G.; Defrenaix, P.; Taverna, M. Recent electrokinetic strategies for isolation, enrichment and separation of extracellular vesicles. *TrAC Trends in Analytical Chemistry* 2021, 135, 116179–116191.

(20) Salimian Rizi, F.; Talebi, S.; Manshadi, M. K. D.; Mohammadi, M. Separation of bacteria smaller than 4 μm from other blood components using insulator-based dielectrophoresis: Numerical simulation approach. *Biomechanics and Modeling in Mechanobiology* 2023, 22, 825–836.

(21) di Toma, A.; Brunetti, G.; Chiriacò, M. S.; Ferrara, F.; Ciminelli, C. A novel hybrid platform for live/dead bacteria accurate sorting by on-chip dep device. *International Journal of Molecular Sciences* 2023, 24, 7077–7089.

(22) Chen, K. H.; Lee, S. H.; Kok, L. C.; Ishdorj, T. O.; Chang, H. Y.; Tseng, F. G. A 3D-ACEK/SERS system for highly efficient and selectable electrokinetic bacteria concentration/detection/antibiotic-susceptibility-test on whole blood. *Biosensors and Bioelectronics* 2022, 197, 113740–113749.

(23) Gadish, N.; Voldman, J. High-throughput positive-dielectrophoretic bioparticle microconcentrator. *Analytical Chemistry* 2006, 78 22, 7870–7876.

(24) Yoon, T.; Moon, H. S.; Song, J. W.; Hyun, K. A.; Jung, H. I. Automatically controlled microfluidic system for continuous separation of rare bacteria from blood. *Cytometry, Part A* 2019, 95, 1135–1144.

(25) Zhu, J.; Xuan, X. Curvature-induced dielectrophoresis for continuous separation of particles by charge in spiral microchannels. *Biomicrofluidics* 2011, 5, 024111–024123.

(26) Gallo-Villanueva, R.; Perez-Gonzalez, V.; Cardenas-Benitez, B.; Jind, B.; Martinez-Chapa, S. O.; Lapizco-Encinas, B. Joule heating effects in optimized insulator-based dielectrophoretic devices: An interplay between post geometry and temperature rise. *Electrophoresis* 2019, 40, 1408–1416.

(27) Mantri, D.; Wymenga, L.; Turnhout, J.; Zeijl, H. W.; Zhang, G. Manipulation, sampling and inactivation of the SARS-CoV-2 virus using nonuniform electric fields on micro-fabricated platforms: A review. *Micromachines* 2023, 14, 345–365.

(28) Wu, M.; Ozcelik, A.; Rufo, J.; Wang, Z.; Fang, R.; Jun Huang, T. Acoustofluidic separation of cells and particles. *Microsystems & Nanoengineering* 2019, 5, 32–49.

(29) Nilghaz, A.; Lee, S. M.; Su, H.; Yuan, D.; Tian, J.; Guijt, R. M.; Wang, X. Development of a pumpless acoustofluidic device for rapid food pathogen detection. *Analytica Chimica Acta* 2023, 1275, 341581–341587.

(30) Ning, S.; Liu, S.; Xiao, Y.; Zhang, G.; Cui, W.; Reed, M. A microfluidic chip with a serpentine channel enabling high-throughput cell separation using surface acoustic waves. *Lab on a Chip* 2021, 21, 4608–4617.

(31) Van Assche, D.; Reithuber, E.; Qiu, W.; Laurell, T.; Henriques-Normark, B.; Mellroth, P.; Ohlsson, P.; Augustsson, P. Gradient acoustic focusing of sub-micron particles for separation of bacteria from blood lysate. *Scientific Reports* 2020, 10, 3670–3682.

(32) Hussain, M.; Liu, X.; Tang, S.; Zou, J.; Wang, Z.; Ali, Z.; He, N.; Tang, Y. Rapid detection of pseudomonas aeruginosa based on lab-on-a-chip platform using immunomagnetic separation, light scattering, and machine learning. *Analytica Chimica Acta* 2022, 1189, 339223–339235.

(33) Freitas, T. A.; Proença, C. A.; Baldo, T. A.; Materón, E. M.; Wong, A.; Magnani, R. F.; Faria, R. C. Ultrasensitive immunoassay for detection of Citrus tristeza virus in citrus sample using disposable microfluidic electrochemical device. *Talanta* 2019, 205, 120110–120119.

(34) Modh, H.; Scheper, T.; Walter, J. Aptamer-modified magnetic beads in biosensing. *Sensors* 2018, 18, 1041–1062.

(35) Su, C. H.; Tsai, M. H.; Lin, C. Y.; Ma, Y. D.; Wang, C. H.; Chung, Y. D.; Lee, G. B. Dual aptamer assay for detection of Acinetobacter baumannii on an electromagnetically-driven microfluidic platform. *Biosensors and Bioelectronics* 2020, 159, 112148–112154.

(36) Wang, J.; Jiang, H.; Pan, L.; Gu, X.; Xiao, C.; Liu, P.; Tang, Y.; Fang, J.; Li, X.; Lu, C. Rapid on-site nucleic acid testing: On-chip sample preparation, amplification, and detection, and their integration into all-in-one systems. *Frontiers in Bioengineering and Biotechnology* 2023, 11, 1020430–1020473.

(37) Han, X.; Liu, Y.; Yin, J.; Yue, M.; Mu, Y. Microfluidic devices for multiplexed detection of foodborne pathogens. *Food Research International* 2021, 143, 110246–110259.

(38) So, H.; Lee, K.; Seo, Y. H.; Murthy, N.; Pisano, A. P. Hierarchical silicon nanospikes membrane for rapid and high-throughput mechanical cell lysis. *ACS Applied Materials & Interfaces* 2014, 6, 6993–6997.

(39) Kaba, A. M.; Jeon, H.; Park, A.; Yi, K.; Baek, S.; Park, A.; Kim, D. Cavitation-microstreaming-based lysis and DNA extraction using a laser-machined polycarbonate microfluidic chip. *Sensors and Actuators B: Chemical* 2021, 346, 130511–130526.

(40) Dignan, L. M.; Woolf, M. S.; Tomley, C. J.; Nauman, A. Q.; Landers, J. P. Multiplexed centrifugal microfluidic system for dynamic solid-phase purification of polynucleic acids direct from buccal swabs. *Analytical Chemistry* 2021, 93, 7300–7309.

(41) Jeong, G. S.; Chung, S.; Kim, C. B.; Lee, S. H. Applications of micromixing technology. *The Analyst* 2010, 135, 460–473.

(42) Naik, P.; Jaitpal, S.; Shetty, P.; Paul, D. An integrated one-step assay combining thermal lysis and loop-mediated isothermal DNA amplification (LAMP) in 30 min from *E. coli* and M. smegmatis cells on a paper substrate. *Sensors and Actuators B: Chemical* 2019, 291, 74–80.

(43) Baffou, G.; Cichos, F.; Quidant, R. Applications and challenges of thermoplasmonics. *Nature Materials* 2020, 19, 946–958.

(44) Yu, E. S.; Kang, B. H.; Ahn, M. S.; Jung, J. H.; Park, J. H.; Jeong, K. H. Highly efficient on-chip photo-thermal cell lysis for nucleic acid extraction using localized plasmonic heating of strongly absorbing Au nanoislands. *ACS Applied Materials & Interfaces* 2023, 15, 34323–34331.

(45) Yan, H.; Zhu, Y.; Zhang, Y.; Wang, L.; Chen, J.; Lu, Y.; Xu, Y.; Xing, W. Multiplex detection of bacteria on an integrated centrifugal disk using bead-beating lysis and loop-mediated amplification. *Scientific Reports* 2017, 7, 1460–1470.

(46) Liu, X.; Li, J.; Zhang, L.; Huang, X.; Farooq, U.; Pang, N.; Zhou, W.; Qi, L.; Xu, L.; Niu, L.; et al. Cell lysis based on an oscillating microbubble array. *Micromachines* 2020, 11, 288–299.

(47) Kim, J.; Johnson, M.; Hill, P.; Gale, B. K. Microfluidic sample preparation: Cell lysis and nucleic acid purification. *Integrative Biology* 2009, 1, 574–586.

(48) Banovetz, J. T.; Manimaran, S.; Schelske, B. T.; Anand, R. K. Parallel dielectrophoretic capture, isolation, and electrical lysis of individual breast cancer cells to assess variability in enzymatic activity. *Analytical Chemistry* 2023, 95, 7880–7887.

(49) Emaus, M. N.; Varona, M.; Eitzmann, D. R.; Hsieh, S. A.; Zeger, V. R.; Anderson, J. L. Nucleic acid extraction: Fundamentals of sample preparation methodologies, current advancements, and future endeavors. *TrAC Trends in Analytical Chemistry* 2020, 130, 115985–115996.

(50) Huang, Y.; Gao, Z.; Ma, C.; Sun, Y.; Huang, Y.; Jia, C.; Zhao, J.; Feng, S. An integrated microfluidic chip for nucleic acid extraction and continued cdPCR detection of pathogens. *Analyst* 2023, 148, 2758–2766.

(51) Paul, S.; Moon, H. Drop-to-drop liquid-liquid extraction of DNA in an electrowetting-on-dielectric digital microfluidics. *Biomicrofluidics* 2021, 15, 034110–034119.

(52) Ali, N.; Rampazzo, R.; Costa, A.; Krieger, M. Current nucleic acid extraction methods and their implications to point-of-care diagnostics. *BioMed Research International* 2017, 2017, 1–13.

(53) Fan, Y.; Dai, R.; Guan, X.; Lu, S.; Yang, C.; Lv, X.; Li, X. Rapid automatic nucleic acid purification system based on gas-liquid immiscible phase. *Journal of Separation Science* 2023, 46, 46–57.

(54) Zai, Y.; Min, C.; Wang, Z.; Ding, Y.; Zhao, H.; Su, E.; He, N. A sample-to-answer, quantitative real-time PCR system with low-cost, gravity-driven microfluidic cartridge for rapid detection of SARS-CoV-2, influenza A/B, and human papillomavirus 16/18. *Lab on a Chip* 2022, 22, 3436–3452.

(55) Reynolds, J.; Loeffler, R. S.; Leigh, P. J.; Lopez, H. A.; Yoon, J.-Y. Recent uses of paper microfluidics in isothermal nucleic acid amplification tests. *Biosensors* 2023, 13, 885–912.

(56) Jeong, I. H.; Kim, H. K.; Kim, H. R.; Kim, J.; Kim, B. C. Development of aptamers for rapid airborne bacteria detection. *Analytical and Bioanalytical Chemistry* 2022, 414, 7763–7771.

(57) Kalita, J. J.; Sharma, P.; Bora, U. Recent developments in application of nucleic acid aptamer in food safety. *Food Control* 2023, 145, 109406–109421.

(58) Pham, T. T. D.; Phan, L. M. T.; Park, J.; Cho, S. Review-electrochemical aptasensor for pathogenic bacteria detection. *Journal of the Electrochemical Society* 2022, 169, 087501–087516.

(59) Wang, Z.; Zhang, Z.; Zhang, Y.; Yu, H.; Gao, H.; Chang, D. Research progress of aptamer sensor in pathogenic bacteria detection. *Fudan University Journal of Medical Sciences* 2022, 49, 790–797.

(60) Wu, T.; Wang, C. C.; Wu, M. S.; Wang, P.; Feng, Q. M. Novel integrating polymethylene blue nanoparticles with dumbbell hybridization chain reaction for electrochemical detection of pathogenic bacteria. *Food Chemistry* 2022, 382, 132501–132507.

(61) Zeng, Y.; Qi, P.; Zhou, Y. N.; Wang, Y. W.; Xin, Y.; Sun, Y.; Zhang, D. Multi pathogenic microorganisms determination using DNA composites-encapsulated DNA silver nanocluster/graphene oxide-based system through rolling cycle amplification. *Microchimica Acta* 2022, 189, 403–411.

(62) Euler, M.; Wang, Y. J.; Nentwich, O.; Piepenburg, O.; Hufert, F. T.; Weidmann, M. Recombinase polymerase amplification assay for rapid detection of Rift Valley fever virus. *Journal of Clinical Virology* 2012, 54, 308–312.

(63) Lei, R.; Wang, X. Y.; Zhang, D.; Liu, Y. Z.; Chen, Q. J.; Jiang, N. Rapid isothermal duplex real-time recombinase polymerase amplification (RPA) assay for the diagnosis of equine piroplasmosis. *Scientific Reports* 2020, 10, 4096–4106.

(64) Rani, A.; Ravindran, V. B.; Surapaneni, A.; Shahsavari, E.; Haleyur, N.; Mantri, N.; Ball, A. S. Evaluation and comparison of recombinase polymerase amplification coupled with lateral-flow bioassay for *Escherichia coli* O157:H7 detection using different genes. *Scientific Reports* 2021, 11, 1881–1892.

(65) Clancy, E.; Higgins, O.; Forrest, M. S.; Boo, T. W.; Cormican, M.; Barry, T.; Piepenburg, O.; Smith, T. J. Development of a rapid recombinase polymerase amplification assay for the detection of Streptococcus pneumoniae in whole blood. *BMC Infectious Diseases* 2015, 15, 481–491.

(66) Li, J.; Macdonald, J.; von Stetten, F. Review: A comprehensive summary of a decade development of the recombinase polymerase amplification. *Analyst* 2019, 144, 31–67.

(67) Li, Q.; Duan, L. J.; Jin, D. S.; Chen, Y. X.; Lou, Y. R.; Zhou, Q. J.; Xu, Z. J.; Chen, F. J.; Chen, H. X.; Xu, G. Z.; et al. A real-time fluorogenic recombinase polymerase amplification microfluidic chip (on-chip RPA) for multiple detection of pathogenic microorganisms of penaeid shrimp. *Aquaculture* 2024, 578, 740017–740026.

(68) Shang, Y. T.; Xing, G. W.; Lin, H. F.; Sun, Y. C.; Chen, S. L.; Lin, J. M. Development of nucleic acid extraction and real-time recombinase polymerase amplification (RPA) assay integrated microfluidic biosensor for multiplex detection of foodborne bacteria. *Food Control* 2024, 155, 110047–110053.

(69) Gan, Z. X.; Roslan, M. A. M.; Abd Shukor, M. Y.; Halim, M.; Yasid, N. A.; Abdullah, J.; Yasin, I. S. M.; Wasoh, H. Advances in aptamer-based biosensors and cell-internalizing SELEX technology for diagnostic and therapeutic application. *Biosensors-Basel* 2022, 12, 922–938.

(70) Nie, Z.; Zhao, Y. N.; Shu, X.; Li, D. F.; Ao, Y. S. Q.; Li, M. X.; Wang, S.; Cui, J.; An, X. M.; Zhan, X. Y.; et al. Recombinase polymerase amplification with lateral flow strip for detecting Babesia microti infections. *Parasitology International* 2021, 83, 102351–102356.

(71) Petrucci, S.; Costa, C.; Broyles, D.; Kaur, A.; Dikici, E.; Daunert, S.; Deo, S. K. Monitoring pathogenic viable *E. coli* O157:H7 in food matrices based on the detection of RNA using isothermal amplification and a paper-based platform. *Analytical Chemistry* 2022, 94, 2485–2492.

(72) Yin, W. H.; Zhuang, J. J.; Li, J. L.; Xia, L. P.; Hu, K.; Yin, J. X.; Mu, Y. Digital recombinase polymerase amplification, digital loop-mediated isothermal amplification, and digital CRISPR-Cas assisted assay: Current status, challenges, and perspectives. *Small* 2023, 95, 2303398–2303416.

(73) Kalsi, S.; Valiadi, M.; Turner, C.; Sutton, M.; Morgan, H. Sample pre-concentration on a digital microfluidic platform for rapid AMR detection in urine. *Lab on a Chip* 2019, 19, 168–177.

(74) Xu, P.; Zheng, X.; Tao, Y.; Du, W. B. Cross-interface emulsification for generating size-tunable droplets. *Analytical Chemistry* 2016, 88, 3171–3177.

(75) Yin, J. X.; Zou, Z. Y.; Hu, Z. M.; Zhang, S.; Zhang, F. P.; Wang, B.; Lv, S. W.; Mu, Y. A "sample-in-multiplex-digital-answer-out" chip for fast detection of pathogens. *Lab on a Chip* 2020, 20, 979–986.

(76) Li, R. X.; Su, N.; Ren, X. D.; Sun, X. E.; Li, W. M.; Li, Y. W.; Li, J.; Chen, C.; Wang, H.; Lu, W. P.; et al. Centrifugal microfluidic-based multiplex recombinase polymerase amplification assay for rapid detection of SARS-CoV-2. *Iscience* 2023, 26, 106245–106265.

(77) Liu, D.; Shen, H. C.; Zhang, Y. Q.; Shen, D. Y.; Zhu, M. Y.; Song, Y. L.; Zhu, Z.; Yang, C. Y. A microfluidic-integrated lateral flow recombinase polymerase amplification (MI-IF-RPA) assay for rapid COVID-19 detection. *Lab on a Chip* 2021, 21, 2019–2026.

(78) Bender, A. T.; Sullivan, B. P.; Zhang, J. Y.; Juergens, D. C.; Lillis, L.; Boyle, D. S.; Posner, J. D. HIV detection from human serum with paper-based isotachophoretic RNA extraction and reverse transcription recombinase polymerase amplification. *Analyst* 2021, 146, 2851–2861.

(79) Seder, I.; Coronel-Tellez, R.; Helalat, S. H.; Sun, Y. Fully integrated sample-in-answer-out platform for viral detection using digital reverse transcription recombinase polymerase amplification (dRT-RPA). *Biosensors & Bioelectronics* 2023, 237, 115487–115494.

(80) Soroka, M.; Wasowicz, B.; Rymaszewska, A. Loop-mediated isothermal amplification (LAMP): The better sibling of PCR? *Cells* 2021, 10, 1931–1950.

(81) Trinh, T. N. D.; Lee, N. Y. Advances in nucleic acid amplification-based microfluidic devices for clinical microbial detection. *Chemosensors* 2022, 10, 123–139.

(82) Liu, D. C.; Zhu, Y. Z.; Li, N.; Lu, Y.; Cheng, J.; Xu, Y. C. A portable microfluidic analyzer for integrated bacterial detection using visible loop-mediated amplification. *Sensors and Actuators B: Chemical* 2020, 310, 127834–127841.

(83) Shang, Y. T.; Sun, J. D.; Ye, Y. L.; Zhang, J. M.; Zhang, Y. Z.; Sun, X. L. Loop-mediated isothermal amplification-based microfluidic chip for pathogen detection. *Critical Reviews in Food Science and Nutrition* 2020, 60, 201–224.

(84) Xiao, B.; Zhao, R. M.; Wang, N.; Zhang, J.; Sun, X. Y.; Chen, A. L. Recent advances in centrifugal microfluidic chip-based loop-mediated isothermal amplification. *TrAC Trends in Analytical Chemistry* 2023, 158, 116836–116854.

(85) Jin, J. L.; Duan, L. J.; Fu, J. L.; Chai, F. C.; Zhou, Q. J.; Wang, Y. H.; Shao, X. B.; Wang, L.; Yan, M. C.; Su, X. R.; et al. A real-time LAMP-based dual-sample microfluidic chip for rapid and simultaneous detection of multiple waterborne pathogenic bacteria from coastal waters. *Analytical Methods* 2021, 13, 2710–2721.

(86) Xiao, B.; Zhao, R. M.; Wang, N.; Zhang, J.; Sun, X. Y.; Huang, F. C.; Chen, A. L. Integrating microneedle DNA extraction to hand-held microfluidic colorimetric LAMP chip system for meat adulteration detection. *Food Chemistry* 2023, 411, 135508–135516.

(87) Wang, W.; Lee, H.; Jankelow, A. M.; Hoang, T.-H.; Bacon, A.; Sun, F.; Chae, S.; Kindratenko, V.; Koprowski, K.; Stavins, R. A.; et al. Smartphone clip-on instrument and microfluidic processor for rapid sample-to-answer detection of Zika virus in whole blood using spatial RT-LAMP. *Analyst*, 2022, 147, 3838–3853.

(88) Chaouch, M. Loop-mediated isothermal amplification (LAMP): An effective molecular point-of-care technique for the rapid diagnosis of coronavirus SARS-CoV-2. *Reviews in Medical Virology* 2021, 31, 2215–2223.

(89) Lalli, M. A.; Langmade, S. J.; Chen, X. H.; Fronick, C. C.; Sawyer, C. S.; Burcea, L. C.; Wilkinson, M. N.; Fulton, R. S.; Heinz, M.; Buchser, W. J.; et al. Rapid and extraction-free detection of SARS-CoV-2 from saliva by colorimetric reverse-transcription loop-mediated isothermal amplification. *Clinical Chemistry* 2021, 67, 415–424.

(90) Donia, A.; Shahid, M. F.; Sammer-Ul, H.; Shahid, R.; Ahmad, A.; Javed, A.; Nawaz, M.; Yaqub, T.; Bokhari, H. Integration of RT-LAMP and microfluidic technology for detection of SARS-CoV-2 in wastewater as an advanced point-of-care platform. *Food and Environmental Virology* 2022, 14, 364–373.

(91) Wu, C.; Liu, L. B.; Ye, Z. Z.; Gong, J. J.; Hao, P.; Ping, J. F.; Ying, Y. B. TriD-LAMP: A pump-free microfluidic chip for duplex droplet digital loop-mediated isothermal amplification analysis. *Analytica Chimica Acta* 2022, 1233, 340513–340520.

(92) Mumtaz, Z.; Rashid, Z.; Ali, A.; Arif, A.; Ameen, F.; AlTami, M. S.; Yousaf, M. Z. Prospects of microfluidic technology in nucleic acid detection approaches. *Biosensors-Basel* 2023, 13, 584–606.

(93) Srivastava, P.; Prasad, D. Isothermal nucleic acid amplification and its uses in modern diagnostic technologies. *3 Biotech* 2023, 13, 200–222.

(94) Ishigaki, Y.; Sato, K. Effects of microchannel shape and ultrasonic mixing on microfluidic padlock probe rolling circle amplification (RCA) reactions. *Micromachines* 2018, 9, 272–282.

(95) Garbarino, F.; Minero, G. A. S.; Rizzi, G.; Fock, J.; Hansen, M. F. Integration of rolling circle amplification and optomagnetic detection on a polymer chip. *Biosensors & Bioelectronics* 2019, 142, 111485–111491.

(96) Ciftci, S.; Neumann, F.; Hernández-Neuta, I.; Hakhverdyan, M.; Bálint, A.; Herthnek, D.; Madaboosi, N.; Nilsson, M. A novel mutation tolerant padlock probe design for multiplexed detection of hypervariable RNA viruses. *Scientific Reports* 2019, 9, 2872–2881.

(97) Ciftci, S.; Neumann, F.; Abdurahman, S.; Appelberg, K. S.; Mirazimi, A.; Nilsson, M.; Madaboosi, N. Digital rolling circle amplification-based detection of Ebola and other tropical viruses. *Journal of Molecular Diagnostics* 2020, 22, 272–283.

(98) Zhang, Z. Q.; Ali, M. M.; Eckert, M. A.; Kang, D. K.; Chen, Y. Y.; Sender, L. S.; Fruman, D. A.; Zhao, W. A. A polyvalent aptamer system for targeted drug delivery. *Biomaterials* 2013, 34, 9728–9735.

(99) Li, S. Y.; Jiang, Y. Q.; Yang, X. Y.; Lin, M.; Dan, H. H.; Zou, S.; Cao, X. D. In situ rolling circle amplification surface modifications to improve *E. coli* O157:H7 capturing performances for rapid and sensitive microfluidic detection applications. *Analytica Chimica Acta* 2021, 1150, 338229–338239.

(100) Jiang, Y. Q.; Qiu, Z. Y.; Le, T.; Zou, S.; Cao, X. D. Developing a dual-RCA microfluidic platform for sensitive *E. coli* O157:H7 whole-cell detections. *Analytica Chimica Acta* 2020, 1127, 79–88.

(101) Bhunia, A. K. One day to one hour: How quickly can foodborne pathogens be detected? *Future Microbiology* 2014, 9, 935–946.

(102) Doyle, M. P. *Escherichia coli* O157: H7 and its significance in foods. *International Journal of Food Microbiology* 1991, 12, 289–302.

(103) Li, Y.; Liu, H. M.; Huang, H.; Deng, J.; Fang, L. C.; Luo, J.; Zhang, S.; Huang, J.; Liang, W. B.; Zheng, J. S. A sensitive electrochemical strategy via multiple amplification reactions for the detection of *E. coli* O157: H7. *Biosensors & Bioelectronics* 2020, 147, 111752–111759.

(104) Liu, L.; Liu, J.; Huang, H.; Li, Y. X.; Zhao, G. Y.; Dou, W. C. A quantitative foam immunoassay for detection of *Escherichia coli* O157:H7 based on bimetallic nanocatalyst-gold platinum. *Microchemical Journal* 2019, 148, 702–707.

(105) Thompson, R.; Perry, J. D.; Stanforth, S. P.; Dean, J. R. Rapid detection of hydrogen sulfide produced by pathogenic bacteria in focused growth media using SHS-MCC-GC-IMS. *Microchemical Journal* 2018, 140, 232–240.

(106) Simmel, F. C.; Yurke, B.; Singh, H. R. Principles and applications of nucleic acid strand displacement reactions. *Chemical Reviews* 2019, 119, 6326–6369.

(107) Sun, D. L.; Fan, T. T.; Liu, F.; Wang, F. X.; Gao, D.; Lin, J. M. A microfluidic chemiluminescence biosensor based on multiple signal amplification for rapid and sensitive detection of *E. coli* O157:H7. *Biosensors & Bioelectronics* 2022, 212, 114390–114397.

(108) Cassoli, L. D.; Lima, W. J. F.; Esguerra, J. C.; Da Silva, J.; Machado, P. F.; Mourao, G. B. Do different standard plate counting (IDF/ISSO or AOAC) methods interfere in the conversion of individual bacteria counts to colony forming units in raw milk? *Journal of Applied Microbiology* 2016, 121, 1052–1058.

(109) Voitoux, E.; Lafarge, V.; Collette, C.; Lombard, B. Applicability of the draft standard method for the detection of *Escherichia coli* O157 in dairy products. *International Journal of Food Microbiology* 2002, 77, 213–221.

(110) Liu, X.; Bu, S. J.; Feng, J. Q.; Wei, H. G.; Wang, Z.; Li, X.; Zhou, H. Y.; He, X. X.; Wan, J. Y. Electrochemical biosensor for detecting pathogenic bacteria based on a hybridization chain reaction and CRISPR-Cas12a. *Analytical and Bioanalytical Chemistry* 2022, 414, 1073–1080.

(111) Zhao, L. Z.; Fu, Y. Z.; Ren, S. W.; Cao, J. T.; Liu, Y. M. A novel chemiluminescence imaging immunosensor for prostate specific antigen detection based on a multiple signal amplification strategy. *Biosensors & Bioelectronics* 2021, 171, 112729–112734.

(112) Luo, F. F.; Li, Z.; Dai, G.; Lu, Y. Q.; He, P. G.; Wang, Q. J. Ultrasensitive biosensing pathogenic bacteria by combining aptamer-induced catalysed hairpin assembly circle amplification with microchip electrophoresis. *Sensors and Actuators B: Chemical* 2020, 306, 127577–127582.

(113) Zhang, Y.; Hu, X. Z.; Wang, Q. J. Review of microchip analytical methods for the determination of pathogenic *Escherichia coli*. *Talanta* 2021, 232, 122410–122419.

(114) Bi, S.; Yue, S. Z.; Zhang, S. S. Hybridization chain reaction: A versatile molecular tool for biosensing, bioimaging, and biomedicine. *Chemical Society Reviews* 2017, 46, 4281–4298.

(115) Rutten, I.; Daems, D.; Leirs, K.; Lammertyn, J. Highly sensitive multiplex detection of molecular biomarkers using hybridization chain reaction in an encoded particle microfluidic platform. *Biosensors-Basel* 2023, 13, 100–112.

(116) Dirks, R. M.; Pierce, N. A. Triggered amplification by hybridization chain reaction. *Proceedings of the National Academy of Sciences of the United States of America* 2004, 101, 15275–15278.

(117) Ang, Y. S.; Yung, L. Y. L. Rational design of hybridization chain reaction monomers for robust signal amplification. *Chemical Communications* 2016, 52, 4219–4222.

(118) Ge, Z. L.; Lin, M. H.; Wang, P.; Pei, H.; Yan, J.; Sho, J. Y.; Huang, Q.; He, D. N.; Fan, C. H.; Zuo, X. L. Hybridization chain reaction amplification of microRNA detection with a tetrahedral DNA nanostructure-based electrochemical biosensor. *Analytical Chemistry* 2014, 86, 2124–2130.

(119) Ikbal, J.; Lim, G. S.; Gao, Z. Q. The hybridization chain reaction in the development of ultrasensitive nucleic acid assays. *TrAC Trends in Analytical Chemistry* 2015, 64, 86–99.

(120) Ying, N.; Wang, Y.; Song, X.; Yang, L.; Qin, B.; Wu, Y.; Fang, W. Lateral flow colorimetric biosensor for detection of Vibrio parahaemolyticus based on hybridization chain reaction and aptamer. *Microchimica Acta* 2021, 188, 381–389.

(121) Tang, L.; Yang, J.; Liu, Z.; Mi, Q.; Niu, L.; Zhang, J. Design nanoprobe based on DNA tetrahedron supported hybridization chain reaction and its application to in situ analysis of bacteria. *Chemical Engineering Journal* 2023, 466, 143099–143106.

(122) Li, T.; Ou, G. Z.; Chen, X. L.; Li, Z. Y.; Hu, R.; Li, Y.; Yang, Y. H.; Liu, M. L. Naked-eye based point-of-care detection of *E. coli* O157: H7 by a signal-amplified microfluidic aptasensor. *Analytica Chimica Acta* 2020, 1130, 20–28.

(123) Li, Y.; Uddayasankar, U.; He, B. S.; Wang, P.; Qin, L. D. Fast, sensitive, and quantitative point-of-care platform for the assessment of drugs of abuse in urine, serum, and whole blood. *Analytical Chemistry* 2017, 89, 8273–8281.

(124) Li, Y.; Xuan, J.; Song, Y. J.; Wang, P.; Qin, L. D. A microfluidic platform with digital readout and ultra-low detection limit for quantitative point-of-care diagnostics. *Lab on a Chip* 2015, 15, 3300–3306.

(125) Song, Y. J.; Zhang, Y. Q.; Bernard, P. E.; Reuben, J. M.; Ueno, N. T.; Arlinghaus, R. B.; Zu, Y. L.; Qin, L. D. Multiplexed volumetric bar-chart chip for point-of-care diagnostics. *Nature Communications* 2012, 3, 1283–1291.

(126) Chen, H.; Li, Z.; Zhang, L. Z.; Sawaya, P.; Shi, J. B.; Wang, P. Quantitation of femtomolar-level protein biomarkers using a simple microbubbling digital assay and bright-field smartphone imaging. *Angewandte Chemie, International Edition* 2019, 58, 13922–13928.

(127) Dou, M. W.; Sanchez, J.; Tavakoli, H.; Gonzalez, J. E.; Sun, J. J.; Bard, J. D.; Li, X. J. A low-cost microfluidic platform for rapid and instrument-free detection of whooping cough. *Analytica Chimica Acta* 2019, 1065, 71–78.

(128) Liu, D.; Jia, S. S.; Zhang, H. M.; Ma, Y. L.; Guan, Z. C.; Li, J. X.; Zhu, Z.; Ji, T. H.; Yang, C. J. Integrating target-responsive hydrogel with pressuremeter readout enables simple, sensitive, user-friendly, quantitative point-of-care testing. *ACS Applied Materials & Interfaces* 2017, 9, 22252–22258.

(129) Wang, P.; Kricka, L. J. current and emerging trends in point-of-care technology and strategies for clinical validation and implementation. *Clinical Chemistry* 2018, 64, 1439–1452.

(130) Scida, K.; Li, B. L.; Ellington, A. D.; Crooks, R. M. DNA detection using origami paper analytical devices. *Analytical Chemistry* 2013, 85, 9713–9720.

(131) Xiao, Y.; Pavlov, V.; Niazov, T.; Dishon, A.; Kotler, M.; Willner, I. Catalytic beacons for the detection of DNA and telomerase activity. *Journal of the American Chemical Society* 2004, 126, 7430–7431.

(132) Demirkol, D. O.; Timur, S. A sandwich-type assay based on quantum dot/aptamer bioconjugates for analysis of *E. coli* O157:H7 in microtiter plate format. *International Journal of Polymeric Materials and Polymeric Biomaterials* 2016, 65, 85–90.

(133) Sheng, H.; Ye, B. C. Different strategies of covalent attachment of oligonucleotide probe onto glass beads and the hybridization properties. *Applied Biochemistry and Biotechnology* 2009, 152, 54–65.

(134) Huang, F. C.; Zhang, Y. C.; Lin, J. H.; Liu, Y. J. Biosensors coupled with signal amplification technology for the detection of pathogenic bacteria: A review. *Biosensors-Basel* 2021, 11, 190–222.

(135) Shao, Y. N.; Duan, H.; Zhou, S.; Ma, T. T.; Guo, L.; Huang, X. L.; Xiong, Y. H. Biotin-streptavidin system-mediated ratiometric multiplex immunochromatographic assay for simultaneous and accurate quantification of three mycotoxins. *Journal of Agricultural and Food Chemistry* 2019, 67, 9022–9031.

(136) Guo, Q.; Han, J. J.; Shan, S.; Liu, D. F.; Wu, S. S.; Xiong, Y. H.; Lai, W. H. DNA-based hybridization chain reaction and biotin-streptavidin signal amplification for sensitive detection of *Escherichia coli* O157:H7 through ELISA. *Biosensors & Bioelectronics* 2016, 86, 990–995.

(137) Kolb, H. C.; Finn, M. G.; Sharpless, K. B. Click chemistry: Diverse chemical function from a few good reactions. *Angewandte Chemie, International Edition* 2001, 40, 2004–2021.

(138) Franc, G.; Kakkar, A. Dendrimer design using Cu(I)-catalyzed alkyne-azide "click-chemistry". *Chemical Communications* 2008, 42, 5267–5276.

(139) Mou, X. Z.; Chen, X. Y.; Wang, J. H.; Zhang, Z. T.; Yang, Y. M.; Shou, Z. X.; Tu, Y. X.; Du, X. C.; Wu, C.; Zhao, Y.; et al. Bacteria-instructed click chemistry between functionalized gold nanoparticles for point-of-care microbial detection. *ACS Applied Materials & Interfaces* 2019, 11, 23093–23101.

(140) Liong, M.; Fernandez-Suarez, M.; Issadore, D.; Min, C.; Tassa, C.; Reiner, T.; Fortune, S. M.; Toner, M.; Lee, H.; Weissleder, R. Specific pathogen detection using bioorthogonal chemistry and diagnostic magnetic resonance. *Bioconjugate Chemistry* 2011, 22, 2390–2394.

(141) Zheng, Y.; Ma, Z. F. Multifunctionalized ZIFs nanoprobe-initiated tandem reaction for signal amplified electrochemical immunoassay of carbohydrate antigen 24–2. *Biosensors & Bioelectronics* 2019, 129, 42–49.

(142) Wang, H. Q.; Ma, Z. F. A cascade reaction signal-amplified amperometric immunosensor platform for ultrasensitive detection of tumour marker. *Sensors and Actuators B: Chemical* 2018, 254, 642–647.

(143) Zhag, Y.; Tan, C.; Fei, R. H.; Liu, X. X.; Zhou, Y.; Chen, J.; Chen, H. C.; Zhou, R.; Hu, Y. G. Sensitive chemiluminescence immunoassay for *E. coli* O157:H7 detection with signal dual-amplification using glucose oxidase and laccase. *Analytical Chemistry* 2014, 86, 1115–1122.

(144) Gao, B.; Chen, X. R.; Huang, X. L.; Pei, K.; Xiong, Y.; Wu, Y. Q.; Duan, H.; Lai, W. H.; Xiong, Y. H. Urease-induced metallization of gold nanorods for the sensitive detection of Salmonella enterica Choleraesuis through colorimetric ELISA. *Journal of Dairy Science* 2019, 102, 1997–2007.

(145) Urmann, K.; Reich, P.; Walter, J. G.; Beckmann, D.; Segal, E.; Scheper, T. Rapid and label-free detection of protein a by aptamer-tethered porous silicon nanostructures. *Journal of Biotechnology* 2017, 257, 171–177.

(146) Xu, L. Q.; Dai, Q. Q.; Shi, Z. Y.; Liu, X. T.; Gao, L.; Wang, Z. Z.; Zhu, X. Y.; Li, Z. Accurate MRSA identification through dual-functional aptamer and CRISPR-Cas12a assisted rolling circle amplification. *Journal of Microbiological Methods* 2020, 173, 105917–105922.

(147) Dong, N. N.; Jiang, N.; Zhao, J. W.; Zhao, G. M.; Wang, T. W. Sensitive and enzyme-free pathogenic bacteria detection through self-circulation of molecular beacon. *Applied Biochemistry and Biotechnology* 2022, 194, 3668–3676.

(148) Ge, J.; Lei, J. D.; Zare, R. N. Protein-inorganic hybrid nanoflowers. *Nature Nanotechnology* 2012, 7, 428–432.

(149) Ye, R. F.; Zhu, C. Z.; Song, Y.; Song, J. H.; Fu, S. F.; Lu, Q.; Yang, X.; Zhu, M. J.; Du, D.; Li, H.; et al. One-pot bioinspired synthesis of all-inclusive protein-protein nanoflowers for point-of-care bioassay: Detection of *E. coli* O157:H7 from milk. *Nanoscale* 2016, 8, 18980–18986.

(150) Amanzholova, M.; Shaizadinova, A.; Bulashev, A.; Abeldenov, S. Genetic identification of Staphylococcus aureus isolates from cultured milk samples of bovine mastitis using isothermal amplification with CRISPR/Cas12a-based molecular assay. *Veterinary Research Communications* 2023, 48, 291–300.

(151) Xia, L. P.; Yin, J. X.; Zhuang, J. J.; Yin, W. H.; Zou, Z. Y.; Mu, Y. Adsorption-free self-priming direct digital dual-crRNA CRISPR/Cas12a-assisted chip for ultrasensitive detection of pathogens. *Analytical Chemistry* 2023, 95, 4744–4752.

(152) Xu, J. H.; Ma, J. F.; Li, Y. W.; Kang, L.; Yuan, B.; Li, S. Q.; Chao, J.; Wang, L. H.; Wang, J. L.; Su, S.; et al. A general RPA-CRISPR/Cas12a sensing platform for Brucella spp. detection in blood and milk samples. *Sensors and Actuators B: Chemical* 2022, 364, 131864–131872.

(153) Hu, M. L.; Qiu, Z. Q.; Bi, Z. R.; Tian, T.; Jiang, Y. Z.; Zhou, X. M. Photocontrolled crRNA activation enables robust CRISPR-Cas12a diagnostics. *Proceedings of the National Academy of Sciences of the United States of America* 2022, 119, e2202034119–e2202034128.

(154) Luo, Z.; Ye, C. H.; Xiao, H.; Yin, J. L.; Liang, Y. C.; Ruan, Z. H.; Luo, D. J.; Gao, D. L.; Tan, Q. P.; Li, Y. K.; et al. Optimization of loop-mediated isothermal amplification (LAMP) assay for robust visualization in SARS-CoV-2 and emerging variants diagnosis. *Chemical Engineering Science* 2022, 251, 117430–117439.

(155) Hu, A. T.; Kong, L. Y.; Lu, Z. X.; Zhou, H. B.; Bie, X. M. Construction of a LAMP-CRISPR assay for the detection of Vibrio parahaemolyticus. *Food Control* 2023, 149, 109728–109740.

(156) Li, Z. H.; Zhao, W. C.; Ma, S. X.; Li, Z. X.; Yao, Y. J.; Fei, T. A chemical-enhanced system for CRISPR-based nucleic acid detection. *Biosensors & Bioelectronics* 2021, 192, 113493–113502.

(157) Ma, L.; Wang, J. J.; Li, Y. R.; Liao, D.; Zhang, W. L.; Han, X.; Man, S. L. A ratiometric fluorescent biosensing platform for ultrasensitive detection of Salmonella typhimurium via CRISPR/Cas12a and silver nanoclusters. *Journal of Hazardous Materials* 2023, 443, 130234–130242.

(158) Koscianska, E.; Starega-Roslan, J.; Sznajder, L. J.; Olejniczak, M.; Galka-Marciniak, P.; Krzyzosiak, W. J. Northern blotting analysis of microRNAs, their precursors and RNA interference triggers. *BMC Molecular Biology* 2011, 12, 14–20.

(159) Wienholds, E.; Kloosterman, W. P.; Miska, E.; Alvarez-Saavedra, E.; Berezikov, E.; de Bruijn, E.; Horvitz, H. R.; Kauppinen, S.; Plasterk, R. H. A. MicroRNA expression in zebrafish embryonic development. *Mechanisms of Development* 2005, 309, 310–311.

(160) Pasquinelli, A. E. Non-coding RNA microRNAs and their targets: Recognition, regulation and an emerging reciprocal relationship. *Nature Reviews Genetics* 2012, 13, 271–282.

(161) Cui, Y.; Fan, S. J.; Yuan, Z.; Song, M. H.; Hu, J. W.; Qian, D.; Zhen, D. S.; Li, J. H.; Zhu, B. D. Ultrasensitive electrochemical assay for microRNA-21 based on CRISPR/Cas13a-assisted catalytic hairpin assembly. *Talanta* 2021, 224, 121878–121884.

(162) Chen, J. S.; Ma, E. B.; Harrington, L. B.; Da Costa, M.; Tian, X. R.; Palefsky, J. M.; Doudna, J. A. CRISPR-Cas12a target binding unleashes indiscriminate single-stranded DNase activity. *Science* 2018, 360, 436–439.

(163) Peng, S.; Tan, Z.; Chen, S. Y.; Lei, C. Y.; Nie, Z. Integrating CRISPR-Cas12a with a DNA circuit as a generic sensing platform for amplified detection of microRNA. *Chemical Science* 2020, 11, 7362–7368.

(164) Dai, Y. F.; Somoza, R. A.; Wang, L.; Welter, J. F.; Li, Y.; Caplan, A. I.; Liu, C. C. Exploring the trans-cleavage activity of CRISPR-Cas12a (cpf1) for the development of a universal electrochemical biosensor. *Angewandte Chemie, International Edition* 2019, 58, 17399–17405.

(165) Shan, Y. Y.; Zhou, X. M.; Huang, R.; Xing, D. High-fidelity and rapid quantification of miRNA combining crRNA programmability and CRISPR/Cas13a trans-cleavage activity. *Analytical Chemistry* 2019, 91, 5278–5285.

(166) Chen, P. R.; Wang, L. Y.; Qin, P. P.; Yin, B. C.; Ye, B. C. An RNA-based catalytic hairpin assembly circuit coupled with CRISPR-Cas12a for one-step detection of microRNAs. *Biosensors & Bioelectronics* 2022, 207, 114152–114158.

(167) Rothe, C.; Schunk, M.; Sothmann, P.; Bretzel, G.; Froeschl, G.; Wallrauch, C.; Zimmer, T.; Thiel, V.; Janke, C.; Guggemos, W.; et al. Transmission of 2019-nCoV infection from an asymptomatic contact in Germany. *New England Journal of Medicine* 2020, 382, 970–971.

(168) Corman, V. M.; Landt, O.; Kaiser, M.; Molenkamp, R.; Meijer, A.; Chu, D. K. W.; Bleicker, T.; Brünink, S.; Schneider, J.; Schmidt, M. L.; et al. Detection of 2019 novel coronavirus (2019-nCoV) by real-time RT-PCR. *Eurosurveillance* 2020, 25, 23–30.

(169) Yourik, P.; Fuchs, R. T.; Mabuchi, M.; Curcuru, J. L.; Robb, G. B. Staphylococcus aureus Cas9 is a multiple-turnover enzyme. *RNA* 2019, 25, 35–44.

(170) Song, D.; Han, X. Z.; Xu, W. J.; Liu, J. Y.; Zhuo, Y. X.; Zhu, A. N.; Long, F. Target nucleic acid amplification-free detection of *Escherichia coli* O157:H7 by CRISPR/Cas12a and hybridization chain reaction based on an evanescent wave fluorescence biosensor. *Sensors and Actuators B: Chemical* 2023, 376, 133005–133013.

(171) Chu, Y. X.; Deng, A. P.; Wang, W. J.; Zhu, J. J., Concatenated Catalytic Hairpin Assembly/Hyperbranched Hybridization Chain Reaction Based Enzyme-Free Signal Amplification for the Sensitive Photoelectrochemical Detection of Human Telomerase RNA. *Analytical Chemistry* 2019, 91 (5), 3619–3627.

(172) Chu, Y. X.; Deng, A. P.; Wang, W. J.; Zhu, J. J., Concatenated catalytic hairpin assembly/hyperbranched hybridization chain reaction based enzyme-free signal amplification for the sensitive photoelectrochemical detection of human telomerase RNA. *Analytical Chemistry* 2019, 91 (5), 3619–3627.

(173) Ji, P. P.; Han, G. M.; Huang, Y.; Jiang, H. X.; Zhou, Q. W.; Liu, X. W.; Kong, D. M., Ultrasensitive ratiometric detection of $Pb^{2+}$ using DNA tetrahedron-mediated hyperbranched hybridization chain reaction. *Analytica Chimica Acta* 2021, 1147, 170–177.

(174) Wang, J.; Wang, D. X.; Ma, J. Y.; Wang, Y. X.; Kong, D. M., Three-dimensional DNA nanostructures to improve the hyperbranched hybridization chain reaction. *Chemical Science* 2019, 10 (42), 9758–9767.

(175) Cai, Q. Q.; Shi, H. X.; Sun, M. N.; Ma, N.; Wang, R.; Yang, W. G.; Qiao, Z. H., Sensitive detection of salmonella based on CRISPR-Cas12a and the tetrahedral DNA nanostructure-mediated hyperbranched hybridization chain reaction. *Journal of Agricultural and Food Chemistry* 2022, 70 (51), 16382–16389.

(176) Li, F.; Ye, Q. H.; Chen, M. T.; Zhou, B. Q.; Zhang, J. M.; Pang, R.; Xue, L.; Wang, J.; Zeng, H. Y.; Wu, S.; Zhang, Y. X.; Ding, Y.; Wu, Q. P., An ultrasensitive CRISPR/Cas12a based electrochemical biosensor for Listeria monocytogenes detection. *Biosensors & Bioelectronics* 2021, 179.

# 3 Paper-Based Microfluidic Devices in Food Processing

*Soja Saghar Soman, Shafeek Abdul Samad,*
*Priyamvada Venugopalan, Nityanand Kumawat,*
*and Sunil Kumar*

## 3.1 INTRODUCTION

The availability of adequate amount of safe and nutritious food is the key to sustaining life and promoting good health. Unsafe food causes a vicious cycle of diseases and malnutrition, particularly affecting infants, young children, the elderly and the sick (Savelli et al. 2019). Food processing is the term used for the conversion of raw food materials into edible, safe, and palatable food. Quality control in each step of food processing is crucial to ensure food safety. Conventional quality control methods involve taking samples and analyzing them with analytical techniques that require laboratory facilities, trained personnel, and careful sample preparation procedures prior to the detection in order to obtain reliable results. These methods are not always ideal for on-site rapid surveillance. Hence, the demands for developing simpler and cost-effective techniques for rapid monitoring in field conditions are on the rise. The pioneering works of Whitesides' group in 2007 showed the efficacy of microfluidic paper-based analytical devices (µPADs) and paved the way to apply µPADs for food safety monitoring (Martinez et al. 2007). µPADs allow easy, rapid, and cost-effective point-of-need screening of food materials. The fundamental working principle of µPAD technology is to scale down the analytical device on a small piece of paper substrate and reduce the reagent volumes to microunits while maintaining the accuracy of detection. Most of the µPADs developed for food analysis showed high testing reliability with a correlation coefficient ranging from $R^2 = 0.9000$ to 0.9953 (Liu et al. 2018). The use of µPADs in food safety offers the following advantages over the conventional methods.

1. Compatibility: µPADs exhibit superior compatibility with most of the biological and chemical reagents.
2. The ease of use: µPADs use capillary and gravitational forces for the absorption and flow of liquid samples without any additional devices.
3. The porous nature of paper offers high surface-to-volume ratio, which allows the absorption of more reagents and samples and better contact and mixing. This gives the flexibility on assay timings and the assayed µPADs can be stored for later analysis, when required.
4. The ease of fabrication: The paper devices can be made using methods such as cutting, heating, laminating, stamping, photolithography, etching, and printing.
5. Readily available around the world: The main raw material to fabricate µPADs is natural paper and it is available everywhere and widely used for a variety of applications.
6. Less foot space and ease of transportation: Compared to other analytical devices, µPADs require less storage space and less cold chain maintenance.
7. The ease of disposal: The µPADs are environment friendly and can be disposed of easily by incineration.

DOI: 10.1201/9781032632599-3

The following section of this chapter gives a brief description of the different µPAD fabrication techniques, their advantages, and the challenges associated with them. The fluid flow control techniques along the channel length and readout mechanisms for the detection of analytes are described in the subsequent section. It is followed by a detailed section on the application of µPADs for the analysis of common food additives, adulterants, toxicants, food-borne pathogens, and other food contaminants. In the end, conclusions and future directions are discussed.

## 3.2 µPADs FABRICATION TECHNIQUES

Following the invention of µPADs by Whiteside's group and the rapid demand in multiple domains, several techniques have been developed for the mass production of µPADs. These devices require isolated hydrophilic micro-zones and channels in the hydrophilic paper matrix. It is achieved by the creation of hydrophobic boundaries using micro-fabrication or printing techniques, or by physical isolation of the required hydrophilic zones. This section summarizes some of the most commonly used techniques for the production of µPADs and their scopes.

### 3.2.1 WAX PATTERNING

Wax is hydrophobic and malleable at normal temperature and is insoluble in water. In wax-printing technique, molten wax impregnated into hydrophilic paper substrate serves as hydrophobic barriers. This direct, mask-less technique requires a commercial printing machine or a wax pen and hot plate (Lu et al. 2009; Carrilho, Martinez, and Whitesides 2009). First, solid wax patterns are printed on the paper surface using a wax-printing machine. In the following step, the wax-printed paper substrate is placed on a hotplate. The printed wax melts and slowly penetrates through the thickness of the paper matrix and forms the hydrophobic barrier. This method substantially reduces the process complexity, cost, and time and does not involve the use of organic solvents. However, wax-barrier-based devices are prone to thermal and chemical instabilities and cannot be used for processes involving high temperatures or organic solvents. With the limitation of low resolution, this technique is still promising for its rapid, inexpensive mass production capabilities (Cate et al. 2015). The eventual discontinuation of the printing apparatus by manufacturers would affect the future of wax printing as a mass-production method for µPADs.

### 3.2.2 INKJET PATTERNING

In this non-contact, direct printing method, ink in the commercial inkjet printer is conveniently replaced with a solvent. The inkjet is then used either to selectively etch patterns in a pre-hydrophobized paper or to create hydrophobic barriers in the paper matrix, see Figure 3.1a. The hydrophobization of paper is generally accomplished by soaking in a polystyrene solution (Abe, Suzuki, and Citterio 2008; Abe et al. 2010). Alternatively, the hydrophilic paper is impregnated with a hydrophobic ink by inkjet printing. Most commonly used inks include alkyl ketene dimer (ADK), UV-curable ink, and hydrophobic sol–gel. Inkjet printing offers a potential simple way of creating immuno-chemical and conductive circuits directly on paper. These printed circuits on paper, which are scalable, foldable, low-cost, and disposable, extend the applications from healthcare to diagnostics and to high-performance electronics. Inkjet-based methods allow direct printing of devices on the paper substrate. The major drawbacks of this method are the involvement of hazardous solvents, multiple printing steps, and relatively slower process, which are not suitable for rapid, mass production (Yamada et al. 2015).

### 3.2.3 PHOTOLITHOGRAPHY

Photolithography is a mask-based, indirect patterning technique, where photoresist materials are used to create hydrophobic barriers on paper substrate. It involves the treatment of the paper substrate with a photoresist material followed by exposure to UV light source using a patterned

# Paper-Based Microfluidic Devices in Food Processing

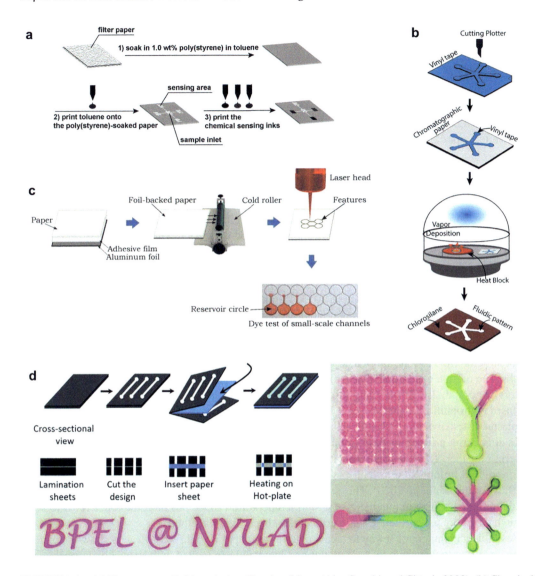

**FIGURE 3.1** (a) The process of inkjet printing. Reprinted from (Abe, Suzuki, and Citterio 2008), (b) Chemical vapor deposition. Reprinted from (Lam et al. 2017), (c) Laser cutting. Reprinted from (Mahmud et al. 2018), (d) Cut and heat (CH)-laminated µPADs. Reprinted from (Kumawat et al. 2022).

photomask. In the subsequent development step, the unexposed resist is dissolved and removed. This method yields micron-resolution patterns at the cost of complex and expensive fabrication process. The first µPADs were fabricated on a chromatographic paper coated with SU-8 2010 photoresist and using a quartz mask (Martinez et al. 2007). During the photolithography process, paper substrate undergoes spin coating, UV exposure, development, resist removal, and multiple baking steps, which demands greater preparation time. The need of costly reagents, sophisticated equipment, and multiple wet-process steps make the lithography technique unattractive for simple, low-cost mass production of the µPADs (Qin et al. 2021; He et al. 2015).

## 3.2.4 Screen Printing

In screen printing, the desired model is first printed on a screen using conventional lithography process. Solid wax is rubbed through the constructed screen which is placed over the paper substrate.

The wax is then melted and impregnated into the paper sheet by heat treatment (Dungchai, Chailapakul, and Henry 2011; Sameenoi et al. 2014). Screen-printing method is a simple, inexpensive, and universal printing technique. It eliminates the requirement of sophisticated instrumentation and skilled labor. Once the screen is created, the screen-printing method offers the flexibility to experiment with a wide range of inks or patterning materials. The poor resolution of printed patterns, lack of automation, and requirement of multiple screens to print different layers are some factors limiting the use of screen printing for mass production of μPADs (Akyazi, Basabe-Desmonts, and Benito-Lopez 2018).

### 3.2.5 OTHER TECHNIQUES

Chemical vapor deposition (CVD)-based technique involves the surface modification of paper substrate by vapor-phase deposition of hydrophobic materials (Figure 3.1b). CVD provides a dry, solventless, substrate-independent way for the production of μPADs. This method has been useful to deposit functional polymers such as poly (o-nitrobenzyl methacrylate) [PoNBMA] and poly (1H,1H,2H,2H-perfluorodecyl acrylate) [PPFDA], pure polymers such as poly(chloro-p-xylene) [PPX], and inorganic compounds such as Trichlorosilane [TCS] on paper substrate (Kwong and Gupta 2012; Chen, Kwong, and Gupta 2013; Demirel and Babur 2014; Lam et al. 2017). On the down side, this fabrication process is relatively complex and requires costly reagents, equipment, and pre-defined masks (He et al. 2015).

Selective plasma treatment is a dry, chemical-free method to alter the hydrophilic nature of the paper structure. μPADs are fabricated by exposing the paper to a plasma discharge through metal masks. The plasma generation is accomplished by vacuum plasma reactors or portable corona generators (Kao and Hsu 2014; Jiang et al. 2016). The technique has enabled the creation of fully enclosed hydrophilic channels using a combination of plasma-assisted fluorocarbon deposition and $O_2$ plasma-assisted etching (Raj, Breedveld, and Hess 2019).

In direct laser printing method, solid toner ink from a laser printer is used to create hydrophobic patterns on paper. The two-step fabrication process involves the automated printing of the patterns, followed by a heating step to melt the toner ink through the thickness of the paper (Ghosh et al. 2019; Ng and Hashimoto 2020). This provides the way for rapid production of high precision, high-resolution μPADs. The requirement of a high temperature (165°C) to melt the toner for it to seep through the entire paper thickness is a critical drawback of this method.

A simple and imaginative way to create patterns on paper is to directly write or draw the pattern using a plotting tool such as an X–Y plotter or a handheld pen loaded with special hydrophobic ink (Li et al. 2017; Li, Liu, et al. 2018b; Sousa, Duarte, and Coltro 2020). The key task behind this process lies in the precise engineering of the fluidic properties of the ink and mechanical system for the ink-ejecting head. The plotting process is inexpensive and does not require any pre-manufactured mask or mold. However, it is not suitable for mass production of μPADs since it requires high manual labor involvement and compromises the precision and resolution of the channels. Physically isolated hydrophilic channels can be created by cutting the paper into desired shapes using a knife, scissor, or cutting plotter (Thuo et al. 2014). This method provides dry, chemical-free, and mask-less production of μPADs.

Further, laser-based cutting systems provide the speed and precision required for mass production capabilities, see Figure 3.1c. $CO_2$ laser is often used for controlled ablation of the cellulose structure or cut through the paper thickness (Nie et al. 2013; Spicar-Mihalic et al. 2013; Mahmud et al. 2016, 2018). The cutting-based techniques create stand-alone hydrophilic patterns, which need a supporting platform to provide mechanical strength. In another method, where $CO_2$ laser cutter is used for cutting/creating patterns in lamination sheets. A filter paper sheet is then placed between the cut lamination sheets and heated on the hot plate for the barrier formation, see Figure 3.1d. The fabricated devices have high mechanical strength against bending, folding, and tearing and are

Paper-Based Microfluidic Devices in Food Processing

highly robust against strong acids, bases and solvents (Kumawat et al. 2022, 2023). The technique presents a simple and low-cost fabrication process using widely used consumables and commercial tools and techniques for automated mass fabrication of µPADs.

## 3.3 LIQUID MANIPULATION OR FLOW CONTROL IN µPADs

Traditional paper-based microfluidic devices (µPADs) utilize capillary action for sample transport to the reaction zones, where the samples react with the pre-deposited reagents. However, proper fluid handling is required for temporally dependent transformations such as sample separation, sample mixing, and the timed addition of multiple reagents. Hence, to expand µPADs applications and performance, studies have been reported to incorporate flow control functionality for the development of accurate, predictable, and programmable µPADs. In this section, the theory of flow control in porous substrates is discussed.

Fluid transport or flow control in µPADs is often enabled by controlling wicking-based flow or capillary action within the paper matrix. The modeling of the capillary action in porous media can be described by Lucas–Washburn equation, Eq. (3.1) (Washburn 1921; Lucas 1918), where $L$ is the distance traveled by the fluid under capillary pressure in paper channel, $t$ is the time, $D$ is the average pore radius of the paper matrix, $\mu$ is the dynamic viscosity of fluid, $\gamma$ is the surface tension, and $\theta$ is the contact angle.

$$L = \sqrt{\frac{\gamma Dt cos\theta}{4\mu}} \tag{3.1}$$

The distance the fluid front travels in a porous media is proportional to the square root of time (Jahanshahi-Anbuhi et al. 2012). Additionally, the fluid flow in a pre-wetted channel with a fixed cross section can be described by Darcy's law (Fu et al. 2011),

$$Q = \frac{kA}{\mu L}\Delta P \tag{3.2}$$

Here $Q$ is the volumetric flow rate, $k$ is the material permeability, $A$ is the cross-sectional area of the paper substrate, and $\Delta P$ is the pressure difference over the medium length.

Evaporation plays an important role in understanding the liquid flow in paper substrates, but both Lucas–Washburn equation and Darcy's law ignore the effect of evaporation on the capillary action. Liu et al. (2015) took into consideration the evaporation effect on the liquid flow length and summarized it as,

$$h_{ev} = \frac{m_e}{\rho\varepsilon w\delta} \tag{3.3}$$

Here $m_e$ is the predicted wicking liquid mass with evaporation, $\rho$ is the density of the fluid, $\varepsilon$ is the substrate's effective porosity, $w$ is the channel width, and $\delta$ is the thickness of the substrate.

The corresponding liquid speed can then be defined as,

$$S_{ev} = \frac{dh_{ev}}{dt} \tag{3.4}$$

With this model, it has been shown that the liquid flow length is proportional to $t^{1/3}$ rather than $t^{1/2}$, as predicted by Lucas–Washburn equation. These studies also revealed that evaporation during liquid flow is significantly different in open and enclosed paper channels, with a faster flow rate for the later (Jahanshahi-Anbuhi et al. 2012). For applications involving multi step reactions, a faster flow rate is desired, while in some applications like reaction incubation in paper reactors, a slower flowing rate is required. Thus, flow rate control in paper devices is essential in the field of microfluidics (Akyazi, Basabe-Desmonts, and Benito-Lopez 2018) and several methods were developed that permitted a controlled fluid flow. These techniques can be categorized into geometry-, chemical-, and mechanical-based methods.

### 3.3.1 Geometry-Based Methods

Flow rate control can be achieved by geometry-based methods that involve change in channel length, channel width, or flow path. The use of baffle design to increase the length of the flow channel was demonstrated by Apilux et al. (2013) that created a time delay resulting in the sequential reagent flow to the detection region. Automation of sequential multistep reaction of ELISA technique was possible in a single step by this simple method (Apilux et al. 2013). Fu et al. (2010) demonstrated controlling the length of the fluid travel by using three inlets with different reagents spaced at different distances from the detection zone, thereby generating sequential multiple flows to the detection zone. Acceleration of the fluid flow rate was shown by simply sandwiching paper channels between two flexible films, wherein the process dynamics obeys the law that the height of the liquid is dependent on time to the power of 1/3 (Jahanshahi-Anbuhi et al. 2012). Toley et al. (2013) demonstrated the use of an absorption pad to produce delay in the fluid flow, wherein the delay can be varied by controlling the dimensions of the pad, see Figure 3.2a. It was shown that the wax-printed barriers can delay the flow time on the device by changing the path length of the liquid flow (Preechakasedkit et al. 2018). Another geometry-based flow control strategy is to pressurize a specific region of the porous channel surface, thereby decreasing the size of the pores, this increases the fluid resistance creating a fluid delay (Shin et al. 2014).

### 3.3.2 Mechanical-Based Methods

Mechanical methods involve physical motion of components that permit connection, disconnection, or close proximity to the channel surface. Toley et al. (2015) demonstrated for the first time mechanically actuated valves for fluid control in designing portable devices, Figure 3.2b. Matsuda et al. (2015) developed an interesting idea of employing an inkjet-printed heater as a valve. The heating caused the fluid to evaporate, stopping the fluid flow at the point. It was also useful for providing heat for chemical reactions. Phillips et al. (2016) utilized tunable wax ink valves that are actuated by heating for fluid flow control. Kong et al. proposed an actuator device made out of a folded chromatography paper, which was actuated by fluid addition at either the crest or trough, engaging or breaking the fluidic contact between the channels (Kong, Flanigan, et al. 2017b). Another simple and easy valve manipulation process was demonstrated using plastic comb binding spines, using ring-shaped binders that can be flipped over and turned back (Han et al. 2018). Integrated paper-based rotation valve was demonstrated to control the connection or disconnection between the detecting zones and fluid channel, thereby improving the detection accuracy in colorimetric detection (Sun et al. 2018).

### 3.3.3 Chemical-Based Methods

Chemical methods utilize various chemicals embedded in the channel that cause variation in flow rates or disconnection of the supply from reservoir to stop the flow. Houghtaling et al. (2013)

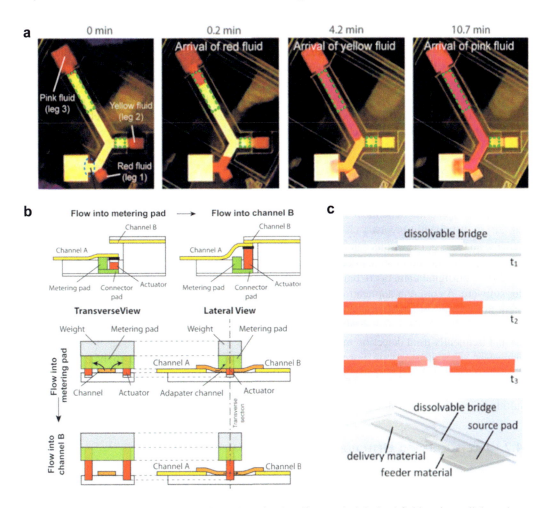

**FIGURE 3.2** (a) Images of sequentially delivered red, yellow, and pink dyed fluids using cellulose shunts. Reprinted from (Toley et al. 2013), (b) Schematic of a volume-metered valve with a cantilever channel and movable metering pad. In a cantilever channel, when the actuator is expanded by the delivered fluid from the metering and connector pads, channels A and B are connected. In a movable metering pad, as the solution flows through the metering pad, the actuator swells up to form space; then, the fluid flows into channel B. Reprinted from (Toley et al. 2015), (c) Schematic of operation mechanism and set-up of a dissolvable bridge. The bridge dissolves at a certain time to shut off the fluid flow. Reprinted from (Houghtaling et al. 2013).

demonstrated the use of dissolvable bridges for the manipulation of fluid volumes and flow in paper networks in limited source settings, see Figure 3.2c. Lutz et al. (2013) used dissolvable sugar to create programmable flow delays resulting in a multistep fluidic protocol. Similarly, water-soluble pullulan films were formatted into the paper device, permitting a slow controlled release of fluid (Jahanshahi-Anbuhi et al. 2014). Chen et al. (2012) developed a one-directional fluidic diode in which pre-embedded surfactants have been used to bridge a hydrophobic gap in the flow path. However, the disadvantage of these methods is that these dissolvable materials are composed of chemicals, which may limit the use of certain reagents, affecting the assay performance. Strong et al. (2019) attempted to overcome this limitation by introducing wax-printed fluid time delays on the top and bottom of pre-fabricated μPAD channels. Similarly, Chen et al. (2019) developed a wax-based valve that can be manually opened by organic solvents without influencing the fluid flow.

## 3.4 DETECTION METHODS IN MPADS

The main applications of μPADs are in disease diagnosis, food safety inspection and environmental analysis. The detection of analytes in these potential areas is an important challenge which includes miniaturizing the sensing equipment for conducting the assays, generate portable devices, and cost-effective detection methods. A large variety of methods have been employed for the detection of analytes in μPADs. Commonly used methods are colorimetric, electrochemical, chemiluminescence, fluorescence, and nanoparticle-based detection techniques.

### 3.4.1 COLORIMETRIC DETECTION

Colorimetric detection is the most widely used detection method for μPADs because it provides easy read-out of the generated chemical signal, paving the way for instrument-free measurements (Dai et al. 2017). It typically involves visually observing the color change during a chemical reaction, where the analysis of the results can be evaluated by naked eyes or in a semiquantitative way (Sechi et al. 2013). This method can also be quantified by handheld scanners, cell phone cameras (Yang et al. 2016), or digital cameras (Chaiyo et al. 2016) and further analysis through the aid of a computer software. A common application of colorimetry is the simultaneous detection of glucose and protein in artificial urine (Martinez et al. 2007), see Figure 3.3a. In the areas of protein and drug analysis, as well as in environmental monitoring and food safety, colorimetric detection is the simplest and frequently used detection method.

However, achieving accurate detection with the naked eye is challenging due to the inhomogeneity of the color distribution on the paper assay (Bruzewicz, Reches, and Whitesides 2008). High background noise of the paper or the sample is another disadvantage of colorimetric detection resulting in low detection limits.

### 3.4.2 ELECTROCHEMICAL DETECTION

One of the major challenges with colorimetric μPADs is achieving low detection limits with high sensitivity and selectivity. To circumvent these issues, Dungchai, Chailapakul, and Henry (2009) developed the first electrochemical μPADs using electrochemical detection, involving the direct conversion of a biological or chemical signal to an electrical one, to enhance the analytical performance of μPADs. Electrochemical detection is low-cost and simple, has high specificity and sensitivity, and can be characterized by low-power consumption and minimal instrumentation (Bhardwaj et al. 2017; Pavithra, Muruganand, and Parthiban 2018; Fan, Hao, and Kan 2018; Wang, Jian, et al. 2018a; Wang, Cheng, et al. 2018b). Furthermore, electrochemical detection-based μPADs are in general insensitive to lighting conditions and contaminants, leading to more stable output unlike colorimetric-based μPADs (Nie et al. 2010), see Figure 3.3b.

A paper-based electrochemical sensor comprises a three-electrode system: a counter, working, and reference electrode. The fabrication process for electrochemical sensors involves the deposition of electrodes in the form of conductive inks (silver or carbon inks) on the paper matrix (Dungchai, Chailapakul, and Henry 2009). The conductive electrode pads are screen-printed in the hydrophilic region of the paper in which enzymes are spotted. The wicking property of the paper is used to flow the sample to the sensing zones with electrodes, and a precise signal is obtained in a short time. Creative methods for fabricating electrodes have allowed for an array of materials to be incorporated with a resulting increase in applications and performance (Adkins, Boehle, and Henry 2015). The electrochemical detection mechanism can be based on several modes, including cyclic voltammetry (Li, Li, Ge, et al. 2014c), chronoamperometry, square wave voltammetry (Shiroma et al. 2012), and potentiometric measurements (Lisak, Cui, and Bobacka 2015).

Paper-Based Microfluidic Devices in Food Processing

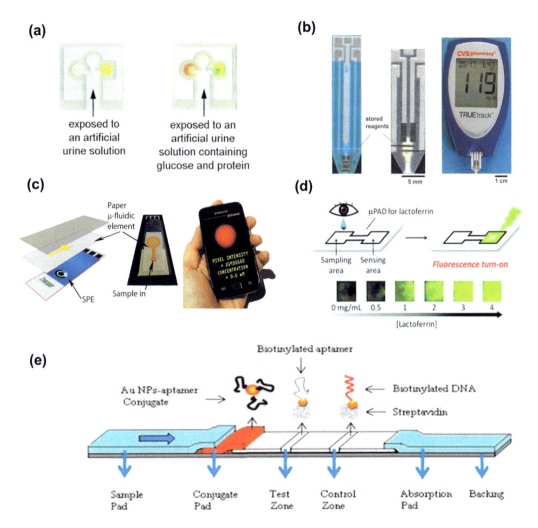

**FIGURE 3.3** (a) Colorimetric detection for simultaneously detecting glucose and protein in an artificial urine sample. Reprinted from (Martinez et al. 2007), (b) Components of an electrochemical µPAD-based system that uses a commercial glucometer as an electrochemical reader. A photograph of a commercial test strip made from plastic (left) and an electrochemical µPAD made from a single layer of paper (center), and (right), the glucometer used as a reader. Reprinted from (Nie et al. 2010), (c) ECL detection of a sample solution (2-(dibutylamino)-ethanol (DBAE)). The device was filled with a 10 mM Ru(bpy)3 2fl solution before drying and was then aligned and fixed onto the face of the screen-printed electrode (SPE) by laminating with transparent plastic. A drop of sample was introduced through a small aperture in the plastic at the base of the channel. After the detection zone being fully wetted, a potential of 1.25 V was applied and the resulting emission was captured and analyzed. Reprinted from (Delaney et al. 2011), (d) Fluorescence-based lactoferrin assay on µPADs. Reproduced from (Yamada et al. 2014), (e) Lateral flow strips with aptamers conjugated with gold NPs. Reprinted from (Quesada-González and Merkoçi 2015).

Electrochemical method-based µPADs have been employed for the detection of analytes such as glucose, lactate, uric acid, and cholesterol in blood and urine (Dungchai, Chailapakul, and Henry 2009; Nie et al. 2010), neurotransmitters and drugs (Rattanarat et al. 2012; Shiroma et al. 2012; Feng et al. 2015; Su et al. 2014; Liu, Ge, et al. 2014a), DNA and nucleosides (Lu et al. 2012; Cunningham, Brenes, and Crooks 2014), heavy metals (Apilux et al. 2010; Ruecha et al. 2015;

Chaiyo et al. 2016), ions (Mensah et al. 2014; Hu, Stein, and Bühlmann 2016; Rattanarat et al. 2012), and immunoassays (Zang et al. 2012; Wu et al. 2012; Li, Li, Yang, et al. 2014a**; Li, Xu, et al. 2014b; Wang et al. 2016). High potential of this detection technique is demonstrated in the commercial success story of detection of glucose with the combination of simple electrochemical µPADs with a handheld glucometer (Nie et al. 2010), see Figure 3.3b. Regardless of the attractive features of electrochemical detection such as high sensitivity, miniaturization possibilities, and low levels of background noise from paper, the need for a detection instrument makes the device complex and costly, which is a significant drawback for point-of-care devices (Yetisen, Akram, and Lowe 2013). There is still room for fundamental improvements in the devices by focusing on combining the unique advantages of self-pumping microfluidic network with the power of electrochemistry to provide impacts for areas like home medical diagnostics (Adkins, Boehle, and Henry 2015).

### 3.4.3 Chemiluminescence and Electrochemiluminescence Detection

Chemiluminescence (CL) mechanism is based on the emission of light generated by a chemical reaction. The luminescence in this method is controlled by the mixing of fluid flow reagents. The major advantage of the CL method is that it doesn't require a light source and it requires only a single light detector. It uses inexpensive reagents and is characterized by a high signal-to-noise ratio and low limits of detection (Luo et al. 2017). However, the measurement using this method needs to be done in the dark making the detection process complicated (Yu et al. 2011). This sensing method was employed for the detection of DNA (Liu and Zhang 2015), cancer markers (Wang et al. 2015), metals (Alahmad et al. 2016), and food safety analysis (Liu, Kou, et al. 2014b).

Electrochemiluminescence (ECL) sensing methods are based on generating luminescence during electrochemical reactions. This method can combine CL reactions and electrochemical detection methods via screen-printed electrodes (Delaney et al. 2011), see Figure 3.3c. ECL is better compared to CL methods in terms of lower background, temporally controlled signal capture, and improved selectivity via potential control (Hu and Xu 2010). This technique is used in the detection of tumor markers (Wang et al. 2012) and cancer cells (Li et al. 2015). Nevertheless, in this technique, readouts must be performed in the dark and require a photomultiplier tube which is costly in a miniature form, making it less viable in resource limited settings (Yetisen, Akram, and Lowe 2013)

### 3.4.4 Fluorescence Detection

µPADs based on fluorometric detection exploit the interaction between the target molecules and fluorescent dyes and measure the emission intensity during analysis under an excitation illumination. The two major flaws of this detection technique are the need for additional instrumentation and the significant background noise from the additives (to improve the whiteness) in commercial paper (Pelton 2009). This detection technique requires more equipment like light source, and detectors, which makes it less suitable for paper-based devices that aim to be facile and low cost. However, many fluorescent-based µPADs have been developed with low detection limits (Thom et al. 2012), for detecting bacteria (Rosa et al. 2014), drugs (Caglayan et al. 2016), cancer cells (Liang et al. 2016), heavy metal ions (Qi et al. 2017), immunoassays (Liang, Wang, and Liu 2012), and eye disorders (Yamada et al. 2014), see Figure 3.3d. Even though fluorescence sensing provides high specificity and selectivity, further improvement in the technique can only be done with the cost-effective size reduction of fluorescence readers.

### 3.4.5 Nanoparticle-Based Detection

Another sensing method for the detection of µPADs makes use of nanoparticles (NPs). The optical properties of NPs, due to SPR (surface plasmon resonance) (Mulvaney 1996), enable naked eye observation of any bio-recognition event. Gold NPs are used with aptamers for the detection of

# Paper-Based Microfluidic Devices in Food Processing

thrombin (Quesada-González and Merkoçi 2015), see Figure 3.3e. Multicolored silver NPs have been produced for the differential detection of dengue, yellow fever, and Ebola viruses (Yen et al. 2015). Non-metallic NPs like latex and carbon dots (CDs) are also employed in the detection of analytes in multiplexed assays. Latex NPs are widely used for the detection of antigens or antibodies due to their low cost and biocompatibility with multiple molecules (Quesada-González and Merkoçi 2015). Pedro et al. (2014) used CDs for colorimetric pH detection. The strong dark suspension color of CDs provides a high contrast to the cellulose fibers of a paper matrix. Besides these conventional detection methods, there are applications that employ spectroscopy-based mechanisms.

## 3.5 APPLICATION OF μPADs IN FOOD PROCESSING

The prerequisite for the application of μPAD technology in food safety analysis is the conversion of food samples into liquid form for optimum assay performance. Detecting hazardous substances and pathogens in food is quite different from detecting those substances in clinical samples. For example, the food sample could be a piece of steak composed of muscle fibers with raw and denatured proteins. It contains lipids, insoluble fat, meat juices, salts, and spices added during cooking, also rich in many compounds produced in the Maillard reaction during cooking (Tamanna and Mahmood 2015; Hua et al. 2018). This shows that, depending upon the nature of the analyte in the food and the μPAD matrix, strong physical and chemical interactions may interfere with the availability of the analyte to the detecting agent.

Sampling and sample preparation are crucial steps in food analysis, which influences the reliability and accuracy of results. It is crucial to use a representative sample of a defined whole, e.g., a single manufacturing batch, or a particular lot in a production batch. The method of sampling and pretreatment depends on the size and nature of this whole. Homogenization and pretreatment are carried out by cutting, chopping, blending, concentrating, drying, and diluting in an appropriate liquid media to make it compatible with the μPADs. This typically involves mixing the food with a solvent, an extraction step to free the analyte from the food matrix, followed by a clean-up step to remove the interfering substances, and finally a concentration step or dilution step to be added, if required. Generally, the μPAD technology is compatible with the common food sample preparation methods for analysis. Table 3.1 sums up the major applications of μPAD technology in the food processing field.

### 3.5.1 Detection of Food Preservatives and Additives

Food industry widely uses food preservatives and additives to enhance the flavor, color, and shelf life of the food. Most of these food additives are safe and non-toxic in limited amount but can cause unhealthy conditions if consumed in large quantities. Many types of μPAD devices have been developed for detecting food preservatives and additives (Soman et al. 2024), and some of the leading methods are described in this section.

Researchers have developed inkjet-printed paper-based sensors to detect food additives such as food coloring agents (e.g., sunset yellow), food preservatives (e.g., nitrite), and vitamins (e.g., vitamin C). The device uses fluorescent CDs for the detection purpose. CDs are nanomaterials that can emit or quench florescence when binding with specific substances such as food additives and preservatives (Deng et al. 2021; Gan et al. 2020). In this method, an optimized ink is prepared using common organic solvents (e.g., absolute ethanol), a retarding medium to slow the drying (e.g., polyethylene glycol), and surfactants (e.g., FS3100 and SE-F) to control the viscosity and surface tension. In one example, the optimized ink was filled into a clean printer cartridge (manufactured by Hewlett Packard or HP) connected to a computer, and a pre-designed pattern was printed using an HP Deskjet 2628 printer on the No. 2 medium-speed flow qualitative filter paper. After patterning, the microfluidic chips were prepared by drying in temperatures ranging from 60 to 100°C. The method demonstrates the fluorescence-based qualitative detection of food additives at various

## TABLE 3.1
## Applications of μPAD Technology in Food Processing

| No. | Application | Fabrication method | Detection scheme |
| --- | --- | --- | --- |
| 1 | Neurotoxic pesticides – bendiocarb, carbaryl, malathion Hossain et al. (2009) | Inkjet printing | Colorimetric assays |
| 2 | Food-borne pathogens – *Escherichia coli, Salmonella spp.*, and *Listeria monocytogenes* Jokerst et al. (2012) | Wax printing | Colorimetric assays |
| 3 | Quality control – food additives – nitrite in food He et al. (2013) | DUV lithography | Colorimetric assays |
| 4 | Quality control – amylose in rice Hu et al. (2015) | Plasma treatment | Colorimetric assays |
| 5 | Food-borne pathogens – *Salmonella* Jin et al. (2015)) | Wax printing | Colorimetric assays |
| 6 | Pesticide residue – DDV in vegetables Liu, Kou, et al. (2014b) | Cutting/masking with adhesive tapes | Chemiluminescence |
| 7 | Quality control – preservative additives – nitrite in ham, sausage, and water Cardoso, Garcia, and Coltro (2015) | Stamping | Colorimetric assays |
| 8 | Neurotoxic pesticides – MPO and CPO in cabbage and green mussel Nouanthavong et al. (2016) | Screen printing | Colorimetric assays |
| 9 | Adulteration – caramel in whiskey Cardoso et al. (2017) | Wax printing | Colorimetric assays |
| 10 | Chemical residue – bisphenol A in food packets Kong, Wang, et al. (2017a) | Screen printing | Colorimetric assays |
| 11 | Adulteration – ketamine in beverages Narang et al. (2017) | Wax printing | Electrochemical |
| 12 | Quality control – benzoic acid in food Liu et al. (2018) | Wax printing | Colorimetric assays |
| 13 | Pesticide residue – thiram, thiabendazole, and methyl parathion in fruits and vegetables Ma, Wang, et al. (2018b) | Screen printing | SERS |
| 14 | Mycotoxin – DON, ZEN, T-2, and HT-2 in cereals Li, Chen, et al. (2018a) | Photolithography | Chemiluminescence |
| 15 | Drug residue – clenbuterol in milk Ma, Nilghaz, et al. (2018a) | Wax printing | Colorimetric assays |
| 16 | Food allergens and toxins – egg white lysozyme, ß-conglutin lupin, okadaic acid, and brevetoxin Weng and Neethirajan (2018) | Photolithography | Fluorescence |
| 17 | Mycotoxin – DON in food/feed (Jiang et al. 2019) | Wax printing | Hybrid/colorimetric assays |
| 18 | Quality control – food additives – colorants from the skin of dals and vegetables Kumar and Santhanam (2019) | Inkjet printing | SERS |
| 19 | Quality control – food additives – tartrazine and indigo carmine Gharaghani, Akhond, and Hemmateenejad (2020) | Laser printing | Colorimetric assays |
| 20 | Antibiotic residue – norfloxacin in meat Trofimchuk, Nilghaz, et al. (2020a) | Wax printing | Colorimetric assays |
| 21 | Antioxidant residue – gallic acid or oenotannin in fruits Martínez-Pérez-Cejuela et al. (2023) | Hybrid – cutting/lamination | Hybrid/colorimetric assays |
| 22 | Adulteration – syrups in natural honey Masoomi, Sharifi, and Hemmateenejad (2024) | Laser printing | Colorimetric assays |
| 23 | Pesticide residue – butachlor in mung beans Wu et al. (2024) | Wax printing | Colorimetric assays |

concentrations using a UV lamp at 365 nm, see Figure 3.4a. On the downside, the performance of the inkjet printing is highly reliant on the seamless formulation of the printing inks, which requires high-level optimization.

Low-cost, sustainable biosensors using cellulose paper substrate have been developed to measure glucose in commercial beverage samples with a limit of detection comparable to high-performance liquid chromatography (Kuek et al. 2014). The device was made up of hydrophilic cellulose paper disk with adsorbed immobilized glucose oxidase enzyme placed on top of a screen-printed carbon electrode. The device used amperometric biosensing. An amperometric biosensor functions based on the oxidation and reduction of an electroactive species on a biosensor surface using immobilized

Paper-Based Microfluidic Devices in Food Processing

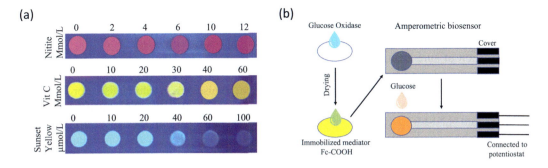

**FIGURE 3.4** (a) A paper-based chip was used for visual determination of food additives at different concentrations. Fluorescent carbon dots (CDs), namely, red-CDs, blue-CDs, and yellow-CDs, were visualized under UV spectrum to detect nitrite, vitamin C, and sunset yellow in a sample, (b) Amperometric biosensing of glucose in commercial beverages in cellulose paper-based μPADs.

enzymes. Amperometry measures the electric current versus time $(i - t)$ when a constant electric potential is maintained. The glucose biosensor could accurately analyze very low sample volumes (5 μL) of commercial beverages using the reaction chemistry illustrated in Figure 3.4b.

Nitrite is a preservative used to extend the shelf life of meat and to give meat a fresh appearance. High nitrate and nitrite content in meat and vegetables can cause methemoglobinemia and gastric cancers. The colorimetric determination of nitrite was performed through Griess reaction carried on μPADs (Cardoso, Garcia, and Coltro 2015; Giustarini et al. 2004). In this reaction, nitrite reacts with sulfanilamide and produces positively charged diazonium salt. This salt couples with $N$-α-naphthyl-ethylenediamine to produce a magenta azo dye compound. The formation of azo compound is directly proportional to the nitrite concentration in the sample. The image of the color developed was captured using a phone camera and analyzed using ImageJ. The test can be carried out quickly within ~15 min, with a sensitivity of 1.1 mg-kg$^{-1}$ (see Figure 3.5a). The microfluidic reaction on μPADs could detect nitrite in meat samples including processed meat, pork, ham, sausages, and drinking water (Trofimchuk, Hu, et al. 2020b; Jayawardane et al. 2014).

Benzoic acid and its salts are frequently used to preserve pickled food and beverages. A Whatman qualitative filter paper-based microfluidic chip device was fabricated for the detection of benzoic acid in food using the Janovsky reaction theory. Janovsky reaction was performed in the following manner in the reaction zones of the μPAD. The circular reaction zones of the μPADs were implanted with 5 N sodium hydroxide and dried at 30°C for 20 min. The benzoic acid sample was converted to 3,5-dinitrobenzoic acid using KNO$_3$ and H$_2$SO$_4$ reagents at 40°C for 40 min and was added to the reaction zones. Then, the μPAD device was transferred to a portable detection system and heated at a temperature of 45°C for 20 min on a hot plate to carry out the chemical reaction. The resulting color change from light to dark brownish-orange shade in the detection zone is imaged using a complementary metal oxide semiconductor (CMOS) camera. The color change was analyzed using an RGB analysis software after transferring to a smartphone. The color change was proportional to the benzoic acid concentration in the sample (Liu et al. 2018; Bansal et al. 2017). Researchers have analyzed 21 different types of commercial food samples, including sauces, processed fruits, and dried and pickled vegetables using this method. The standard deviation of the results was not more than 6.6% compared to the HPLC method and the coefficient of correlation across the tests was equal to $R^2 = 0.9953$.

### 3.5.2 Detection of Food Adulteration with Chemical Toxicants, Pesticides, and Herbicides

Numerous μPAD developments have been published to detect adulterants such as hazardous chemicals, toxicants, pesticides, and herbicides in food processing field, in the last decade. Metallic compounds such as mercury, arsenic, cobalt, and cadmium are commonly found in food adulterants. The most common method of food adulteration in solid food is the addition of substances such as sand, ground stones, and pebbles to food grains and used tea leaves to tea. Moreover, deliberate contamination of pure forms of food products with low-quality foreign substances such as rancid oils, and altered species of meat are emerging concerns.

Researchers developed μPAD devices to detect mercury poisoning in salmon fish. For the detection, the μPADs were coated with modified gold nanoparticles (AuNPs). The AuNPs have been modified with N,N'-bis (2-dihydroxyethyl) dithiooxamide (HEDTO) and coated on a hydrophobic–hydrophilic barrier created by triethoxymethylsilane. Upon adding mercury-containing samples, HEDTO-AuNPs formed aggregates and showed a color change from red to blue on μPADs. The detection was facilitated by capturing the images of color change using a smartphone camera. The images were analyzed using Adobe Photoshop CS6 image processing software. The technique could offer a limit of detection of 15 nM of mercury from food samples (Shariati and Khayatian 2020).

One modification in the μPADs technology is the development of electrochemical micro fluidic paper-based analytical devices integrated with nanotechnology, called EμPADs. EμPADs report the detection and quantification of anesthetic drugs such as ketamine in alcoholic and non-alcoholic drinks using electrochemical sensing. Ketamine is a criminally abused drug in party beverages, and in incidences of robbery to sedate the victims. In this method, the EμPAD surface was coated with zeolite-nanoflakes and graphene-oxide nanocrystals (Zeo-GO). When the test beverage (alcoholic drink or fruit juice) was applied to the circular working zone of the EμPADs, the nanocrystals of ZeO–GO EμPAD showed electro-oxidation of ketamine presented in the beverages. Thus, causing a change in the signal intensity while sensing with a sensor. The change in the intensity was correlated to the drug concentration in the beverages. This method showed a swift response time of 2 seconds, with a limit of detection of 0.001 nM/ml and a wide range of detection from 0.001-to 5nM/mL(Narang et al. 2016). Similar types of μPADs were devised for the detection of misused drugs such as Estazolam and Clenbuterol (Sha et al. 2020; Ma, Nilghaz, et al. 2018a).

Melamine is a nitrogen-containing substance illegally added to milk, infant formula, and pet food to artificially increase the protein value of the food (Gao, Huang, and Wu 2018; Devries et al. 2016; Alam et al. 2017). The addition of 1% melamine in food causes an artificial elevation of food protein content to 4% (Song et al. 2015). Melamine in excess amount can form insoluble melamine-cyanurate crystals in the kidney and cause renal damage (Kumar, Seth, and Kumar 2014; Manzoori, Amjadi, and Hasanzadeh 2011; Wang et al. 2010). Colorimetric detection of melamine adulteration in milk was detected on a Whatman filter paper-based visual sensor using Triton X-100 modified, stabilized AuNPs. The chemical reaction happens between the melamine and the AuNPs through the ligand exchange with citrate ions on the surface of AuNPs, leading to the Triton X-100 to be removed. The AuNPs aggregate and produce a color change from a wine red to blue depending on the concentration of the melamine. This color change causes a shift in absorption peak while measuring using UV–vis absorption spectroscopy. The method could accurately detect melamine as low as 5.1 nM in milk samples, which is a lower value of melamine concentration as per food safety regulations (Trapiella-Alfonso et al. 2013). The naked eye could detect 1.0 μM Melamine in milk samples. The same technique was modified to suit the rapid testing in field conditions using smartphone-based detection of color change.

A paper-based microfluidic device was fabricated and tested for 10–50% (v/v) palm oil adulteration in sunflower oil. The devices were prepared in the form of circular discs and rectangular channel strips using simple cutting and crafting techniques and used for visible colorimetric detection of oil adulteration. The data is analyzed using digitally acquired images along with UV–vis spectrophotometry. The digital imaging of spot tests of oil samples was performed and the average grayscale intensity data for different oil mixture compositions was plotted in a graph and interpreted. The coefficient of determination ($R^2$) of the tests was 0.9464, indicating a good empirical fit and suitability of the method for quantitative detection. The assay device was evaluated for the stability of signal over a time span of six days after the test. The images were captured every day and analyzed

Paper-Based Microfluidic Devices in Food Processing     **69**

for grayscale intensity values using ImageJ software. The signal remained fairly constant over the test duration, allowing the flexibility of data analysis after the test (Muthukumar et al. 2021).

The use of organophosphorus (OP) pesticides such as carbamates is common in agriculture and forms a major chemical hazard in food. Carbamate is listed as an endocrine disruptor compound, capable to cause hormonal abnormalities if consumed (Moreira et al. 2022). Analyzing these compounds in food materials is important in the food processing step. The use of μPADs technology has been developed for the colorimetric determination of carbamate pesticides based on the inhibition of acetylcholinesterase (AChE) enzyme by the pesticide while reacting with a specific chemical substrate (Xie et al. 2020; Aidil et al. 2013; Ellman et al. 1961). The reaction chemistry is as follows: acetylthiocholine iodide substrate is hydrolyzed into thiocholine and acetic acid by AChE. The thiocholine base reacts with 5,5′-dithiobis-(2-nitrobenzoic acid) (DTNB) to generate a yellow color detectable at 405 nm. The intensity of the yellow color developed is inversely proportional to the pesticide concentration. The images were captured using a desktop scanner and analyzed for color intensity using ImageJ software. The device was designed as a spot-array assay with a well diameter of 10 mm detection zone on the patterned paper. The device was distinguished with color codes on hydrophobic and hydrophilic zones, blue color was identified as hydrophobic zone and yellow for the reaction developing zone. The design was printed on the sheet of Whatman chromatography paper using a wax printer. The hydrophobic barrier was created on the paper by heating the paper at 150°C for 2 min in an oven to melt the wax to impregnate the wax throughout the thickness of the paper in the defined design. The AChE inhibition was determined by plotting a calibration curve of the pesticide concentration against percentage inhibition (Beshana et al. 2022; Nouanthavong et al. 2016). The results were compared with no pesticide controls and different concentrations of pesticides. Another, important point to be noted in assay using enzyme kinetics is that high concentration of pesticides may inhibit the complete activity of AChE, shown as the absence of yellow color. Therefore, a standard detection range is to be followed while conducting these assays.

Researchers have developed a method to detect methyl parathion on fruit surface. They developed a simple, flexible, and sensitive surface-enhanced Raman scattering (SERS) method using AuNPs. The filter paper substrate was immersed in the prepared AuNPs solution. The spectroscopic probe molecule 4-mercaptobenzoic acid (4-MBA) was used to evaluate the performance of the paper substrate for an optimized signal using a portable Raman spectrometer coupled with 785 nm laser. Then, the substrate was applied to detect methyl parathion standard solutions with a linear range between 0.018 $\mu g/cm^2$ and 0.354 $\mu g/cm^2$ with a limit of detection of 0.011 $\mu g/cm^2$. After standardization, the actual fruit peel sample spiked with methyl parathion was used to verify the test. The test recovery rate was 94.09–98.72%, indicating high reliability in testing fruit samples without rigorous pretreatment. The method showed excellent reproducibility and stability (Xie et al. 2020).

Lead (Pb) is a toxic metal of greatest public health concern as per the World Health Organization (WHO). Pb exposure accounts for more than one million deaths each year and causes the loss of 24.4 million disability-adjusted life years. Pb causes irreversible acute and chronic neurotoxicity and immune dysregulations (Mishra 2009; O'Connor et al. 2020). Pb toxicity occurs through food crops, preserved eggs, and drinking water. A distance measuring-paper-based analytical device (dPAD) has been developed for the detection of Pb in preserved century eggs. The assay uses the competitive binding chemistry between carminic acid (CA) and polyethyleneimine (PEI) to detect Pb in food samples. The device was patterned with a sample reservoir and a hydrophilic channel containing a colorimetric indicator reactive to the sample. When the sample is added and flows through the channel, the indicator generates a colored band, whose length can be measured by keeping a ruler along the length of the channel. The length of the band will be inversely proportional to the concentration of the sample added. The method allows to read and interpret the results without using any sophisticated read-out system. dPAD showed good linear correlation of $R^2$ value 0.974 for measuring Pb, within the ranges of 5–100 $\mu g \cdot mL^{-1}$. The results from the dPAD were comparable to the measurement obtained by atomic absorption spectroscopy (Katelakha et al. 2021). dPAD has the potential to be applied in food processing in developing countries, and it does not require laborious instruments and procedures.

A colorimetric determination method for analyzing the widely used herbicide, glyphosate in food grains was developed on µPADs. Glyphosate is linked to disease conditions such as organ toxicity and cancer (Marino et al. 2021). The glyphosate detecting system showed improved selectivity and sensitivity by using Mn–ZnS quantum dot (QD) embedded molecularly imprinted polymers (MIP) on µPADs. The detection of glyphosate is based on the oxidation of 2,2′-azino-bis(3-ethylbenzothiazoline)-6-sulfonic acid (ABTS) by $H_2O_2$ in the presence of Mn–ZnS QD-MIP. Glyphosate non-binding Mn–ZnS QD-MIP generates ˙OH from $H_2O_2$, producing a dark green color of the test zone. The binding of glyphosate to the Mn–ZnS QD-MIP turns off the generation of ˙OH, resulting light green color (Sawetwong et al. 2021).

### 3.5.3 DETECTION OF FOOD-BORNE PATHOGENS AND MICROBIAL TOXINS

Every year, around 600 million people become sick by consuming unsafe food, as per the reports of the World Health Organization. Food-borne pathogenic bacteria causes 420,000 deaths worldwide (Mazur et al. 2023). The major microorganisms associated with food poisoning are: bacteria, viruses, parasites, fungi, and mycotoxins. µPAD-based spot assays were developed for the detection of *Escherichia coli* O157:H7, *Salmonella Typhimurium*, and *Listeria monocytogenes* in food samples, using wax printing on filter paper (Jokerst et al. 2012). The detection was based on the measurement of color change of a chromogenic substance by the bacterial-specific enzymes (Table 3.2).

A novel µPAD-based single-input channel microfluidic device with aptasensors could detect *Escherichia coli* O157:H7 and *Salmonella typhimurium* simultaneously. Aptasensors are single-stranded nucleic acid (DNA or RNA) probes that can be artificially synthesized in the lab. Aptasensors specific to a certain bacterial species interact with the surface of the cells directly and have the ability to detect whole bacteria in a sample without complex sample preparation procedures (Somvanshi et al. 2022). The µPAD device was fabricated by polystyrene (PS) microparticles assembled with AuNPs, combined with a salt-based aggregation mechanism to produce stable colorimetric signals. The µPAD devices were completely dried after adding samples and used for image acquisition and analysis. Images were captured using an iPhone camera and grayscale images were analyzed using a python algorithm. The colorimetric results show linearity over a wide concentration range of bacterial culture ($10^2$CFU/mL to $10^8$CFU/mL). The assay showed a limit of detection (LOD) of $10^3$CFU/mL *Escherichia coli* O157:H7 and that of *S. typhimurium* was $10^2$ CFU/mL. This method shows the robustness of multiplexing and rapid testing on µPADs, which has the potential to screen pathogenic bacteria in water and food products in field conditions.

### TABLE 3.2
### Chromogenic Detection Systems on µPADs Using Bacterial Specific Enzymes

| Bacteria | Enzyme | Chemical substrate | Colour change |
|---|---|---|---|
| *Escherichia coli* | β-Galactosidase | Chlorophenol red | Yellow to red |
| *Salmonella enterica* | Esterase | 5-Bromo-6-chloro-3-indolyl caprylate | No color to purple |
| *Salmonella typhimurium* | β-Galactosidase (used as an antibody–enzyme complex, for more specificity) | Chlorophenol red-β-d-galactopyranoside | Yellow to purple |
| *Listeria monocytogenes* | Phospholipase | 5-Bromo-4-chloro-3-indolyl myo-inositol-1-phosphate | No color to blue |
| *Staphylococcus aureus* | Protease | Proteolysis of magnetic nanobead–peptide probes | Black to golden |

Coliforms form the most common bacterial pathogen responsible for water contamination. If consumed, it may cause extreme illness to fatality. An open-channel, µPAD microfluidic device was fabricated for coliform lysis and detection. The device was created by direct printing of omniphilic channels on an omniphobic, fluorinated paper. This µPAD device demonstrated the flow and control of both high- and low-surface tension liquids such as cell lysing agents. The µPAD device could detect *Escherichia coli*, at a concentration as low as ~$10^4$ CFU-mL$^{-1}$, using the bacteria-specific, β-galactosidase enzyme (Snyder et al. 2020).

Indirect detection of bacteria using bacterial-released proteases is also possible on µPADs. Paper strip containing magnetic nanobead–peptide probes integrated with a gold sensing platform detected the presence of *Staphylococcus aureus* in food samples such as ground beef, turkey sausage, lettuce, and milk. The reaction detects the cleavage by *Staphylococcus aureus*-specific proteases and dissociation of the magnetic nanobeads from the sensor surface. A paper-based microfluidic platform combining advanced molecule techniques using loop-mediated isothermal amplification (LAMP) combined with clustered regularly interspaced short palindromic repeats (CRISPR) technology for fluorometric detection of nucleic acids is reported. This technology has the potential to detect food-borne pathogen with high sensitivity and specificity (Yang et al. 2020; Anindita et al. 2023).

An inexpensive µPAD-based chromogenic array, named paper chromogenic array (PCA) to identify individual pathogens (*Listeria monocytogenes*, *Salmonella enteritidis*, and *Escherichia coli* O157:H7) or multiple pathogens in the presence of nonpathogenic background microflora on food was developed (Jia et al. 2021). The array is integrated with a machine-learning approach. Grade 1 cellulose chromatography paper PCA was fabricated into which 22 chromogenic dye spots were infused along with standard color dots (Figure 3.5b). When the array is exposed to volatile organic compounds emitted by the pathogens of interest, the dye spots exhibited remarkable color changes and pattern shifts. The pattern was analyzed digitally and used to construct an advanced deep feed-forward neural network. After training, the network demonstrated excellent performance in identifying pathogens with an accuracy of up to 93%.

The presence of microbial toxins in food materials is a serious problem worldwide. Aflatoxin B1 is a common milk contaminant and carcinogen, which can cause liver diseases. A microfluidic paper device was developed for the rapid detection of aflatoxin B1, using specific aptasensors in a colorimetric

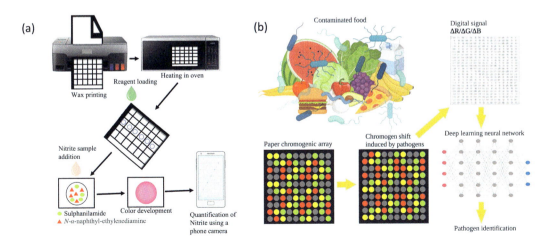

**FIGURE 3.5** (a) The colorimetric determination of nitrite, a food preservative, was performed through Griess reaction carried out on µPADs. The intensity of color developed was analyzed using a mobile device. (b) Pathogen detection array (PCA). The array was developed and integrated with a machine-learning approach. PCA was fabricated using Grade 1 cellulose chromatography paper into which 22 chromogenic dye spots were infused along with standard color dots. Pathogen-specific volatile organic compounds caused remarkable color changes and pattern shifts. The pattern is analyzed using a neural network to identify the pathogen.

assay. Aptameric–AuNP conjugate was physically adsorbed on a μPAD and aflatoxin B1 was allowed to flow over the μPAD. The nanoconjugate was characterized using UV–vis spectroscopy, and dynamic light scattering for measuring hydrodynamic diameter and zeta potential. The assay could detect 1 μM to 1 pM of aflatoxin B1 with a limit of detection of 10 nM in standard samples (Kasoju et al. 2020). In another study, a paper-based microfluidics chip to measure the mycotoxin, deoxynivalenol was devised using a colorimetric competitive immunoassay, using gold nanoparticles, with a detection range of 0.01-20 ppm (Jiang et al. 2019).

### 3.5.4 Detection of Food Allergens

Food allergy is an important food-associated health concern affecting up to 10% of the population, causing mild to severe symptoms, or may cause fatality even. Rapid and accurate detection of allergens in food is most critical for effective curbing of grave outcomes. Most of the available allergen tests are time consuming and costly. μPADs demonstrate the potential to address these challenges using cost-effective materials. A lateral flow immunoassay (LFI) μPAD device for quick detection of allergic protein in food samples was fabricated using modified cellulose material. The device could detect allergens within 15 min including sample preparation. The sample flow rate was optimized by defined geometric patterns. This μPAD could detect as low as 1 ppm ovalbumin – a major egg allergen, in different food samples, in various stages of food processing (Hua and Lu 2020). A sandwich immunoassay strip test with nitrocellulose membrane was developed to detect Bowman–Birk inhibitor, a type of antinutritional factor present in soybean, causing indigestion and stunted growth in humans and animals (Wang et al. 2019). The test could detect allergen concentrations up to 0.5 μg/mL for visual and 0.23 μg/mL for strip reader. The assay was read by a TSR3000 Membrane Strip Reader (BioDot, USA) for quantitative analysis.

A trichromatic multiplex lateral flow immunoassay (xLFIA) to detect milk casein, ovalbumin, and hazelnut proteins was devised. The antibodies directed against the proteins were individually adsorbed onto gold and silver NPs to produce specific-colored probes on nitrocellulose membrane (Hi-flow plus 180) strips. These were inserted in an LFIA comprising three lines, each responsive for one allergen. The xLFIA easily identified allergens in commercial biscuits as low as 0.1 mg/l (Anfossi et al. 2019). This paper-based immunoassay integrated with microfluidic device for allergen detection offers advantages in fluid control and multiplexing capabilities.

### 3.5.5 Detection of Antibiotics Residues in Food

Antimicrobial resistance is a global threat due to the non-judicious use of antibiotics to treat bacterial infections in food producing animals. Predictive statistical models estimated 4·95 million deaths associated with bacterial antimicrobial resistance in 2019 (Mazur et al. 2023). A number of cost-effective μPAD-based diagnostic devices have been developed to detect antibiotics residues in milk, egg, and meat (Taghizadeh-Behbahani, Shamsipur, and Hemmateenejad 2022; Nilghaz and Lu 2019; Ma, Nilghaz, et al. 2018a).

## 3.6 CONCLUSIONS AND PERSPECTIVES

The screening of foodborne pathogens, allergens, and chemical contaminants in food processing industry is a major concern in the food safety sector. If the food quality control is not followed stringently, these factors can cause mild to serious illnesses and may lead to mortality. Conventional analytical methods require well-equipped laboratories, trained personnel, costly reagents, and tedious sample preparation procedures prior to the detection in order to obtain reliable results. Hence, they are not always ideal for on-site rapid surveillance. Therefore, there is a high demand in developing low-cost, robust, reliable, and rapid detection methods for food safety and quality analysis.

µPAD-based diagnostic devices and microfluidic technology are a fast-growing field in food processing to produce safer food. This chapter describes the assay principles, state-of-the-art designs, and various technology in µPAD devices to apply in food processing. Prominent research trends and future research directions for maximizing the value of microfluidic technology in food sector are identified. One key area for improvement in the food sector is to develop robust µPADs compatible methods to prepare complex food samples to load on the µPADs. The µPAD technology is promising in the food sector by using novel materials and improved methods to enhance the sensitivity and specificity of the assays, at low cost.

Another significant area for improvement is detection since most of the detection methods still depend on traditional techniques and instruments. The most common method is colorimetric detection but it has a low detection limit. All the other methods require expensive reagents or miniaturized equipment limiting their potential for mass commercialization. For the advancement of µPAD field, development of systems free of detection equipment is necessary. Research toward this direction is expected to be dominated in the foreseeable future.

For the successful commercialization of µPADs in food processing and diagnostics, simple, and low-cost fabrication techniques that address process parameters with minimal human intervention are ideally suited. Some of the recent techniques have shown potential for low-cost automated mass fabrication which will make their widespread adoption for point-of-use applications.

## REFERENCES

Abe, K., K. Kotera, K. Suzuki, and D. Citterio. 2010. "Inkjet-printed paperfluidic immuno-chemical sensing device." *Analytical and Bioanalytical Chemistry* 398 (2):885–893. https://doi.org/10.1007/s00216-010-4011-2

Abe, K., K. Suzuki, and D. Citterio. 2008. "Inkjet-printed microfluidic multianalyte chemical sensing paper." *Analytical Chemistry* 80 (18):6928–6934. https://doi.org/10.1021/ac800604v

Adkins, J., K. Boehle, and C. Henry. 2015. "Electrochemical paper-based microfluidic devices." *Electrophoresis* 36 (16):1811–1824. https://doi.org/10.1002/elps.201500084

Aidil, M., M. Sabullah, M. E. Halmi, R. Sulaiman, M. S. A. Shukor, Y. Shukor, N. Shaharuddin, M. Syed, and A. Syahir. 2013. "Assay for heavy metals using an inhibitive assay based on the acetylcholinesterase from pangasius hypophthalmus (Sauvage, 1878)." *Fresenius Environmental Bulletin* 22:3572–3576.

Akyazi, T., L. Basabe-Desmonts, and F. Benito-Lopez. 2018. "Review on microfluidic paper-based analytical devices towards commercialisation." *Analytica Chimica Acta* 1001:1–17. https://doi.org/10.1016/j.aca.2017.11.010

Alahmad, W., K. Uraisin, D. Nacapricha, and T. Kaneta. 2016. "A miniaturized chemiluminescence detection system for a microfluidic paper-based analytical device and its application to the determination of chromium(iii)." *Analytical Methods* 8 (27):5414–5420. https://doi.org/10.1039/C6AY00954A

Alam, M. F., A. A. Laskar, S. Ahmed, M. A. Shaida, and H. Younus. 2017. "Colorimetric method for the detection of melamine using in-situ formed silver nanoparticles via tannic acid." *Spectrochim Acta A Mol Biomol Spectrosc* 183:17–22. https://doi.org/10.1016/j.saa.2017.04.021

Anfossi, L., F. Di Nardo, A. Russo, S. Cavalera, C. Giovannoli, G. Spano, S. Baumgartner, K. Lauter, and C. Baggiani. 2019. "Silver and gold nanoparticles as multi-chromatic lateral flow assay probes for the detection of food allergens." *Analytical and Bioanalytical Chemistry* 411 (9):1905–1913. https://doi.org/10.1007/s00216-018-1451-6

Anindita, S., M. Calum, P. Aashish, and B. Murray. 2023. "A paper based microfluidic platform combining LAMP-CRISPR/Cas12a for fluorometric detection of nucleic acids." *bioRxiv*:2023.2003.2002.530841. https://doi.org/10.1101/2023.03.02.530841

Apilux, A., W. Dungchai, W. Siangproh, N. Praphairaksit, C. S. Henry, and O. Chailapakul. 2010. "Lab-on-Paper with Dual Electrochemical/Colorimetric Detection for Simultaneous Determination of Gold and Iron." *Analytical Chemistry* 82 (5):1727–1732. https://doi.org/10.1021/ac9022555

Apilux, A., Y. Ukita, M. Chikae, O. Chailapakul, and Y. Takamura. 2013. "Development of automated paper-based devices for sequential multistep sandwich enzyme-linked immunosorbent assays using inkjet printing." *Lab on a Chip* 13 (1):126–135. https://doi.org/10.1039/C2LC40690J

Bansal, S., A. Singh, M. Mangal, A. K. Mangal, and S. Kumar. 2017. "Food adulteration: Sources, health risks, and detection methods." *Crit Rev Food Sci Nutr* 57 (6):1174–1189. https://doi.org/10.1080/1040839 8.2014.967834

Beshana, S., A. Hussen, S. Leta, and T. Kaneta. 2022. "Microfluidic paper based analytical devices for the detection of carbamate pesticides." *Bulletin of Environmental Contamination and Toxicology* 109:1–8. https://doi.org/10.1007/s00128-022-03533-3

Bhardwaj, J., S. Devarakonda, S. Kumar, and J. Jang. 2017. "Development of a paper-based electrochemical immunosensor using an antibody-single walled carbon nanotubes bio-conjugate modified electrode for label-free detection of foodborne pathogens." *Sensors and Actuators B: Chemical* 253:115–123. https://doi.org/10.1016/j.snb.2017.06.108

Bruzewicz, D. A., M. Reches, and G.M. Whitesides. 2008. "Low-Cost Printing of Poly(dimethylsiloxane) Barriers To Define Microchannels in Paper." *Analytical Chemistry* 80 (9):3387–3392. https://doi.org/10.1021/ac702605a

Caglayan, M. G., S. Sheykhi, L. Mosca, and P. Anzenbacher. 2016. "Fluorescent zinc and copper complexes for detection of adrafinil in paper-based microfluidic devices." *Chemical Communications* 52 (53):8279–8282. https://doi.org/10.1039/C6CC03640F

Cardoso, T. M. G., R. B. Channon, J. A. Adkins, M. Talhavini, W. K. T. Coltro, and C. S. Henry. 2017. "A paper-based colorimetric spot test for the identification of adulterated whiskeys." *Chemical Communications* 53 (56):7957–7960. https://doi.org/10.1039/c7cc02271a

Cardoso, T. M. G., P. T. Garcia, and W. K. T. Coltro. 2015. "Colorimetric determination of nitrite in clinical, food and environmental samples using microfluidic devices stamped in paper platforms." *Analytical Methods* 7 (17):7311–7317. https://doi.org/10.1039/c5ay00466g

Carrilho, E., A. W. Martinez, and G. M. Whitesides. 2009. "Understanding wax printing: A simple micropatterning process for paper-based microfluidics." *Analytical Chemistry* 81 (16):7091–7095. https://doi.org/10.1021/ac901071p

Cate, D. M., J. A. Adkins, J. Mettakoonpitak, and C. S. Henry. 2015. "Recent developments in paper-based microfluidic devices." *Analytical Chemistry* 87 (1):19–41. https://doi.org/10.1021/ac503968p

Chaiyo, S., A. Apiluk, W. Siangproh, and O. Chailapakul. 2016. "High sensitivity and specificity simultaneous determination of lead, cadmium and copper using µPAD with dual electrochemical and colorimetric detection." *Sensors and Actuators B: Chemical* 233:540–549. https://doi.org/10.1016/j.snb.2016.04.109

Chen, B., P. Kwong, and M. Gupta. 2013. "Patterned fluoropolymer barriers for containment of organic solvents within paper-based microfluidic devices." *Acs Applied Materials & Interfaces* 5 (23):12701–12707. https://doi.org/10.1021/am404049x

Chen, H., J. Cogswell, C. Anagnostopoulos, and M. Faghri. 2012. "A fluidic diode, valves, and a sequential-loading circuit fabricated on layered paper." *Lab on a Chip* 12 (16):2909–2913. https://doi.org/10.1039/C2LC20970E

Cunningham, J. C., N. J. Brenes, and R. M. Crooks. 2014. "Paper electrochemical device for detection of DNA and thrombin by target-induced conformational switching." *Analytical Chemistry* 86 (12):6166–6170. https://doi.org/10.1021/ac501438y

Dai, G., J Hu, X. Zhao, and P. Wang. 2017. "A colorimetric paper sensor for lactate assay using a cellulose-Binding recombinant enzyme." *Sensors and Actuators B: Chemical* 238:138–144. https://doi.org/10.1016/j.snb.2016.07.008

Delaney, J. L., C. F. Hogan, J. Tian, and W. Shen. 2011. "Electrogenerated chemiluminescence detection in paper-based microfluidic sensors." *Analytical Chemistry* 83 (4):1300–1306. https://doi.org/10.1021/ac102392t

Demirel, G., and E. Babur. 2014. "Vapor-phase deposition of polymers as a simple and versatile technique to generate paper-based microfluidic platforms for bioassay applications." *Analyst* 139 (10):2326–2331. https://doi.org/10.1039/c4an00022f

Deng, Y., Q. Li, Y. Zhou, and J. Qian. 2021. "Fully inkjet printing preparation of a carbon dots multichannel microfluidic paper-based sensor and its application in food additive detection." *ACS Applied Materials & Interfaces* 13 (48):57084–57091. https://doi.org/10.1021/acsami.1c14435

Devries, J., G. Greene, A. Payne, S. Zbylut, P. Scholl, P. Wehling, J. Evers, and J. Moore. 2016. "Non-protein nitrogen determination: A screening tool for nitrogenous compound adulteration of milk powder." *International Dairy Journal* 68. https://doi.org/10.1016/j.idairyj.2016.12.003

Dungchai, W., O. Chailapakul, and C. S. Henry. 2011. "A low-cost, simple, and rapid fabrication method for paper-based microfluidics using wax screen-printing." *Analyst* 136 (1):77–82. https://doi.org/10.1039/c0an00406e

Dungchai, W., O. Chailapakul, and C. S. Henry. 2009. "Electrochemical detection for paper-based microfluidics." *Analytical Chemistry* 81 (14):5821–5826.

Ellman, G. L., K. D. Courtney, V. Andres, Jr., and R. M. Feather-Stone. 1961. "A new and rapid colorimetric determination of acetylcholinesterase activity." *Biochem Pharmacol* 7:88–95. https://doi.org/10.1016/0006-2952(61)90145-9

Fan, L., Q. Hao, and X. Kan. 2018. "Three-dimensional graphite paper based imprinted electrochemical sensor for tertiary butylhydroquinone selective recognition and sensitive detection." *Sensors and Actuators B: Chemical* 256:520–527. https://doi.org/10.1016/j.snb.2017.10.085

Feng, Q.-M., M. Cai, C.-G. Shi, N. Bao, and H.-Y. Gu. 2015. "Integrated paper-based electroanalytical devices for determination of dopamine extracted from striatum of rat." *Sensors and Actuators B: Chemical* 209:870–876. https://doi.org/10.1016/j.snb.2014.12.062

Fu, E., B. Lutz, P. Kauffman, and P. Yager. 2010. "Controlled reagent transport in disposable 2D paper networks." *Lab on a Chip* 10 (7):918–920. https://doi.org/10.1039/B919614E

Fu, E., S. A. Ramsey, P. Kauffman, B. Lutz, and P. Yager. 2011. "Transport in two-dimensional paper networks." *Microfluidics and Nanofluidics* 10 (1):29–35. https://doi.org/10.1007/s10404-010-0643-y

Gan, L., Q. Su, Z. Chen, and X. Yang. 2020. "Exploration of pH-responsive carbon dots for detecting nitrite and ascorbic acid." *Applied Surface Science* 530:147269. https://doi.org/10.1016/j.apsusc.2020.147269

Gao, N., P. Huang, and F. Wu. 2018. "Colorimetric detection of melamine in milk based on Triton X-100 modified gold nanoparticles and its paper-based application." *Spectrochim Acta A Mol Biomol Spectrosc* 192:174–180. https://doi.org/10.1016/j.saa.2017.11.022

Gharaghani, F. M., M. Akhond, and B. Hemmateenejad. 2020. "A three-dimensional origami microfluidic device for paper chromatography: Application to quantification of Tartrazine and Indigo carmine in food samples." *Journal of Chromatography A* 1621. https://doi.org/ARTN46104910.1016/j.chroma.2020.461049

Ghosh, R., S. Gopalakrishnan, R. Savitha, T. Renganathan, and S. Pushpavanam. 2019. "Fabrication of laser printed microfluidic paper-based analytical devices (LP-mu PADs) for point-of-care applications." *Scientific Reports* 9. https://doi.org/ARTN 789610.1038/s41598-019-44455-1

Giustarini, D., I. Dalle-Donne, R. Colombo, A. Milzani, and R. Rossi. 2004. "Adaptation of the Griess reaction for detection of nitrite in human plasma." *Free Radic Res* 38 (11):1235–1240. https://doi.org/10.1080/10715760400017327

Han, J., A. Qi, J. Zhou, G. Wang, B. Li, and L. Chen. 2018. "Simple way to fabricate novel paper-based valves using plastic comb binding spines." *ACS Sensors* 3 (9):1789–1794. https://doi.org/10.1021/acssensors.8b00518

He, Q. H., C. C. Ma, X. Q. Hu, and H. W. Chen. 2013. "Method for fabrication of paper-based microfluidic devices by alkylsilane self-assembling and UV/O3-Patterning." *Analytical Chemistry* 85 (3):1327–1331. https://doi.org/10.1021/ac303138x

He, Y., Y. Wu, J. Z. Fu, and W. B. Wu. 2015. "Fabrication of paper-based microfluidic analysis devices: a review." *Rsc Advances* 5 (95):78109–78127. https://doi.org/10.1039/c5ra09188h

Hossain, S. M. Z., R. E. Luckham, A. M. Smith, J. M. Lebert, L. M. Davies, R. H. Pelton, C. D. M. Filipe, and J. D. Brennan. 2009. "Development of a bioactive paper sensor for detection of neurotoxins using piezoelectric inkjet printing of sol-gel-derived bioinks." *Analytical Chemistry* 81 (13):5474–5483. https://doi.org/10.1021/ac900660p

Houghtaling, J., T. Liang, G. Thiessen, and Elain Fu. 2013. "Dissolvable bridges for manipulating fluid volumes in paper networks." *Analytical Chemistry* 85 (23):11201–11204. https://doi.org/10.1021/ac4022677

Hu, J., A. Stein, and P. Bühlmann. 2016. "A disposable planar paper-based potentiometric ion-sensing platform." *Angewandte Chemie International Edition* 5d5 (26):7544–7547. https://doi.org/10.1002/anie.201603017

Hu, L., and G. Xu. 2010. "Applications and trends in electrochemiluminescence." *Chemical Society Reviews* 39 (8):3275–3304. https://doi.org/10.1039/B923679C

Hu, X. Q., L. Lu, C. Y. Fang, B. W. Duan, and Z. W. Zhu. 2015. "Determination of apparent amylose content in rice by using paper-based microfluidic chips." *Journal of Agricultural and Food Chemistry* 63 (44):9863–9868. https://doi.org/10.1021/acs.jafc.5b04530

Hua, M. Z., S. Li, S. Wang, and X. Lu. 2018. "Detecting Chemical Hazards in Foods Using Microfluidic Paper-Based Analytical Devices (µPADs): The real-world application." *Micromachines (Basel)* 9 (1). https://doi.org/10.3390/mi9010032

Hua, M. Z., and X. Lu. 2020. "Development of a microfluidic paper-based immunoassay for rapid detection of allergic protein in foods." *ACS Sensors* 5 (12):4048–4056. https://doi.org/10.1021/acssensors.0c02044

Jahanshahi-Anbuhi, S., P. Chavan, C. Sicard, S. M. Vincent Leung Z. Hossain, R. Pelton, J. D. Brennan, and C. D. M. Filipe. 2012. "Creating fast flow channels in paper fluidic devices to control timing of sequential reactions." *Lab on a Chip* 12 (23):5079–5085. https://doi.org/10.1039/C2LC41005B

Jahanshahi-Anbuhi, S., A. Henry, V.incent Leung, C. Sicard, K. Pennings, R. Pelton, J. D. Brennan, and C. D. M. Filipe. 2014. "Paper-based microfluidics with an erodible polymeric bridge giving controlled release and timed flow shutoff." *Lab on a Chip* 14 (1):229–236. https://doi.org/10.1039/C3LC50762A

Jayawardane, B. M., S. Wei, I. D. McKelvie, and S. D. Kolev. 2014. "Microfluidic paper-based analytical device for the determination of nitrite and nitrate." *Analytical Chemistry* 86 (15):7274–7279. https://doi.org/10.1021/ac5013249

Jia, Z., Y. Luo, D. Wang, Q. N. Dinh, S. Lin, A. Sharma, E. M. Block, et al. 2021. "Nondestructive multiplex detection of foodborne pathogens with background microflora and symbiosis using a paper chromogenic array and advanced neural network." *Biosens Bioelectron* 183:113209. https://doi.org/10.1016/j.bios.2021.113209

Jiang, Q., J. D. Wu, K. Yao, Y. L. Yin, M. M. Gong, C. B. Yang, and F. Lin. 2019. "Paper-Based Microfluidic Device (DON-Chip) for rapid and low-cost deoxynivalenol quantification in food, feed, and feed ingredients." *Acs Sensors* 4 (11):3072–3079. https://doi.org/10.1021/acssensors.9b01895

Jiang, Y., Z. X. Hao, Q. H. He, and H. W. Chen. 2016. "A simple method for fabrication of microfluidic paper-based analytical devices and on-device fluid control with a portable corona generator." *Rsc Advances* 6 (4):2888–2894. https://doi.org/10.1039/c5ra23470k

Jin, S. Q., S. M. Guo, P. Zuo, and B. C. Ye. 2015. "A cost-effective Z-folding controlled liquid handling microfluidic paper analysis device for pathogen detection via ATP quantification." *Biosensors & Bioelectronics* 63:379–383. https://doi.org/10.1016/j.bios.2014.07.070

Jokerst, J. C., J. A. Adkins, B. Bisha, M. M. Mentele, L. D. Goodridge, and C. S. Henry. 2012. "Development of a paper-based analytical device for colorimetric detection of select foodborne pathogens." *Analytical Chemistry* 84 (6):2900–2907. https://doi.org/10.1021/ac203466y

Kao, P. K., and C. C. Hsu. 2014. "One-step rapid fabrication of paper-based microfluidic devices using fluorocarbon plasma polymerization." *Microfluidics and Nanofluidics* 16 (5):811–818. https://doi.org/10.1007/s10404-014-1347-5

Kasoju, A., N. S. Shrikrishna, D. Shahdeo, A. A. Khan, A. M. Alanazi, and S. Gandhi. 2020. "Microfluidic paper device for rapid detection of aflatoxin B1 using an aptamer based colorimetric assay." *RSC Advances* 10 (20):11843–11850. https://doi.org/10.1039/D0RA00062K

Katelakha, K., V. Nopponpunth, W. Boonlue, and W. Laiwattanapaisal. 2021. "A simple distance paper-based analytical device for the screening of lead in food matrices." *Biosensors* 11 (3):90.

Kong, Q. K., Y. H. Wang, L. N. Zhang, S. G. Ge, and J. H. Yu. 2017a. "A novel microfluidic paper-based colorimetric sensor based on molecularly imprinted polymer membranes for highly selective and sensitive detection of bisphenol A." *Sensors and Actuators B-Chemical* 243:130–136. https://doi.org/10.1016/j.snb.2016.11.146

Kong, T., S. Flanigan, M. Weinstein, U. Kalwa, C. Legner, and S. Pandey. 2017b. "A fast, reconfigurable flow switch for paper microfluidics based on selective wetting of folded paper actuator strips." *Lab on a Chip* 17 (21):3621–3633. https://doi.org/10.1039/C7LC00620A

Kuek, L., C. Shiong, S. N. Tan, and C. Z. Floresca. 2014. "A "green" cellulose paper based glucose amperometric biosensor." *Sensors and Actuators B: Chemical* 193:536–541. https://doi.org/10.1016/j.snb.2013.11.054

Kumar, A., and V. Santhanam. 2019. "Paper swab based SERS detection of non-permitted colourants from dals and vegetables using a portable spectrometer." *Analytica Chimica Acta* 1090:106–113. https://doi.org/10.1016/j.aca.2019.08.073

Kumar, N., R. Seth, and H. Kumar. 2014. "Colorimetric detection of melamine in milk by citrate stabilized gold nanoparticles." *Analytical biochemistry* 456. https://doi.org/10.1016/j.ab.2014.04.002

Kumawat, N., S. S. Soman, S. Vijayavenkataraman, and S. Kumar. 2022. "Rapid and inexpensive process to fabricate paper based microfluidic devices using a cut and heat plastic lamination process." *Lab Chip* 22 (18):3377–3389. https://doi.org/10.1039/d2lc00452f

Kumawat, N., S. S. Soman, S. Vijayavenkataraman, and S. Kumar. 2023. Patent application reference number: US20230285960A1. https://patents.google.com/patent/US20230285960A1/en

Kwong, P., and M. Gupta. 2012. "Vapor phase deposition of functional polymers onto paper-based microfluidic devices for advanced unit operations." *Analytical Chemistry* 84 (22):10129–10135. https://doi.org/10.1021/ac302861v

Lam, T., J. P. Devadhasan, R. Howse, and J. Kim. 2017. "A chemically patterned microfluidic paper-based analytical device (C-mu PAD) for Point-of-Care Diagnostics." *Scientific Reports* 7. https://doi.org/ARTN118810.1038/s41598-017-01343-w

Li, L., H. P. Chen, X. L. Lv, M. Wang, X. Z. Jiang, Y. F. Jiang, H. Y. Wang, Y. F. Zhao, and L. R. Xia. 2018a. "Paper-based immune-affinity arrays for detection of multiple mycotoxins in cereals." *Analytical and Bioanalytical Chemistry* 410 (8):2253–2262. https://doi.org/10.1007/s00216-018-0895-z

Li, L., Y. Zhang, F. Liu, M. Su, L. Liang, S. Ge, and J. Yu. 2015. "Real-time visual determination of the flux of hydrogen sulphide using a hollow-channel paper electrode." *Chemical Communications* 51 (74):14030–14033. https://doi.org/10.1039/c5cc05710h

Li, L., W. Li, H. Yang, C. Ma, J. Yu, M. Yan, and X. Song. 2014a. "Sensitive origami dual-analyte electrochemical immunodevice based on polyaniline/Au-paper electrode and multi-labeled 3D graphene sheets." *Electrochimica Acta* 120:102–109. https://doi.org/10.1016/j.electacta.2013.12.076

Li, L., J. Xu, X. Zheng, C. Ma, X. Song, S. Ge, J. Yu, and M. Yan. 2014b. "Growth of gold-manganese oxide nanostructures on a 3D origami device for glucose-oxidase label based electrochemical immunosensor." *Biosensors and Bioelectronics* 61:76–82. https://doi.org/10.1016/j.bios.2014.05.012

Li, W., L. Li, S. Ge, X. Song, L. Ge, M. Yan, and J. Yu 2014c. "Multiplex electrochemical origami immunodevice based on cuboid silver-paper electrode and metal ions tagged nanoporous silver–chitosan." *Biosensors and Bioelectronics* 56:167–173. https://doi.org/10.1016/j.bios.2014.01.011

Li, Z. D., F. Li, Y. Xing, Z. Liu, M. L. You, Y. C. Li, T. Wen, Z. G. Qu, X. L. Li, and F. Xu. 2017. "Pen-on-paper strategy for point-of-care testing: Rapid prototyping of fully written microfluidic biosensor." *Biosensors & Bioelectronics* 98:478–485. https://doi.org/10.1016/j.bios.2017.06.061

Li, Z. D., H. Liu, X. C. He, F. Xu, and F. Li. 2018b. "Pen-on-paper strategies for point-of-care testing of human health." *Trac-Trends in Analytical Chemistry* 108:50–64. https://doi.org/10.1016/j.trac.2018.08.010

Liang, J., Y. Wang, and B. Liu. 2012. "Paper-based fluoroimmunoassay for rapid and sensitive detection of antigen." *RSC Advances* 2 (9):3878–3884. https://doi.org/10.1039/C2RA20156A

Liang, L., M. Su, L. Li, F. Lan, G. Yang, S. Ge, J. Yu, and X. Song. 2016. "Aptamer-based fluorescent and visual biosensor for multiplexed monitoring of cancer cells in microfluidic paper-based analytical devices." *Sensors and Actuators B: Chemical* 229:347–354. https://doi.org/10.1016/j.snb.2016.01.137

Lisak, G., J. Cui, and J. Bobacka. 2015. "Paper-based microfluidic sampling for potentiometric determination of ions." *Sensors and Actuators B: Chemical* 207:933939. https://doi.org/10.1016/j.snb.2014.07.044

Liu, C. C., Y. N. Wang, L. M. Fu, and K. L. Chen. 2018. "Microfluidic paper-based chip platform for benzoic acid detection in food." *Food Chemistry* 249:162–167. https://doi.org/10.1016/j.foodchem.2018.01.004

Liu, F., S. Ge, J. Yu, M. Yan, and X. Song. 2014a. "Electrochemical device based on a Pt nanosphere-paper working electrode for in situ and real-time determination of the flux of H2O2 releasing from SK-BR-3 cancer cells." *Chemical Communications* 50 (71):10315–10318. https://doi.org/10.1039/C4CC04199B

Liu, F., and C. Zhang. 2015. "A novel paper-based microfluidic enhanced chemiluminescence biosensor for facile, reliable and highly-sensitive gene detection of Listeria monocytogenes." *Sensors and Actuators B: Chemical* 209:399–406. https://doi.org/10.1016/j.snb.2014.11.099

Liu, W., J. Kou, H. Z. Xing, and B. X. Li. 2014b. "Paper-based chromatographic chemiluminescence chip for the detection of dichlorvos in vegetables." *Biosensors & Bioelectronics* 52:76–81. https://doi.org/10.1016/j.bios.2013.08.024

Liu, Z., J. Hu, Y. Zhao, Z. Qu, and F. Xu. 2015. "Experimental and numerical studies on liquid wicking into filter papers for paper-based diagnostics." *Applied Thermal Engineering* 88:280–287. https://doi.org/10.1016/j.applthermaleng.2014.09.057

Lu, J., S. Ge, L. Ge, M. Yan, and J. Yu. 2012. "Electrochemical DNA sensor based on three-dimensional folding paper device for specific and sensitive point-of-care testing." *Electrochimica Acta* 80:334–341. https://doi.org/10.1016/j.electacta.2012.07.024

Lu, Y., W. W. Shi, L. Jiang, J. H. Qin, and B. C. Lin. 2009. "Rapid prototyping of paper-based microfluidics with wax for low-cost, portable bioassay." *Electrophoresis* 30 (9):1497–1500. https://doi.org/10.1002/elps.200800563

Lucas, R. 1918. "Ueber das Zeitgesetz des kapillaren Aufstiegs von Flüssigkeiten." *Kolloid-Zeitschrift* 23 (1):15–22. https://doi.org/10.1007/BF01461107

Luo, M., K. Shao, Z. Long, L. Wang, C. Peng, J. Ouyang, and N. Na. 2017. "A paper-based plasma-assisted cataluminescence sensor for ethylene detection." *Sensors and Actuators B: Chemical* 240:132–141. https://doi.org/10.1016/j.snb.2016.08.156

Ma, L., A. Nilghaz, J. R. Choi, X. Liu, and X. Lu. 2018a. "Rapid detection of clenbuterol in milk using microfluidic paper-based ELISA." *Food Chem* 246:437–441. https://doi.org/10.1016/j.foodchem.2017.12.022

Ma, Y. D., Y. H. Wang, Y. Luo, H. Z. Duan, D. Li, H. Xu, and E. K. Fodjo. 2018b. "Rapid and sensitive on-site detection of pesticide residues in fruits and vegetables using screen-printed paper-based SERS swabs." *Analytical Methods* 10 (38). https://doi.org/10.1039/c8ay01698d

Mahmud, M. A., E. J. M. Blondeel, M. Kaddoura, and B. D. MacDonald. 2016. "Creating compact and microscale features in paper-based devices by laser cutting." *Analyst* 141 (23):6449–6454. https://doi.org/10.1039/c6an02208a

Mahmud, M. A., E. J. M. Blondeel, M. Kaddoura, and B. D. MacDonald. 2018. "Features in microfluidic paper-based devices made by laser cutting: How small can they be?" *Micromachines* 9 (5). https://doi.org/ARTN 22010.3390/mi9050220

Manzoori, J., M. Amjadi, and J. Hasanzadeh. 2011. "Enhancement of the chemiluminescence of permanganate-formaldehyde system by gold/silver nanoalloys and its application to trace determination of melamine." *Microchimica Acta* 175:47–54. https://doi.org/10.1007/s00604-011-0651-y

Marino, M., E. Mele, A. Viggiano, S. L. Nori, R. Meccariello, and A. Santoro. 2021. "Pleiotropic outcomes of glyphosate exposure: From organ damage to effects on inflammation, cancer, reproduction and development." *Int J Mol Sci* 22 (22). https://doi.org/10.3390/ijms222212606

Martinez, A. W., S. T. Phillips, M. J. Butte, and G. M. Whitesides. 2007. "Patterned paper as a platform for inexpensive, low-volume, portable bioassays." *Angewandte Chemie-International Edition* 46 (8):1318–1320. https://doi.org/10.1002/anie.200603817

Martínez-Pérez-Cejuela, H., R. B. R. Mesquita, E. F. Simó-Alfonso, J. M. Herrero-Martínez, and A. O. S. S. Rangel. 2023. "Combining microfluidic paper-based platform and metal-organic frameworks in a single device for phenolic content assessment in fruits." *Microchimica Acta* 190 (4). https://doi.org/ARTN 12610.1007/s00604-023-05702-5

Masoomi, S., H. Sharifi, and B. Hemmateenejad. 2024. "A paper-based optical tongue for characterization of iranian honey: Identification of geographical/botanical origins and adulteration detection." *Food Control* 155. https://doi.org/ARTN11005210.1016/j.foodcont.2023.110052

Matsuda, Y., S. Shibayama, K. Uete, H. Yamaguchi, and T. Niimi. 2015. "Electric conductive pattern element fabricated using commercial inkjet printer for paper-based analytical devices." *Analytical Chemistry* 87 (11):5762–5765. https://doi.org/10.1021/acs.analchem.5b01568

Mazur, F., A. D. Tjandra, Y. Zhou, Y. Gao, and R. Chandrawati. 2023. "Paper-based sensors for bacteria detection." *Nat Rev Bioeng* 1 (3):180–192. https://doi.org/10.1038/s44222-023-00024-w

Mensah, S. T., Y. Gonzalez, P. Calvo-Marzal, and K. Y. Chumbimuni-Torres. 2014. "Nanomolar detection limits of Cd2+, Ag+, and K+ Using paper-strip ion-selective electrodes." *Analytical Chemistry* 86 (15):7269–7273. https://doi.org/10.1021/ac501470p

Mishra, K. P. 2009. "Lead exposure and its impact on immune system: a review." *Toxicol In Vitro* 23 (6):969–972. https://doi.org/10.1016/j.tiv.2009.06.014

Moreira, S., R. Silva, D. F. Carrageta, M. G. Alves, V. Seco-Rovira, P. F. Oliveira, and M. de Lourdes Pereira. 2022. "Carbamate pesticides: shedding light on their impact on the male reproductive system." *Int J Mol Sci* 23 (15). https://doi.org/10.3390/ijms23158206

Mulvaney, P. 1996. "Surface plasmon spectroscopy of nanosized metal particles." *Langmuir* 12 (3):788–800. https://doi.org/10.1021/la9502711

Muthukumar, R., A. Kapoor, S. Balasubramanian, V. Vaishampayan, and M. Gabhane. 2021. "Detection of adulteration in sunflower oil using paper-based microfluidic lab-on-a-chip devices." *Materials Today: Proceedings* 34:496–501. https://doi.org/10.1016/j.matpr.2020.03.099

Narang, J., N. Malhotra, C. Singhal, A. Mathur, D. Chakraborty, A. Anil, A. Ingle, and C. S. Pundir. 2017. "Point of care with micro fluidic paper based device integrated with nano zeolite-graphene oxide nano-flakes for electrochemical sensing of ketamine." *Biosensors & Bioelectronics* 88:249–257. https://doi.org/10.1016/j.bios.2016.08.043

Narang, J., N. Malhotra, C. Singhal, A. Mathur, D. Chakraborty, A. Anil, A. Ingle, and C. Pundir. 2016. "Point of care with micro fluidic paper based device integrated with nano zeolite –graphene oxide nanoflakes for electrochemical sensing of ketamine." *Biosensors and Bioelectronics* 88. https://doi.org/10.1016/j.bios.2016.08.043

Ng, J. S., and M. Hashimoto. 2020. "Fabrication of paper microfluidic devices using a toner laser printer." *Rsc Advances* 10 (50):29797–29807. https://doi.org/10.1039/d0ra04301j

Nie, J. F., Y. Z. Liang, Y. Zhang, S. W. Le, D. N. Li, and S. B. Zhang. 2013. "One-step patterning of hollow microstructures in paper by laser cutting to create microfluidic analytical devices." *Analyst* 138 (2):671–676. https://doi.org/10.1039/c2an36219h

Nie, Z., F. Deiss, X. Liu, O. Akbulut, and G. M. Whitesides. 2010. "Integration of paper-based microfluidic devices with commercial electrochemical readers." *Lab on a Chip* 10 (22):3163–3169. https://doi.org/10.1039/C0LC00237B

Nilghaz, A., and X. Lu. 2019. "Detection of antibiotic residues in pork using paper-based microfluidic device coupled with filtration and concentration." *Anal Chim Acta* 1046:163–169. https://doi.org/10.1016/j.aca.2018.09.041

Nouanthavong, S., D. Nacapricha, C. S. Henry, and Y. Sameenoi. 2016. "Pesticide analysis using nanoceria-coated paper-based devices as a detection platform." *Analyst* 141 (5):1837–1846. https://doi.org/10.1039/c5an02403j

O'Connor, D., D. Hou, Y. Sik Ok, and B. Lanphear. 2020. "The effects of iniquitous lead exposure on health." *Nature Sustainability* 3:77–79. https://doi.org/10.1038/s41893-020-0475-z

Pavithra, M., S. Muruganand, and C. Parthiban. 2018. "Development of novel paper based electrochemical immunosensor with self-made gold nanoparticle ink and quinone derivate for highly sensitive carcinoembryonic antigen." *Sensors and Actuators B: Chemical* 257:496–503. https://doi.org/10.1016/j.snb.2017.10.177

S. Gómez-de Pedro, A. Salinas-Castillo, M. Ariza-Avidad, A. Lapresta-Fernández, C. Sánchez-González, C. S. Martínez-Cisneros, M. Puyol, L. F. Capitan-Vallvey, and J. Alonso-Chamarro. 2014. "Microsystem-assisted synthesis of carbon dots with fluorescent and colorimetric properties for pH detection." *Nanoscale* 6 (11):6018–6024. https://doi.org/10.1039/C4NR00573B

Pelton, Robert. 2009. "Bioactive paper provides a low-cost platform for diagnostics." *TrAC Trends in Analytical Chemistry* 28 (8):925–942. https://doi.org/10.1016/j.trac.2009.05.005

Preechakasedkit, P., W. Siangproh, N. Khongchareonporn, N. Ngamrojanavanich, and O. Chailapakul. 2018. "Development of an automated wax-printed paper-based lateral flow device for alpha-fetoprotein enzyme-linked immunosorbent assay." *Biosensors and Bioelectronics* 102:27–32. https://doi.org/10.1016/j.bios.2017.10.051

Qi, J., B. Li, X. Wang, Z. Zhang, Z. Wang, J. Han, and L. Chen. 2017. "Three-dimensional paper-based microfluidic chip device for multiplexed fluorescence detection of Cu2+ and Hg2+ ions based on ion imprinting technology." *Sensors and Actuators B: Chemical* 251:224–233. https://doi.org/10.1016/j.snb.2017.05.052

Qin, X. X., J. J. Liu, Z. Zhang, J. H. Li, L. Yuan, Z. Y. Zhang, and L. X. Chen. 2021. "Microfluidic paper-based chips in rapid detection: Current status, challenges, and perspectives." *Trac-Trends in Analytical Chemistry* 143. https://doi.org/ARTN11637110.1016/j.trac.2021.116371

Quesada-González, D., and A. Merkoçi. 2015. "Nanoparticle-based lateral flow biosensors." *Biosensors and Bioelectronics* 73:47–63. https://doi.org/10.1016/j.bios.2015.05.050

Raj, N., V. Breedveld, and D. W. Hess. 2019. "Fabrication of fully enclosed paper microfluidic devices using plasma deposition and etching." *Lab on a Chip* 19 (19):3337–3343. https://doi.org/10.1039/c9lc00746f

Rattanarat, P., W. Dungchai, W. Siangproh, O. Chailapakul, and C. S. Henry. 2012. "Sodium dodecyl sulfate-modified electrochemical paper-based analytical device for determination of dopamine levels in biological samples." *Analytica Chimica Acta* 744:1–7. https://doi.org/10.1016/j.aca.2012.07.003

Rosa, A. M. M., A. Filipa Louro, S. A. M. Martins, J. Inácio, A. M. Azevedo, and D. Miguel F. Prazeres. 2014. "Capture and detection of DNA hybrids on paper via the anchoring of antibodies with fusions of carbohydrate binding modules and ZZ-Domains." *Analytical Chemistry* 86 (9):4340–4347. https://doi.org/10.1021/ac5001288

Ruecha, N., N. Rodthongkum, D. M. Cate, J. Volckens, O. Chailapakul, and C. S. Henry. 2015. "Sensitive electrochemical sensor using a graphene–polyaniline nanocomposite for simultaneous detection of Zn(II), Cd(II), and Pb(II)." *Analytica Chimica Acta* 874:40–48. https://doi.org/10.1016/j.aca.2015.02.064

Sameenoi, Y., P. N. Nongkai, S. Nouanthavong, C. S. Henry, and D. Nacapricha. 2014. "One-step polymer screen-printing for microfluidic paper-based analytical device (mu PAD) fabrication." *Analyst* 139 (24):6580–6588. https://doi.org/10.1039/c4an01624f

Savelli, C. J., A. Bradshaw, P. Ben Embarek, and C. Mateus. 2019. "The FAO/WHO international food safety authorities network in review, 2004-2018: Learning from the past and looking to the future." *Foodborne Pathog Dis* 16 (7):480–488. https://doi.org/10.1089/fpd.2018.2582

Sawetwong, P., S. Chairam, P. Jarujamrus, and M. Amatatongchai. 2021. "Enhanced selectivity and sensitivity for colorimetric determination of glyphosate using Mn–ZnS quantum dot embedded molecularly imprinted polymers combined with a 3D-microfluidic paper-based analytical device." *Talanta* 225:122077. https://doi.org/10.1016/j.talanta.2020.122077

Sechi, D., B. Greer, J. Johnson, and N. Hashemi. 2013. "Three-dimensional paper-based microfluidic device for assays of protein and glucose in urine." *Analytical Chemistry* 85 (22):10733–10737. https://doi.org/10.1021/ac4014868

Sha, X., S. Q. Han, H. Zhao, N. Li, C. Zhang, and W. L. Hasi. 2020. "A rapid detection method for on-site screening of estazolam in beverages with Au@Ag Core-shell Nanoparticles Paper-based SERS Substrate." *Anal Sci* 36 (6):667–674. https://doi.org/10.2116/analsci.19P361

Shariati, S., and G. Khayatian. 2020. "Microfluidic paper-based analytical device using gold nanoparticles modified with N,N'-bis(2-hydroxyethyl)dithiooxamide for detection of Hg(ii) in air, fish and water samples." *New Journal of Chemistry* 44 (43):18662–18667. https://doi.org/10.1039/D0NJ03986A

Shin, J. H., J. Park, S. H. Kim, and J. K. Park. 2014. "Programmed sample delivery on a pressurized paper." *Biomicrofluidics* 8 (5):054121. https://doi.org/10.1063/1.4899773

Shiroma, L. Y., M. Santhiago, A. L. Gobbi, and L. T. Kubota. 2012. "Separation and electrochemical detection of paracetamol and 4-aminophenol in a paper-based microfluidic device." *Analytica Chimica Acta* 725:44–50. https://doi.org/10.1016/j.aca.2012.03.011

Snyder, S. A., M. Boban, C. Li, J. Scott VanEpps, G. Mehta, and A. Tuteja. 2020. "Lysis and direct detection of coliforms on printed paper-based microfluidic devices." *Lab on a Chip* 20 (23):4413–4419. https://doi.org/10.1039/D0LC00665C

Soman, S. S., S. A. Samad, P. Venugopalan, N. Kumawat, and S. Kumar. 2024. "Microfluidic paper analytic device (μPAD) technology for food safety applications." *Biomicrofluidics* 18 (3):031501. https://doi.org/10.1063/5.0192295

Somvanshi, S., G. Ana Ulloa, M. Zhao, Q. Liang, A. Barui, A. Lucas, K. M. Jadhav, J. Allebach, and L. Stanciu. 2022. "Microfluidic paper-based aptasensor devices for multiplexed detection of pathogenic bacteria." *Biosensors and Bioelectronics* 207:114214. https://doi.org/10.1016/j.bios.2022.114214

Song, J., Y. Wan, and L. Ma. 2015. "Colorimetric detection of melamine in pretreated milk using silver nanoparticles functionalized with sulfanilic acid." *Food Control* 50:356–361. https://doi.org/10.1016/j.foodcont.2014.08.049

Sousa, L. R., L. C. Duarte, and W. K. T. Coltro. 2020. "Instrument-free fabrication of microfluidic paper-based analytical devices through 3D pen drawing." *Sensors and Actuators B-Chemical* 312. https://doi.org/ARTN12801810.1016/j.snb.2020.128018

Spicar-Mihalic, P., B. Toley, J. Houghtaling, T. Liang, P. Yager, and E. Fu. 2013. "CO2 laser cutting and ablative etching for the fabrication of paper-based devices." *Journal of Micromechanics and Microengineering* 23 (6). https://doi.org/Artn06700310.1088/0960-1317/236/067003

Strong, E. B., C. Knutsen, J. T. Wells, A. R. Jangid, M.L. Mitchell, N. W. Martinez, and A. W. Martinez. 2019. "Wax-printed fluidic time delays for automating multi-step assays in paper-based Microfluidic Devices (MicroPADs)." *Inventions* 4 (1). https://doi.org/10.3390/inventions4010020

Su, M., L. Ge, S. Ge, N. Li, J. Yu, M. Yan, and J. Huang. 2014. "Paper-based electrochemical cyto-device for sensitive detection of cancer cells and in situ anticancer drug screening." *Analytica Chimica Acta* 847:1–9. https://doi.org/10.1016/j.aca.2014.08.013

Sun, X., B. Li, A. Qi, C. Tian, J. Han, Y. Shi, B. Lin, and L. Chen. 2018. "Improved assessment of accuracy and performance using a rotational paper-based device for multiplexed detection of heavy metals." *Talanta* 178:426–431. https://doi.org/10.1016/j.talanta.2017.09.059

Taghizadeh-Behbahani, M., M. Shamsipur, and B. Hemmateenejad. 2022. "Detection and discrimination of antibiotics in food samples using a microfluidic paper-based optical tongue." *Talanta* 241:123242. https://doi.org/10.1016/j.talanta.2022.123242

Tamanna, N., and N. Mahmood. 2015. "Food Processing and Maillard Reaction Products: Effect on Human Health and Nutrition." *Int J Food Sci* 2015:526762. https://doi.org/10.1155/2015/526762

Thom, N. K., K. Yeung, M. B. Pillion, and S. T. Phillips. 2012. ""Fluidic batteries" as low-cost sources of power in paper-based microfluidic devices." *Lab on a Chip* 12 (10):1768–1770. https://doi.org/10.1039/C2LC40126F

Thuo, M. M., R. V. Martinez, W. J. Lan, X. Y. Liu, J. Barber, M. B. J. Atkinson, D. Bandarage, J. F. Bloch, and G. M. Whitesides. 2014. "Fabrication of low-cost paper-based microfluidic devices by embossing or cut-and-stack methods." *Chemistry of Materials* 26 (14):4230–4237. https://doi.org/10.1021/cm501596s

Toley, B. J., B. McKenzie, T. Liang, J. R. Buser, P. Yager, and E. Fu. 2013. "Tunable-Delay shunts for paper microfluidic devices." *Analytical Chemistry* 85 (23):11545–11552. https://doi.org/10.1021/ac4030939

Toley, B. J., J. A. Wang, M. Gupta, J. R. Buser, L. K. Lafleur, B. R. Lutz, E. Fu, and P. Yager. 2015. "A versatile valving toolkit for automating fluidic operations in paper microfluidic devices." *Lab on a Chip* 15 (6):1432–1444. https://doi.org/10.1039/C4LC01155D

Trapiella-Alfonso, L., J. M. Costa-Fernandez, R. Pereiro, and A. Sanz-Medel. 2013. "Synthesis and characterization of hapten-quantum dots bioconjugates: Application to development of a melamine fluorescentimmunoassay." *Talanta* 106:243–248. https://doi.org/10.1016/j.talanta.2013.01.027

Trofimchuk, E., A. Nilghaz, S. Sun, and X. N. Lu. 2020a. "Determination of norfloxacin residues in foods by exploiting the coffee-ring effect and paper-based microfluidics device coupling with smartphone-based detection." *Journal of Food Science* 85 (3):736–743. https://doi.org/10.1111/1750-3841.15039

Trofimchuk, E., Y. Hu, A. Nilghaz, M. Z. Hua, S. Sun, and X. Lu. 2020b. "Development of paper-based microfluidic device for the determination of nitrite in meat." *Food Chemistry* 316:126396. https://doi.org/10.1016/j.foodchem.2020.126396

Wang, C., X. Qin, B. Huang, F. He, and C. Zeng. 2010. "Hemolysis of human erythrocytes induced by melamine-cyanurate complex." *Biochem Biophys Res Commun* 402 (4):773–777. https://doi.org/10.1016/j.bbrc.2010.10.108

Wang, H., Y. Jian, Q. Kong, H. Liu, F. Lan, L. Liang, S. Ge, and J. Yu. 2018a. "Ultrasensitive electrochemical paper-based biosensor for microRNA via strand displacement reaction and metal-organic frameworks." *Sensors and Actuators B: Chemical* 257:561–569. https://doi.org/10.1016/j.snb.2017.10.188

Wang, P., Z. Cheng, Q. Chen, L. Qu, X. Miao, and Q. Feng. 2018b. "Construction of a paper-based electrochemical biosensing platform for rapid and accurate detection of adenosine triphosphate (ATP)." *Sensors and Actuators B: Chemical* 256:931–937. https://doi.org/10.1016/j.snb.2017.10.024

Wang, S., L. Ge, Y. Zhang, X. Song, N. Li, S. Ge, and J. Yu. 2012. "Battery-triggered microfluidic paper-based multiplex electrochemiluminescence immunodevice based on potential-resolution strategy." *Lab on a Chip* 12 (21):4489–4498. https://doi.org/10.1039/C2LC40707H

Wang, Y., H. Xu, J. Luo, J. Liu, L. Wang, Y. Fan, S. Yan, Y. Yang, and X. Cai. 2016. "A novel label-free microfluidic paper-based immunosensor for highly sensitive electrochemical detection of carcinoembryonic antigen." *Biosensors and Bioelectronics* 83:319–326. https://doi.org/10.1016/j.bios.2016.04.062

Wang, Y., H. Liu, P. Wang, J. Yu, S. Ge, and M. Yan. 2015. "Chemiluminescence excited photoelectrochemical competitive immunosensing lab-on-paper device using an integrated paper supercapacitor for signal amplication." *Sensors and Actuators B: Chemical* 208:546–553. https://doi.org/10.1016/j.snb.2014.11.088

Wang, Y., Y. Li, J. Wu, Y. Pei, X. Chen, Y. Sun, M. Hu, et al. 2019. "Development of an immunochromatographic strip test for the rapid detection of soybean Bowman-Birk inhibitor." *Food and Agricultural Immunology* 30 (1):1202–1211. https://doi.org/10.1080/09540105.2019.1680613

Washburn, E. W. 1921. "The dynamics of capillary flow." *Physical Review* 17 (3):273–283. https://doi.org/10.1103/PhysRev.17.273

Weng, X., and S. Neethirajan. 2018. "Paper-based microfluidic aptasensor for food safety." *Journal of Food Safety* 38 (1). https://doi.org/ARTNe1241210.1111/jfs.12412

Wu, X., H. Kuang, C. Hao, C. Xing, L. Wang, and C. Xu. 2012. "Paper supported immunosensor for detection of antibiotics." *Biosensors and Bioelectronics* 33 (1):309–312. https://doi.org/10.1016/j.bios.2012.01.017

Wu, Y., L. Zhang, D. Zhang, and R. Yu. 2024. "A surface molecularly imprinted microfluidic paper based device with smartphone assisted colorimetric detection for butachlor in mung bean." *Food Chemistry* 435:137659. https://doi.org/10.1016/j.foodchem.2023.137659

Xie, J., L. Li, I. M. Khan, Z. Wang, and X. Ma. 2020. "Flexible paper-based SERS substrate strategy for rapid detection of methyl parathion on the surface of fruit." *Spectrochim Acta A Mol Biomol Spectrosc* 231:118104. https://doi.org/10.1016/j.saa.2020.118104

Yamada, K., T. G. Henares, K. Suzuki, and D. Citterio. 2015. "Paper-based inkjet-printed microfluidic analytical devices." *Angewandte Chemie-International Edition* 54 (18):5294–5310. https://doi.org/10.1002/anie.201411508

Yamada, K., S. Takaki, N. Komuro, K. Suzuki, and D. Citterio. 2014. "An antibody-free microfluidic paper-based analytical device for the determination of tear fluid lactoferrin by fluorescence sensitization of Tb3+." *Analyst* 139 (7):1637–1643. https://doi.org/10.1039/C3AN01926H

Yang, K., H. Peretz-Soroka, Y. Liu, and F. Lin. 2016. "Novel developments in mobile sensing based on the integration of microfluidic devices and smartphones." *Lab on a Chip* 16 (6):943–958. https://doi.org/10.1039/c5lc01524c

Yang, K., W. Yu, G. Huang, J. Zhou, X. Yang, and W. Fu. 2020. "Highly sensitive detection of Staphylococcus aureus by a THz metamaterial biosensor based on gold nanoparticles and rolling circle amplification." *RSC Adv* 10 (45):26824–26833. https://doi.org/10.1039/d0ra03116j

Yen, C.-W., H. de Puig, J. O. Tam, J. Gómez-Márquez, I. Bosch, K. Hamad-Schifferli, and L. Gehrke. 2015. "Multicolored silver nanoparticles for multiplexed disease diagnostics: distinguishing dengue, yellow fever, and Ebola viruses." *Lab on a Chip* 15 (7):1638–1641. https://doi.org/10.1039/C5LC00055F

Yetisen, A. K., M. S. Akram, and C. R. Lowe. 2013. "Paper-based microfluidic point-of-care diagnostic devices." *Lab on a Chip* 13 (12):2210–2251. https://doi.org/10.1039/C3LC50169H

Yu, J., L. Ge, J. Huang, S. Wang, and S. Ge. 2011. "Microfluidic paper-based chemiluminescence biosensor for simultaneous determination of glucose and uric acid." *Lab on a Chip* 11 (7):1286–1291. https://doi.org/10.1039/C0LC00524J

Zang, D., L. Ge, M. Yan, X. Song, and J. Yu. 2012. "Electrochemical immunoassay on a 3D microfluidic paper-based device." *Chemical Communications* 48 (39):4683–4685. https://doi.org/10.1039/C2CC16958D

# 4 Microfluidic Nanodetection Systems for Molecular Interactions

*Manman Du\*, Xin Wang\*, Xin Meng, Yaohua Du, and Xinwu Xie*

## 4.1 INTRODUCTION

The essence of life is the process of interaction between biomolecules, and the exploration of physiological phenomena, such as gene expression, signal transduction, and immune responses, can be classified as the study of the interactions between biomolecules. Biomolecular interactions are common in living organisms, such as the immune reaction of mutual recognition between antigens and antibodies, the binding of enzymes or receptors to substrate molecules, and the binding of proteins and gene sequences. Thus, the detection and analysis of biomolecular interactions are of great importance in the field of food processing, for instance, to ensure food safety and for functional food development.

For cells, proteins are the executors and sit at the core position. Most biomolecules (proteins, DNA, RNA, peptides, sugars, lipids, and small molecules) perform their biological functions via interaction with proteins (Asmari et al., 2018). Therefore, the determination of the interaction between proteins and other biomolecules is crucial and indispensable in life science and biomedical research.

Traditional techniques are used to determine biomolecular interactions including electrophoresis, circular dichroism, yeast two-hybrid system, nuclear magnetic resonance (NMR), isothermal titration calorimetry (ITC), mass spectrometry (MS), enzyme-linked immune sorbent assay (ELISA), atomic force microscopy (AFM), fluorescence resonance energy transfer (FRET), and protein chip technology. Traditional detection technologies are not only difficult to employ in a high-throughput manner for high-sensitive detection, but they also require a large amount of sample and are time-consuming and laborious (Jeong et al., 2022). Further, they can only detect the equilibrium states of the biomolecular interaction, with minimal detection of dynamic processes.

With the continuous development of various new materials, technologies, and principles, research on biomolecular detection technology based on microfluidics is booming, and the types of detection processes cover almost all known functional biomolecules. Microfluidics refers to the science and technology involved in systems that process or manipulate miniscule fluid volumes (nL-aL volumes) through microchannels (10–100 μm in size). Microfluidics can integrate multiple processes such as sampling, chemical reaction, separation, dilution, and detection into one chip (Zhou et al., 2019; Fong Lei, 2013). The application of microfluidic devices in bioanalytical field has grown tremendously in the last decade, including to achieve the "sample-in/result-out" analysis mode and the overwhelming cost competitiveness (Yang et al., 2020). Compared to traditional detection technologies, microfluidic systems have unique advantages such as being high throughput, requiring low

---

\* These authors contributed equally to this chapter.

reagent consumption, and enabling rapid analysis, miniaturization, and automation (Zhou et al., 2019; Yang et al., 2020). In addition, microfluidic detection systems can provide dynamic, real-time changes in molecular binding or dissociation processes, helping better understand the process of biomolecular interactions.

The development of nanomaterials has provided powerful new tools and development directions for biological detection. Especially since entering the 21st century, nanotechnology has become a hot research direction, and the research of nanomaterials and the development of nanodevices have gradually matured. Nanomaterials have characteristics such as surface and interface effects, small size effects, quantum size effects, and macroscopic quantum tunneling effects, which make them distinguishable from large-scale materials formed by the same elements, and have electrical, optical, magnetic, thermal, and other desirable physical and chemical properties (Miernicki et al., 2019). Nanomaterials are ideal carriers for immobilization of biomolecules due to their ease of surface modification. Further, due to the large specific surface area of nanomaterials, they can both participate in the immobilization of biomolecules and carry out signal amplification to improve the detection sensitivity (Tavakoli et al., 2022; Fattahi & Hasanzadeh, 2022). Therefore, the combination of nanomaterials and microfluidic detection systems provides advantages for the integration and portability of detection equipment, as well as improves detection sensitivity and shortens detection time.

At present, microfluidic nanodetection systems are widely used in various biomolecular detection and biomolecular interaction analyses, including single-cell detection (Jammes & Maerkl, 2020; Chen et al., 2019), protein–protein interaction (Arter et al., 2020), protein–DNA interaction (Lee et al., 2021), and protein and other molecular interactions (Weng & Neethirajan, 2018). Moreover, commercial instruments based on microfluidic detection systems have been developed for the analysis of biomolecular interactions, such as Biacore equipment based on the principle of surface plasmon resonance (SPR) and the Fortebio/Sartorius Octet® based on the principle of biolayer interferometry (BLI).

Therefore, this chapter will focus on the principles of biomolecular interactions, the main components of microfluidic nanodetection systems, and the research progress and development trends in biomolecular interactions.

## 4.2 MOLECULAR INTERACTIONS IN MICROFLUIDIC SYSTEMS

The inner workings of a cell consist of a series of biochemical processes that are accomplished through molecular interactions. The study of biomolecular interaction is highly valuable for probing the mechanism of these interactions and establishing convenient methods to effectively understand the functions of the molecules. The interaction between molecules can be in the form of short-range bonding forces or non-bonding long-range action, such as hydrogen bonding, hydrophobic action, aromatic $\pi$–$\pi$ accumulation, charge transfer, and *van der Waals* force. At present, the analysis of intermolecular interactions is mainly concerned with the qualitative and quantitative analysis of intermolecular interactions (Yang et al., 2013). Qualitative analysis is mainly to analyze the mode of action, site of action, molecular structure, configuration transformation, and long-range electron transfer properties of biological macromolecules (proteins and nucleic acids), while quantitative analysis is mainly to analyze the binding ratio, binding site number, binding constant, free energy change, diffusion coefficient, material turbulence, and other thermodynamic and kinetic parameters. The biological properties of the molecules are determined by qualitative and quantitative analyses of these effects. Liang et al. (2020) describe the experience of using quantum dot (QD)-based lateral flow immunoassay, combined with a portable fluorescence immunoassay chip detector, to detect serum-specific IgE against *Dermatophagoides pteronyssinus* and *Dermatophagoides farinae*, two common mite allergens in China.

### 4.2.1 MOLECULAR INTERACTION MECHANISM

#### 4.2.1.1 Intermolecular Forces

In the process of intermolecular interaction, the second bond between atoms plays an important role. For example, in proteins, interactions such as *van der Waals* force, hydrogen bonding networks, electrostatic interactions, and hydrophobic effects can affect protein folding and stability.

1. van der Waals force

   The *van der Waals* force is a general term for atomic or intermolecular electric dipole interaction, including electrostatic force (orientation force), induced force, and dispersive force. The *van der Waals* force is a weak intermolecular force between molecules. The bond energy of *van der Waals* bonds is 1–2 orders of magnitude lower than that of other chemical bonds; the action range is 0.3–0.5 nm, with no directionality and saturation (Gao et al., 2022).

2. Hydrogen bonding interactions

   The strength of a hydrogen bond lies between that of a covalent bond and the *van der Waals* force, which is 8.4–168 kJ/mol. Hydrogen bond energy is low, making them easy to form and break, so hydrogen bonds are of great significance for the recognition and reaction of biomolecules (Lin et al., 2023). In addition, although hydrogen bonds are weak, the common effect of a large number of hydrogen bonds in living organisms can stabilize structures. For example, the presence of α helices in proteins, β folding, and the DNA double helix structure all rely on a large number of hydrogen bonds to stabilize the overall structures of the molecules. Bodin-Thomazo et al. (2022) explored the capacity of a single pH-sensitive copolymer, PDMS60-b-PDMAEMA50, and salts to form and stabilize multiple W/O/W emulsions loaded with sucrose or catechin by a one-step mechanical process or a microfluidic method. This study highlighted some of the barriers to break to formulate multiple emulsions stabilized by a PDMS-b-PDMAEMA copolymer or other polymers which can form hydrogen bonds and interact with encapsulated drugs.

3. Electrostatic force

   Electrostatic force refers to the interaction force between stationary and charged bodies. The charged body can be regarded as composed of many point charges, and the interaction force between each pair of stationary point charges follows Coulomb's law, also known as the Coulomb force. The magnitude of the electrostatic interaction force is dependent on the interparticle distance, charge size, and medium properties. Electrostatic interaction forces have important applications in many fields. For example, in biology, electrostatic forces can facilitate the binding between proteins and biological macromolecules such as nucleic acids by forming ion pairs or ion bridges (Zhou et al., 2022).

4. Hydrophobic forces

   Hydrophobic forces (hydrophobic bonds) often play a key role in the spatial folding of protein polypeptide chains, the formation of biofilms, the interaction between biological macromolecules, and the enzyme catalysis of substrate molecules (Parvizi et al., 2023). Nonpolar molecules or the nonpolar groups of molecules attract and gather together in an aqueous environment, while water molecules are originally excluded near the nonpolar groups. The result is that the entropy of the surrounding water molecules increases. This interaction force can cause protein macromolecules to fold close to their surface to form a certain cleft structure, forming an anhydrous microenvironment, which is sometimes necessary for enzyme binding to substrate molecules or receptor binding to ligand molecules (Vloemans et al., 2021).

# Microfluidic Nanodetection Systems for Molecular Interactions

### 4.2.1.2 Spatial Structure

The spatial structure of different biological macromolecules is very different, and changes in the spatial structure have a large influence on the interaction of molecules in three-dimensional (3D) space, such as specific binding between antibodies and antigens, enzymes and substrates, and hormones and receptors (Wang et al., 2023). Molecular recognition is achieved through the respective binding sites of the two molecules. To achieve molecular recognition, one requires that the binding site of two molecules is structurally complementary; the other requires that two binding sites have corresponding groups that can produce enough force between each other that can be combined together. Molecular recognition exists among the glycans, proteins, nucleic acids, and lipids or among each other.

### 4.2.1.3 Electrodynamic and Thermodynamic Effects

Electrodynamics and thermodynamics are two other important factors in the interactions between biological macromolecules (Hossain et al., 2023; Jimenez et al., 2023). The charge attraction, electrostatic Coulomb repulsion, and hydrophobic action between molecules are the dominant factors in intermolecular interactions. These forces have obvious effects on the distance, direction, and effect between molecules. Second, physical quantities such as molecular energy, activity, entropy, and enthalpy play a crucial role in the process of intermolecular interactions.

### 4.2.1.4 Non-specific Adsorption

Non-specific adsorption (non-specific binding) is a type of adsorption caused by non-covalent bond forces. The compounds to be tested in a solution are adsorbed to the solid surface because of electrostatic or hydrophobic interactions with the compounds (Kovalchuk & Simmons, 2023). The adsorption phenomenon and adsorption degree of the compounds to be tested are closely related to the solid surface of the solution, the composition of the solution, and the properties of the compounds to be tested. Three elements of non-specific adsorption determine whether there is adsorption, and some other factors such as the environmental temperature, the solution pH, the contact time between the solution and the solid surface, and even the freezing and melting times of the solution will have a certain impact on the adsorption degree. Electrode functionalization protocols were developed considering a possible charge transfer through the sensing layer, in addition to analyte-specific binding by corresponding antibodies and reduction of non-specific protein adsorption to prevent false-positive signals (Alsabbagh et al., 2021).

In order to better understand intermolecular interactions, several common types of intermolecular interactions in microfluidic systems are listed below.

### 4.2.2 COMMON TYPES OF MOLECULAR INTERACTIONS IN MICROFLUIDIC SYSTEMS

### 4.2.2.1 Protein–Protein Interactions

Current standard techniques for investigating protein–protein interactions require volumes and/or concentrations significantly higher than those relevant under physiological conditions. Such approaches further mainly rely on surface immobilization, thereby potentially interfering with the physiological properties of the interaction investigated. Microfluidic techniques offer an orthogonal method for investigating such interactions under physiological conditions. These low-volume techniques have thus been applied to investigate a wide range of protein properties and functions, as well as to study protein–protein interactions. The types of protein–protein interactions mainly include antigen–antibody interactions, enzyme–substrate interactions, and receptor–protein ligand interactions.

1. Antigen–antibody interactions
   Binding of an antigen to an antibody is a highly specific reaction, where an antibody can only bind to its specific antigen and not to other antigens. This specificity is determined by

the structure of the antibody, each having a specific structure that recognizes and binds a specific antigen. Binding of antigen to antibody is a complex process involving steps such as affinity, specificity, and crosslinking. Schneider et al. (2023) applied microfluidic diffusional sizing demonstrating reliable quantification of alloantibody binding affinity and concentration of alloantibodies binding to human leukocyte antigens, an extensively used clinical biomarker in organ transplantation.

2. Enzyme–substrate interactions

The specificity of the enzymatic reaction is usually imagined as being the result of a steric fit. The "key and lock" or "induced fit" image is evoked to illustrate the underlying principle. Various types of weak and usually non-covalent interactions can be formed between biologicals. Among them, hydrogen bridges, a number of *van der Waals* forces, and electrostatic interactions are the most prominent. Only when several of these interactions can occur simultaneously because the putative interaction points between the two molecules coincide in a 3D framework, and "binding" takes place (Freitag, 1999). The strength of the resulting bound can cover several orders of magnitude and, in some cases (e.g., certain hormone–receptor pairs), approach that of a covalent one. Recently, the integration of microfluidics and enzyme immobilization technology has enabled the development of low-cost, time-efficient, and high-accuracy analytical biosensors for detecting biomolecules (Yamaguchi & Miyazaki, 2022).

3. Receptor–protein ligand interactions

Any extracellular signaling molecule that causes a certain response by the target cell depends on the binding of the signaling molecule to a specific receptor. Receptor proteins are classified as membrane receptors, intracellular receptors, or nuclear receptors according to their cellular location. Signaling molecules, such as hormones, pheromones, or neurotransmitters, are called ligands, which must bind to the characteristic sites of receptor proteins, cause conformational changes of receptor molecules, and then initiate changes in cell function. Goral et al. (2011) developed a label-free optical biosensor with microfluidics for differentiating ligand-directed functional selectivity on trafficking of receptors.

### 4.2.2.2 Protein–Nucleic Acid Interactions

Genome replication, transcription, translation, modification, and other processes are inseparable from the interaction between nucleic acids and proteins. Research on this topic is helpful to understand various processes including epigenetic regulation, such as finding target DNA transcription-related regulatory factors and finding RNA translation regulation-related proteins. The metabolic processes involving protein–RNA interaction mainly include the transcription of RNA molecules, the post-transcriptional processing of precursor RNA, the transport and localization of RNA in cells, and the stabilization of intracellular RNA. Protein–DNA interactions mainly include gene transcription and regulation, DNA replication and repair, DNA recombination and packaging, and the formation of chromatin and ribosomes. Lee et al. (2021) reported a rapid electrokinetic detection of low-molecular-weight thiols using the interaction between Bacillus subtilis-derived protein and its operator DNA element in ion concentration polarization-coupled microfluidic multiple channels.

### 4.2.2.3 Protein–Small Molecule Interactions

Protein–small molecule interaction is a very important concept in biology, which involves many important biochemical processes in organisms. Proteins are one of the most important molecules in living organisms. They play important roles in cells, including catalyzing reactions, transmitting signals, and transporting substances. Small molecules, on the other hand, refer to organic molecules with low molecular weights that can interact with proteins to affect their function.

There are many types of protein–small molecule interactions, the most common of which is the ligand–receptor interaction. In this interaction, a stable complex is formed between a small molecule (ligand) and a protein (receptor), thereby affecting the function of the receptor. This interaction is very important in drug development because many drugs work through ligand–receptor interactions with proteins. In addition, there are interactions between proteins and small organic molecules such as cell metabolites, drugs, and food additives. Amir et al. (2023) reviewed the recent microfluidics developments for modeling and diagnosing common diseases, including cancer, neurological, cardiovascular, respiratory, and autoimmune disorders, and their applications in drug development.

### 4.2.3 MOLECULAR INTERACTION ANALYSIS METHODS

Many methods can be used to detect intermolecular interactions, such as various optical methods (ultraviolet–visible absorption spectroscopy, infrared spectroscopy, Raman spectroscopy, SPR, fluorescence spectroscopy, and dual-polarization interferometry), circular dichroism, AFM, NMR, X-ray crystal diffraction, electrochemical methods, MS, ITC, chromatography, electrophoresis, equilibrium dialysis, ultracentrifugation, and microscale thermophoresis, all of which are widely used in the detection of molecular interactions.

However, traditional molecular interaction analysis methods are often slow, sample-consuming, and sensitivity-limited, making them unsuitable for high-throughput studies of weak transient interactions and requiring a large amount of reagents. Furthermore, methods based on surface analysis are often challenged by non-specific binding and the need for suitable antibody reagents. These drawbacks cannot meet the demands for on-site detection or analysis and raise the need for experimental techniques to study molecular interactions directly in free solution that operate rapidly within short analysis time and require minimal sample consumption.

Microfluidic technology has the advantages of high speed, high sensitivity, strong specificity, and ease of integration. The consumption of samples and reagents by microfluidic technology is very small, usually only a few microliters, which saves on the amount of reagents required. Further, multiple reactions can be run simultaneously for high-throughput analysis, reducing testing costs. The analysis speed of microfluidic chips is very fast, usually only a few minutes or even tens of seconds. Microfluidic technology can integrate multiple processes such as sampling, chemical reaction, separation, dilution, and detection into one chip to enable the miniaturization and automation of the detection platform. Thus, the characteristics of microfluidic technology fully meet the needs of molecular interaction analysis, which will be developed into a portable device for on-site biodetection and analysis of molecular interactions to enable environmental testing, medical research, food and agricultural safety, military medicine, etc.

## 4.3 MICROFLUIDIC NANODETECTION SYSTEM

Microfluidics is an emerging technology with advantages including low reagent consumption, short analysis time, high efficiency, high throughput, integrated detection sample-to-result, low cost, and portability, which has been proven to be useful for studying and controlling biomolecular interactions. The applications of microfluidic technology in food processing have been extended to other fields. For example, in food safety measurements, the current industry practice of using 1.0 mg/L free chlorine requires >1.0 s total contact to achieve a 5-$\log_{10}$ reduction in an *E. coli* O157: H7 population, whereas by using a microfluidic device, 10.0 g/L free chlorine solution is effective for a 5-$\log_{10}$ reduction in as little as 0.25 s. In bio-compound detection, a newly developed microfluidic flow injection analyzer allows the determination of the total polyphenols in white wine with relative errors reduced by >13% compared to the conventional flow-injection analysis method. The applications of microfluidic devices in food processing have been comprehensively studied in laboratory settings, and microfluidic devices have been broadly applied in various food processing sectors for processing applications and food safety control.

Due to the unique physical and chemical properties of nanomaterials, which can enhance signal modulation, such as signal amplification and signal transduction, their applications in the field of biosensing have received extensive attention. Therefore, a large number of microfluidic detection systems that integrate nanobiosensor chips have been developed for biological analyses. The microfluidic nanodetection platform provides a dynamic and real-time technology for testing the interaction between biomolecules. It can reflect the changes in the process of molecular binding or dissociation in real time. Being able to observe the specificity and strength of the binding of two molecules helps to understand the process of biomolecular interaction more realistically.

### 4.3.1 Main Units of Microfluidic Detection Systems

Microfluidic detection systems mainly include a microfluidic chip, a signal detection chip/module, a signal acquisition and amplification module, a core control module, a system software, and a power module. Some microfluidic detection systems also include auxiliary modules to supply exciting energy or stabilize the system conditions, such as light and temperature. However, some detection systems are rather simpler, for example, some sensors use a microfluidic chip that yields results that can be detected with the naked eye.

#### 4.3.1.1 Microfluidic Chip

The basic principle of microfluidic detection is to utilize the flow characteristics and microscale channel structure of microfluidics to achieve rapid manipulation and efficient analysis of samples. The size of microchannels in microfluidic chips usually ranges from nanometers to micrometers. This microscale channel structure can provide a larger specific surface area and shorter diffusion distance, thereby improving the sensitivity and speed of analysis. Microfluidic devices such as micropumps, microvalves, and microreservoirs are connected through a micropipeline network, and microelectrodes, microdetections, and other components are integrated into the chip to realize sample injection, dilution, mixing, reaction, separation, metering, and other functions (Yang & Gijs, 2018).

Micropumps and microvalves are the main parts of microfluidic chips. They are the power source and core components of sample driving in microfluidic chips, which can control the flow rate, flow direction, mixing degree, and other parameters of the sample, so as to achieve accurate sample processing. A micropump is a device that continuously delivers a working fluid (liquid or gas) at precise volumes from a storage chamber to a designated location (Wu et al., 2023). Considering the small volume of the reaction chamber of the microfluidic chip, low consumption of samples and reagents, and the need for easy integration and miniaturization, micropumps with small volumes and low flow rates are generally selected.

Microvalves can effectively control the on-off, speed, and flow direction of different types of microfluidic chips. An ideal microvalve should have the characteristics of low cost, small size, easy integration, high flow control accuracy, no leakage, and fast response (Wu et al., 2023). There are single, double, or multiple microreservoirs in the microfluidic chips, which are used for sample storage, mixing, reaction, cell culturing, etc.

To form a complete microfluidic chip system, each individual microfluidic device needs to be connected. The connection of microfluidic devices needs to prevent liquid leakage and ensure the smooth operation of the whole system. At present, the connection methods for microfluidic devices include the LEGO® connection method, pipe connection method, Luer connection method, O-ring/gasket connection method, plasma connection method, and adhesive connection method (Wu et al., 2023).

#### 4.3.1.2 Signal Detection Chip/Module

The detection Chip/Module of the microfluidic system usually needs to have the characteristics of high sensitivity and signal-to-noise ratio, low cost, miniaturization, fast analysis speed, and easy integration with the microfluidic Chip/Module. The main function of the detection chip is to determine the composition and content of the analyte after separation or processing by the microfluidic chip, which is an important part of the microfluidic detection system. The core component of the Chip/Module is a transducer that converts the biomolecular interactions into another detectable signal and generally detects the signal as an electric output for further processing. Some signal detection modules can be miniaturized into a very small size as a chip. In some cases, the signal detection module is integrated into the microfluidic chip as a very compact device. At present, the main detection methods of microfluidic signals are the electrochemical method, optical method, MS method, and other methods.

1. Electrochemical method

   The principle of electrochemical detection is to use immobilized biomolecules as recognition elements and electrodes as transducer elements to convert the interaction between biomolecules into detectable electrochemical signals. The electrochemical detection methods used on microfluidic chips mainly include voltammetry/amperometry, the potentiometric method, electrical impedance, and the field effect transistor (FET) (Welch et al., 2021). Depending on the way the potential is applied, voltammetry/amperometry is commonly used for biomolecule detection using cyclic voltammetry, chronoamperometry, differential pulse voltammetry, square wave voltammetry, and linear sweep voltammetry, among others. The electrochemical method has the advantages of high sensitivity, label-free, fast response, low cost, and easy miniaturization (Ebrahimi et al., 2022). Moreover, the detection sensitivity is not affected by the diameter of the microchannel, and there is no requirement for light transmission by the detection part, which is very suitable for the integration and miniaturization of the microfluidic analysis system.

2. Optical method

   Optical methods are widely used in biochemical analysis due to their high sensitivity. The principle of optical detection is to use the interaction between matter and light by receiving and processing reflection, transmission, refraction, scattering, energy, fluorescence, and other light signals to obtain the information and characteristics of the object. The optical detection methods used in conjunction with microfluidic technology mainly include fluorescence, absorptiometry, chemiluminescence, SPR, reflection interference spectroscopy, BLI, and surface-enhanced Raman spectroscopy (SERS). Optical methods are highly sensitive, safe, reliable, non-contact, and non-destructive detection methods (Alhalaili et al., 2022). However, when using an optical method for detection, it is necessary to pay attention to optical parameters, such as the light transmittance of the microfluidic chip material, and consider the influence of ambient light sources and particles on the measurement results.

3. Mass spectrometry

   MS detection technology mainly uses a specific ion source to convert the sample to be tested into high-speed ions. These ions are separated under the action of an electric field or a magnetic field due to different mass/charge ratios, and the detector is used to record the relative intensities of various ions, form a mass spectrum for analysis, and provide reliable detection results. At present, the MS techniques used for the detection of biomolecules are mainly gas chromatography-MS, time-of-flight MS, electrospray MS, and capillary electrophoresis-MS (CE-MS) (Hung et al., 2014). The biggest advantage of MS is that it can quickly and sensitively provide molecular spatial structure information and apply it to microfluidic chip analysis, which can greatly improve the detection effect. For example, Khatri et al. (2017) developed a microfluidics-based CE-MS system that can analyze monosaccharides, oligosaccharides, and glycopeptides under the same electrophoretic conditions, which greatly simplifies the task.

4. Other detection methods

   In addition to detection methods based on electrochemical, optical, and MS principles, there are also detection methods based on thermal effects, piezoelectric effects, magnetic properties, acoustic waves, and micromechanical gravity, such as quartz crystal microbalances (QCM), microcantilever (MCL) biosensors, magnetic biosensors, surface acoustic wave sensors, and ring-mediated isothermal amplification technology. Among them, QCM uses quartz crystal as the sensitive element of a microbalance, and electron diffraction, electron beam evaporation, sputtering deposition, and other technologies are used to prepare a thin film on its surface. When biomolecules interact with the surface of the film, the quality of the quartz crystal will change, resulting in change in the crystal vibration frequency, revealing the mass of the sample.

MCL biosensors are mainly measured by the change in the surface stress of the MCL. Biomolecules interact on the surface of the cantilever, which causes the surface of the MCL to bend. The MCL converts the molecular recognition of the biomolecules into a nanoscale mechanical offset, and then the surface of the microorganism can be detected by outputting the signal through the readout system. These methods can also be combined with microfluidic technology for real-time, highly sensitive, and label-free detection of biomolecules and can also be used for the analysis of biomolecular interactions such as cells, proteins, enzymes, and nucleic acids.

### 4.3.1.3 Signal Acquisition and Amplification Module

The signal of the microfluidic detection chip/module is first converted into an electrical signal by the detection circuit, and it then undergoes a series of processing steps such as filtering, rectification, and amplification before finally being collected by an analog-to-digital converter. During the detection process, the microfluidic chip signal will be interfered with by the inherent noise of the system, high-frequency noise, and baseline drift. Among these, the inherent noise of the system can be directly filtered out by the hardware in the microfluidic chip detection equipment, but the high-frequency noise and baseline drift are difficult to filter out by optimizing the hardware design of the instrument, requiring research on corresponding methods for signal processing.

For the signal processing of the microfluidic chip, the most critical issue is to preserve the signal reflecting the characteristics of the detected substance as much as possible. However, for biomolecular detection, the signal detected by the detection chip/module is very weak, at the microampere level or even smaller, which is difficult to detect, so it is necessary to amplify this weak signal. Signal amplification refers to processing the input signal through an amplifier to obtain a larger output signal. The purpose of signal amplification is to enhance the strength of the signal so that it can be better received and processed during transmission. The signal amplification module generally includes a pre-amplification circuit, bias circuit, post-amplification circuit, and the like. Finally, the processed signal is collected by a data acquisition module and sent to a computer or mobile terminal to perform operations such as waveform display, data analysis, and storage.

### 4.3.1.4 Core Control Module

The core control module refers to the part responsible for controlling and adjusting the work of some other modules or components in the electronic equipment. Its function is to control the operating status of the device or implement specific functions through logical judgment and processing based on input signals or conditions. The core control module plays a very important role in the whole system, which is equivalent to the human brain, and needs to be carefully selected. Because of the excellent performance and progress of microcontrollers, they have been widely used in the development of various microfluidic instruments. A microcontroller is an integrated circuit chip that integrates central processing unit, random access memory, read-only memory, interrupt system, timer/counter, a large number of I/O port, and even includes a digital-to-analog conversion module, an analog-to-digital conversion module, and other functions. There are many types of microcontrollers on the market, such as ARM, MIPS, MPC/PPC, and Super H series. According to the actual needs of the detection system, an appropriate microcontroller should be selected as the main controller. The design of the microcontroller control module circuit is a complex and important task. During the design process, functional requirements, stability, reliability, input and output signals, power consumption, maintainability, scalability, and security protection need to be considered. Only by comprehensively considering these factors and making a reasonable circuit design can a high-performance, stable, and dependable single-chip control module circuit be realized.

### 4.3.1.5 System Software

For a practical instrument, the hardware part is its skeleton, and the software part is its soul. The two complement each other to realize their functions together. System software refers to the system that controls and coordinates computers and external devices and supports the development and operation of application software. The main function is to schedule, monitor, and maintain computer systems, responsible for managing various independent hardware in the computer system so that they can work in coordination. The system software mainly includes the program design of the slave computer (Core Control Module) and the software design of the master computer (usually as the user interface). The master computer can be a PC, an embedded terminal, or a mobile terminal. The slave computer mainly uses a microcontroller as the main control chip to control the operation and function of each module. The master computer software enables system data display and user interaction, and its main functions include establishing communication with the slave computer, sending control commands, real-time data display, dynamic curve drawing, data storage, and historical data import. The slave computer can control the microcontroller to collect data and then integrate the data and send it to the master computer until it receives a stop command from the upper computer. When the master computer sends corresponding control commands, such as sampling switch or voltage zero setting, the slave computer program will perform the corresponding operations. The master computer software can be written in a variety of programming languages, such as Java, Kotlin, C/C++, Python, JavaScript, and graphical programming language, by calling the official Bluetooth application program interface (API) and serial port API to communicate with other devices and complete the development of other functions.

### 4.3.2 NANOMATERIALS FOR MICROFLUIDIC DETECTION SYSTEMS

Integrating nanomaterial-based biosensors into microfluidic systems has become a hot research topic. On the one hand, nanomaterials are capable of playing versatile biosensing roles in microfluidic detection systems, such as molecular event recognition, signal amplification, signal transduction, and modulation (such as amplification and quenching); on the other hand, microfluidic system provides a more versatile platform for real-time sensing of biomolecules using nanomaterials (Tavakoli et al., 2022).

Nanomaterials (size ~1–100 nm) have some unique effects due to the particularity of scale and structure, such as the quantum size effect, surface effect, small size effect, and macro quantum tunneling effect, resulting in unique optical, electrical, magnetic, thermal, and other properties that traditional macroscopic materials do not display (Tavakoli et al., 2022). The characteristics of nanomaterials can improve the microfluidic chip detection sensitivity of the sensor, shorten the detection time, and further stabilize the physical and chemical properties of the sensor so that the detection performance of the sensor can be significantly improved. Moreover, nanomaterials have the characteristics of a large specific surface area and easy functionalization of the surface. These unique properties of nanomaterials make them ideal for combining biological interactions with transducers that can generate detectable signals. Nanostructures facilitate electron transfer between immobilized bio-recognition elements (e.g., nucleic acids, proteins, and antibodies) and electrode surfaces, making them widely used in biosensing devices. According to their dimensions, nanomaterials can be divided into zero-dimensional (0D), one-dimensional (1D), two-dimensional (2D), and 3D nanomaterials.

### 4.3.2.1 Zero-Dimensional Nanomaterials

0D nanomaterials refer to materials that are in the nanoscale range (1–100 nm) or are composed of basic units in three dimensions. The surface atoms occupy a significant proportion, which increases the surface state density and various quantum effects are very significant. Such materials have the advantages of narrow emission peaks, high stability, and size-dependent photoluminescence. 0D nanomaterials include nanoparticles (NPs), nanoclusters, and QDs. For example, metal NPs are ideal materials for biomolecular detection due to their ease of synthesis, regulation, and surface modification, as well as their excellent stability, biocompatibility, and high absorption coefficients. Common metal NPs include Au NPs, Ag NPs, and Pt NPs. The quantum size effect and quantum confinement effect of QDs yield unique luminescence characteristics and electronic properties. Their surfaces can be used as a scaffold structure to bind biological macromolecules such as proteins and nucleic acids, which is ideal for fluorescent probes (Resch-Genger et al., 2008).

### 4.3.2.2 One-Dimensional Nanomaterials

1D nanomaterials refer to nanomaterials that have two dimensions at the nanoscale and only one dimension beyond the nanoscale, with an aspect ratio usually >1000. 1D nanomaterials have a high specific surface area, which is conducive to surface modification to obtain receptor sensitivity, while their long longitudinal dimensions are conducive to the construction of devices such as FETs (Welch et al., 2021). 1D nanomaterials are often used as nanocarriers to analyze the interaction or chemical reaction between the sensing electrode and the surface of the analyte. Common 1D nanomaterials include carbon nanotubes (CNTs) and silicon nanowires (SiNWs).

### 4.3.2.3 Two-Dimensional Nanomaterials

2D nanomaterials are nanomaterials that have only one dimension between 1 and 100 nm, and the other two dimensions are not limited to the nanometer size. 2D nanomaterials include graphene and its chemical derivatives, black phosphorus, transition metal dichalcogenides, transition metal carbon/nitride, and metal-organic frameworks (MOFs) (Li et al., 2022). 2D nanomaterials have attracted extensive attention in the field of biosensing due to their photothermal properties, high electrical conductivity, high mechanical flexibility, large specific surface area, many active sites, good chemical stability, biocompatibility, and ease of surface modification.

### 4.3.2.4 Three-Dimensional Nanomaterials

3D nanomaterials refer to composite nanomaterials composed of one or more 0D, 1D, and 2D nanomaterials as structural units, which are not limited to the nanoscale in any dimension but are still regarded as nanomaterials, because they have nanocrystalline structures or involve the existence of nanoscale characteristics. For the nanocrystal structure, the most typical is the multiple arrangement of nanocrystals in different directions to form 3D nanomaterials. In addition, 3D space assembly and arrangement into a system with nanostructure characteristics (e.g., composites of NPs and conventional materials) and assembly systems with porous media are also called 3D nanomaterials, including nanomesoporous materials, metal nanopores, graphene foams, nanoglasses, and nanoceramics. 3D nanomaterials use the defects of low-dimensional materials to assemble them into stable 3D structures, which retain the intrinsic properties of the original materials but have new expansions in structure and properties. 3D nanomaterials are mainly used in the construction of sensing interfaces of electrochemical biosensors, as well as cell scaffolds, and for bioreceptor immobilization due to their good biocompatibility.

### 4.3.3 Typical Microfluidic Nanodetection System

At present, many researchers have developed various microfluidic nanodetection systems for biomolecular detection or biomolecular interaction analyses. For example, Xie et al. (2022) constructed a self-contained integrated microfluidic nanodetection system based on SiNW-FET biosensors for biological detection and analysis, as shown in Figure 4.1. All analysis processes, including liquid sample delivery, light modulation, constant temperature control, signal amplification, data acquisition, and result display, are automatically completed. The detection limit of this device is 1.0 fg/mL using a simulated sample of the typical microorganism *Mycobacterium tuberculosis*, which verifies

# Microfluidic Nanodetection Systems for Molecular Interactions

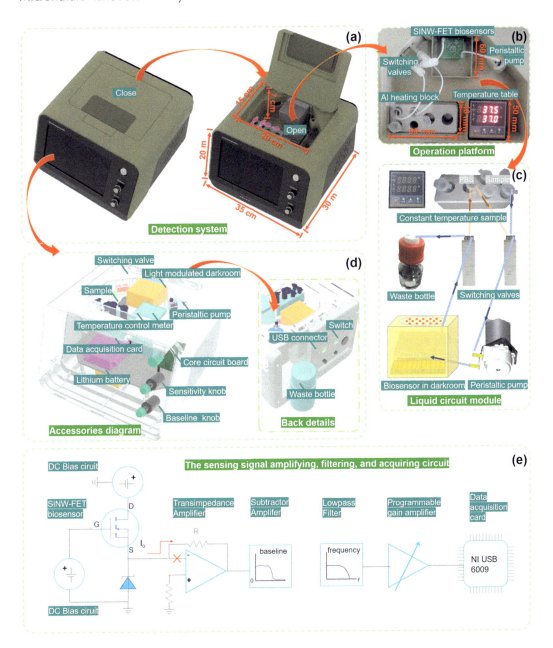

**FIGURE 4.1** Construction of the integrated SiNW-FET detection system. (a) Overview of the integrated detection system. (b) Layout drawing of the operating area of the system. (c) Composition of the liquid circuit module. (d) Internal structures of the system. (e) Sensing signal amplifying, filtering, and acquiring circuit of the system. (Reproduced from Ref. Xie et al. (2022) with permission from the Royal Society of Chemistry. Copyright 2022, The Royal Society of Chemistry.)

the feasibility of the system for biological detection. Furthermore, the association–dissociation process of antibody–protein pairs was analyzed using this system, demonstrating its potential for molecular interaction analysis. The system is highly integrated, small in size, and easy to carry. It will likely be developed into a portable device for on-site biological detection and molecular interaction analysis in the fields of environmental detection, medical research, food and agricultural safety, and military medicine.

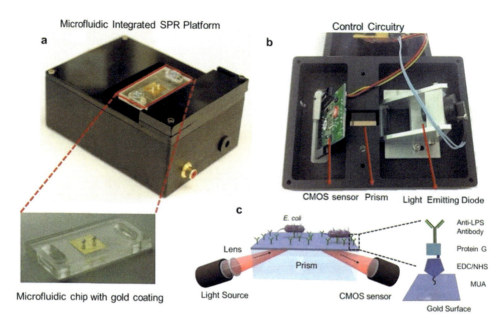

**FIGURE 4.2** Schematic of a quantitative portable plasma platform integrated with microfluidic for pathogen detection and quantification. (a) The surface-activated disposable microfluidic chips were mounted on the top side of the device. (b) The electronic setup of the device is represented from bottom. (c) Schematics of the microfluidic integrated SPR platform. (Tokel et al. (2015). Copyright 2015, The Authors.)

Tokel et al. (2015) proposed a portable, multiplexed, and inexpensive microfluidic-integrated SPR detection platform for the detection and quantification of bacteria (Figure 4.2). The microfluidic chip design consists of a single microchannel with inlet and outlet ports. Two layers of polymethyl methacrylate (PMMA) are assembled using a layer of double-sided adhesive (DSA). A second DSA layer and a gold-plated substrate form the microchannels. A microchannel (12mm × 7mm × 50 μm) is located in the center of the chip. The PMMA-DSA-PMMA-DSA-gold chip is assembled as a single use, disposable microchip. The design of the SPR platform is based on a Kretschmann configuration that uses prism coupling to satisfy the momentum conservation for external light source excitation of plasmons. The collimated point source LED output is focused with a cylindrical lens and illuminates the microchip surface through a glass prism. A rectangular prism is placed on the platform to facilitate the insertion of microfluidic chips and prisms. The reflected light is captured by a complementary metal-oxide-semiconductor (CMOS) sensor whose surface normal is parallel to the direction of light being monitored. The light source, CMOS sensor, and related optical and electronic components are packaged in a portable case with dimensions of $13.5 \times 10 \times 5.2$ cm$^3$. Changes in the resonance angle are monitored using custom-made software. The software captures image frames from the sensor, calculates the resonance angle in real time, and then plots the resonance curve and the sensor map (resonance angle as a function of time) as a readout for kinetic measurements. The platform reliably captures and detects *Escherichia coli* concentrations ranging from ~$10^5$ to $3.2 \times 10^7$ CFUs/mL in phosphate-buffered saline and peritoneal dialysis fluid. Utilizing the multiplexing and specificity of the *Staphylococcus aureus* sample detection platform, this system can potentially be extended to the detection of other pathogens or for immunodiagnostics.

Qi et al. (2021) developed a microfluidic biosensor for rapid, sensitive, and automatic detection of *Salmonella* using MOF NH$_2$-MIL-101(Fe) with mimic peroxidase activity to amplify the biological signal and a Raspberry Pi with self-developed app to analyze color images (Figure 4.3). The microfluidic biosensor mainly consists of two parts: a microfluidic chip and a portable device. The

Microfluidic Nanodetection Systems for Molecular Interactions 95

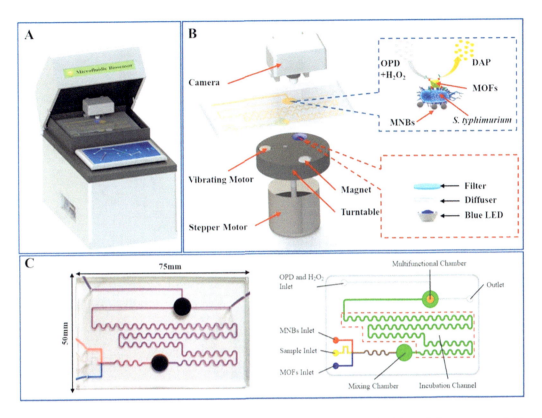

**FIGURE 4.3** (a) The prototype of this microfluidic biosensor. (b) The principle and the structure of this microfluidic biosensor. (c) The diagram of the microfluidic chip. (Qi et al. (2021). Copyright 2021, Elsevier B.V.)

microfluidic chip is largely composed of three portions: (1) an active vibrating mixer for efficient mixing of the immune magnetic nanobeads (MNBs), the sample, and the MOFs; (2) a serpentine incubating channel for sufficient forming of the MNB–*Salmonella*–MOF sandwich complexes; and (3) a multifunctional chamber for magnetic separation of the complexes, MOF catalysis of o-phenylenediamine and $H_2O_2$, and optical detection of the catalysate. The microfluidic chip was fabricated using 3D printing and surface plasma bonding. The portable device was developed based on a Raspberry Pi for automatic control of solutions, rotation of the turntable, and analysis of images. The application program for this device was developed in the Python environment based on the OpenCV function library using PyCharm IDE with three functions: (1) automatically control the stepper motor to rotate the blue LED, the magnet, and the vibrating motor to their designated positions at their designated times; (2) automatically operate the fluids to achieve mixing, incubation, separation, washing, and catalysis; and (3) automatically collect the catalysate image under the blue LED and analyze the image to determine the bacterial concentration. All of the procedures for separation, labeling, catalysis, and detection were automatically performed after the "Start" button on the app was activated. The experimental results showed that the device could detect ~1.5 × $10^1$–1.5 × $10^7$ CFU/mL of *Salmonella typhimurium* within 1 h, and the minimum detection limit was 14 CFU/mL. The mean recovery of *Salmonella* in the spiked chicken sample was ~112%. The sensor integrates mixing, separation, labeling, and detection and has the advantages of automatic operation, fast response, low reagent consumption, and small volume. It has broad application prospects in the field for the detection of foodborne bacteria.

Srikanth et al. (2022) developed a simple, economical, and miniaturized microfluidic lab-on-a-chip platform that can simultaneously culture and detect bacteria. The device integrates microfluidic

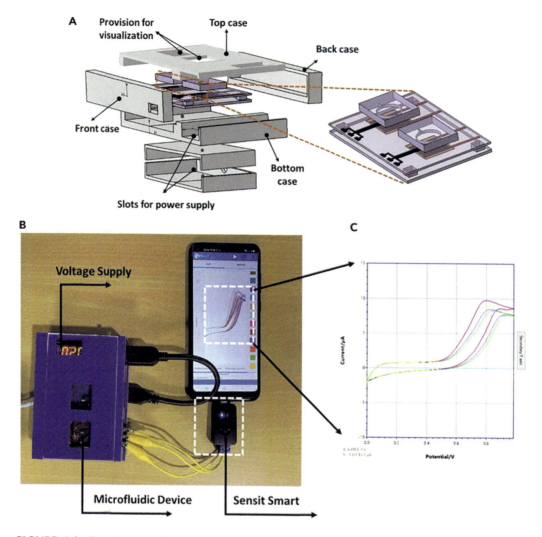

**FIGURE 4.4** Development of portable device: (a) schematic of the handheld device with different parts showcasing integrated microfluidic device. (b) Realized handheld device wherein the integrated microfluidic device is packaged, which is connected to a portable potentiostat. (c) The output curves for bacterial growth in the microfluidic device. (Srikanth et al. (2022). Copyright 2022, The Authors.)

chambers onto multiwalled CNTs-modified screen-printed electrodes for electrochemical detection of bacteria. To enable a miniaturized and an all-in-one platform, 3D-printed packaging was designed and constructed to integrate the microfluidic device with the heater and the detection equipment (Figure 4.4). The voltage to the heater is supplied by a buck booster, and the electrochemical readings are taken using a handheld potentiostat (Sensit smart) connected to a smartphone. Once the setup is ready, the sample is infused into the microfluidic device, and the required voltage is supplied to the heater to provide sufficient temperature for the bacteria for incubation. Within the range of ~$2 \times 10^4$–$1.1 \times 10^9$ CFU/mL, the three-electrode system can accurately quantify the bacterial concentration without the need for biological modification of the electrodes. The viability of the cultured bacteria in the microfluidic device was confirmed by fluorescence imaging.

### 4.3.4 Challenges in Implementing Microfluidic Nanodetection Systems

Although microfluidic nanodetection systems have made great progress in biomolecular detection, there are still some challenges in promoting microfluidic nanodetection systems from the experimental bench to practical applications or commercialization. In particular, for real environmental samples, the efficient capture of targets from complex samples for high-throughput multiple analysis remains a key problem to be solved. This is mainly due to the lack of "biometric elements" with high binding affinity (Mi et al., 2022). In addition, when integrating nanomaterials with microfluidic chips, challenges such as size change, stability, toxicity, and uncontrollable aggregation of nanomaterials must be considered (Fattahi & Hasanzadeh, 2022). Biosensing devices for some microfluidic systems, especially optical sensors, must rely on bulky analytical instruments for quantitative signal reading or external precision accessories for liquid handling. In addition to the high cost of chip manufacturing, the operation of some devices is somewhat complex for non-technical personnel. To address these limitations, the development of new instrument-free signal transduction or readout principles is ideal for working with microfluidic devices (Tavakoli et al., 2022). In addition, it is also a challenge to ensure stable acquisition and output of small signals generated by biomolecular interactions under high background noise. During the entire microfluidic system integration process, the selection of each device, the design and layout of each module, etc. also need to be considered. Therefore, it is necessary to further develop nanotechnologies and microfluidic technologies, as well as cross combination with other disciplines, to achieve the portability, ease of use, multiple reuse, and sensitivity of biomolecular detection, enhance the adaptability to the environment of the instrument, and improve the reliability of the instrument, which are the keys to the commercialization of microfluidic nanodetection systems.

## 4.4 APPLICATION OF MICROFLUIDIC NANODETECTION SYSTEMS IN MOLECULAR INTERACTIONS

As powerful biomolecular analysis tools, microfluidic nanodetection systems have been developed for single-cell, protein–protein interaction, and protein–other molecular interaction analyses. Due to their low sample and reagent consumption, simple analysis process, high analysis speed, high sensitivity, high throughput, portability, and automated sample preparation and analysis potential, these devices are widely applicable in the fields of disease diagnosis, drug development, food safety detection, antibody screening, and pollutant detection.

### 4.4.1 Single-Cell Analysis

Due to the heterogeneity of cells, single-cell-level studies can obtain more accurate and comprehensive information reflecting the physiological state and process of cells, increasing the significance of single-cell analysis. The application of single-cell analysis faces the problem of cell contamination or damage caused by cell sample transportation. Using microfluidic detection systems for single-cell analysis has the characteristics of integration, miniaturization, and high throughput, which can greatly improve the efficiency of single-cell analysis (Sun et al., 2021). Microfluidic detection systems can be used for single-cell culturing, detection, sorting, and monitoring cell activation and differentiation and can also be used to monitor cell viability, adhesion changes, morphology, proliferation, migration, diffusion, and responses to drugs, compounds, and viruses (Szittner et al., 2022).

Wang et al. (2017) combined nanodielectrophoretic microfluidic devices with SERS technology to construct a new type of microfluidic portable biosensor. With three different SERS-tagged molecular probes targeting different epitopes of the same pathogen being deployed simultaneously, the detection of pathogen targets was achieved at single-cell level with sub-species specificity. The integration of microfluidic devices with SERS detection yielded simple and miniaturized instrumentation that was suitable for the detection and characterization of small volume of chemical and biological analytes with high sensitivity and specificity. Li et al. (2018) introduced an innovative label-free optofluidic nanoplasmonic biosensor for real-time single-cell analysis. The system features small-volume microchambers and regulated channels to reliably monitor cytokine secretion from single cells over hours. Distinct interleukin-2 secretion profiles have been detected and distinguished from individual lymphoma cells. Wang et al. (2018) developed a 3D biointerface of graphene-based electrical impedance sensors and integrated them on a microfluidic chip for metastatic cancer diagnosis at single-cell resolution. Compared to traditional 2D interface impedance sensors, the 3D graphene biointerface significantly improves the capture efficiency and sensing sensitivity of single cells, and the impedance signal increases by ~100% at the nodes of cell state changes.

Compared to traditional techniques, microfluidic devices have attracted much attention because of their high efficiency and convenience in the concentration and detection of foodborne pathogens (Mi et al., 2022). For example, An et al. (2020) proposed a single-cell-level analysis method based on droplet microfluidics, which can sensitively and rapidly detect *Salmonella* directly from food samples. The detection limit for *Salmonella* was 50 CFU/mL within 5 h, which was lower than traditional analytical methods for assessing *Salmonella* contamination. Jiang et al. (2020) developed a double-rolling circle amplification (RCA) microfluidic platform for whole-cell detection of *E. coli* O157:H7. This method can be used in different food matrices, including orange juice and milk, with a detection limit of 80 cells/mL. Asgari et al. (2022) developed a sensitive SERS-based microfluidic immunosensor for the isolation and detection of *E. coli* O157:H7 in romaine lettuce. SERS nanoprobes containing Ag NPs and anti-*E. coli* O157:H7 were selectively anchored to *E. coli* O157:H7 cells and isolated from lettuce samples. The detection limit of this method for *E. coli* O157:H7 in romaine lettuce was 0.5 CFU/mL, and this method reduced the analysis time for single-cell detection to only 1 h. The combination of hydrodynamic flow-focusing microfluidic devices with SERS nanoprobes provides a dependable, selective, and sensitive method for the detection of various pathogens in complex food samples.

### 4.4.2 PROTEIN–PROTEIN INTERACTION

The determination of protein–protein interactions can help to reveal the most basic mechanisms of life at the molecular level and understand the most fundamental questions of life science, such as inheritance, reproduction, development, aging, death, and disease. Monitoring protein–protein interactions is essential for the study of complex signaling pathways, drug target binding, oligomerization, and aggregation of pathogenic proteins. Microfluidic-based nanodetection systems can monitor protein-protein interactions in real time with high sensitivity and high selectivity.

Walgama et al. (2016) reported a fast optical microarray imaging method. In real-time microfluidics analysis, the surface plasmonic resonance imager (SPRi) showed that the specificity of the interaction between mouse double micro 2 protein (MDM2) and wild-type p53 was 3.5-fold higher than that of non-specific p53 mutants. For MDM2–p53 interactions and inhibition of the small-molecule Nutlin-3 drug analogue known for its anticancer properties, significant percentage reflectance changes ($\Delta\%$ R) and molecular-level mass changes were detected in the SPRi signal. In addition, it was demonstrated that with array-based SPRi technology, synthetic, inexpensive interacting cancer protein binding domains were sufficient to screen anticancer drugs with excellent specificity and sensitivity. This imaging array, combined with a mass sensor, can be used to quantitatively study any protein–protein interaction and screen small molecules by binding and potency assessment.

Niels Zijlstra et al. (2017) developed a microfluidic detection device based on single-molecule dichroic and tricolor FRET to study the dissociation kinetics and structural properties of low-affinity protein complexes. This work shows that the versatility of the device makes it suitable for studying complexes with dissociation constants ranging from low nanomolar to 10 mm, thus covering a wide range of biomolecular interactions. Javanmard et al. (2009) developed a microfluidic detection device based on electrical impedance, using microelectrodes in microchannels. Specific interactions

Microfluidic Nanodetection Systems for Molecular Interactions

between proteins were detected in real time by detecting changes in solution resistance caused by the obstruction of ionic current due to the binding of functionalized beads on the surface of bioactivated microchannels. Antigen–antibody interactions, glycoprotein–glycoprotein interactions, and antigen–glycoprotein interactions have been successfully detected. The ability of the technique to distinguish between strong and weak interactions was also demonstrated.

Xie et al. (2022) constructed a completely independent and integrated portable microfluidic nanoautomatic detection system based on SiNW-FET biosensors and used this system to analyze the binding–dissociation process of antibody–protein pairs, demonstrating the potential of this system for molecular interaction analysis. Chen et al. (2017) used an indium gallium zinc oxide thin film transistor-based biosensor integrated with a microfluidic channel for transient analysis of streptavidin–biotin complex detection. The electrical and diffusion properties of streptavidin–biotin protein complexes were investigated, and the binding interaction of the streptavidin–biotin complex was studied. The method can be applied to the detection of protein–ligand binding, protein–protein interactions, protein folding and reconfiguration, and other protein characteristics such as denaturation, charge, and diffusivity.

Madeira et al. (2009) described the use of SPR in combination with MS to discover protein–protein interactions. Peptides or proteins were immobilized on the sensor chip and then exposed to brain extract injected through the chip surface by the microfluidic system. Interactions between immobilized ligands and extracts can be monitored in real time. MS was used to recover, trypsinize, and identify proteins that interact with the peptide/protein.

### 4.4.3 PROTEIN–OTHER MOLECULAR INTERACTIONS

Proteins rarely function in the form of monomers in living organisms and often need to interact with other biomolecules to perform the function of regulating life's activities. In addition to protein–protein interactions, the analysis of protein interactions with other biomolecules (such as nucleic acids, enzymes, aptamers, and cell metabolites) is also very important. DNA adenine methyltransferase identification (DamID) measures protein–DNA binding events by methylating adenine bases near each protein–DNA interaction site, followed by selective amplification and sequencing of these methylated regions. Furthermore, these interactions can be visualized using m6A-Tracer, a fluorescent protein that binds methyladenine. Altemose et al. (2020) combined these imaging and sequencing technologies in an integrated microfluidic platform (μDamID) that enables single-cell isolation, imaging, and sorting, followed by DamID. μDamID provides the unique ability to compare paired imaging and sequencing data within each cell and between cells, enabling joint analysis of variability in nuclear localization, sequence identity, and protein–DNA interactions.

Ordinario et al. (2014) developed a CNTs FET-based microfluidic detection platform for the analysis of protein–DNA interactions. The system was able to detect the sequence-specific activity of a DNA-binding protein (*Pvu*II restriction endonuclease) at a concentration as low as 0.5 pM in a volume of 0.025 μL (equivalent to ~7500 proteins). Chiesa et al. (2012) designed a label-free and ultrasensitive biosensor based on monitoring the changes in the resistance of a silicon nanowire transistor, which was integrated into a microfluidic unit. The device was able to detect the biomolecular processes related to the (deoxy)adenosine triphosphate (dATP) hydrolysis required for dissociation of RecA from ssDNA, which defined the early stage of RecA-mediated homologous recombination, and the inhibition of this activity in the presence of a competing protein, SsbA. It has been shown that the interaction of 250 RecA molecules can be detected.

Yan et al. (2021) proposed a fluorescent sensor using CdSe QDs as a fluorescent material in response to the urgent need for monitoring organophosphorus pesticide residues in ecosystems and introduced surface molecularly imprinted polymers and microfluidic technology for rapid detection of dimethoate. The detection platform could effectively respond to dimethoate within 10 min, the detection concentration range was 0.45–80 μmol/L, and the detection limit was 0.13 μmol/L. More importantly, during the molecular imprinting process, the interaction between different functional monomers and dimethoate was investigated. The results show that 3-mercaptopropyl trimethoxysilane has a strong interaction with the dimethoate molecule and is an ideal functional monomer.

Weng and Neethirajan (2018) developed a nanomaterial-enhanced multipurpose paper-based microfluidic aptasensor for accurate detection of food allergens and food toxins. Using graphene oxide and aptamer functionalized QDs as probes, the fluorescence quenching, and recovery of QDs caused by the interaction of graphene oxide and aptamer functionalized QDs, and the target protein were investigated to quantitatively analyze the target concentration. The homogenous assay was performed on the paper-based microfluidic chip, which significantly decreased the sample and reagent consumption and reduced the assay time. Egg white lysozyme, ß-conglutin lupine and food toxins, okadaic acid and brevetoxin standard solutions, and spiked food samples were successfully assayed by the presented aptasensor. Dual-target assay was completed within 5 min.

## 4.5 COMMERCIAL INSTRUMENTATION TO ANALYZE MOLECULAR INTERACTIONS

With the increasing demand for the detection of biomolecular interactions, equipment and technologies for the analysis of biomolecular interactions have also gradually developed. At present, there are many commercialized instruments for biomolecular interaction analysis, including SPR molecular interaction analyzers (represented by the Biacore series) and BLI molecular interaction analyzers (represented by the Fortebio/Sartorius Octet®). These biomolecular interaction analysis techniques can complement and verify each other, as well as provide important technical support for biomedical research such as disease diagnosis, biomolecular detection, drug molecular structure–activity relationship research, target binding verification, affinity quantitative analysis, binding and dissociation kinetic analysis, protein key action site research, and potential drug target discovery.

### 4.5.1 BIACORE SERIES

In 1990, the Swedish Pharmacia company and researchers from Uppsala University jointly invented the world's first Biacore instrument based on SPR technology, enabling the automatic detection of the interaction between different molecules for the first time. Biacore uses the principle of SPR to detect interactions between biomolecules in their native state. Generally, a biomolecule (target molecule) is first immobilized on the surface of the sensor chip and then a solution containing another biomolecule (analyte) that can interact with the target molecule is injected and flows through the surface of the sensor chip. The combination of biomolecules causes an increase in the surface material of the sensor chip, which leads to the enhancement of the refractive index in the same proportion, yielding the detection of the interaction between biomolecules.

Biacore series instruments are composed of a liquid handling system, an optical system, a sensor chip, and a microcomputer. The chip of the Biacore sensor is its core component. It fixes a 100-nm-thick gold film onto a glass sheet, which is embedded in a plastic plate clip, and couples the chip to the glass prism with a polymer whose refractive index matches with that of the prism. In the Biacore technology, a biomolecule must first be coupled to the sensor chip, and then it is used to capture biomolecules that can react specifically with it. The coupling process can be automatically controlled by the instrument. Another distinctive feature of Biacore is the micro-flow cell processing system. BiacoreX has two micro-flow cells, and when it was developed to Biacore1000, the number of flow cells reached four. Therefore, four samples can be analyzed independently at the same time. The volume of the micro-flow cell is very small, each only 60 nL, which is the scribe line engraved on the surface of the sensor chip.

Biacore has a high degree of automation: sample processing, sample injection, and sample recovery are all fully automatic. The sample volume required for the entire analysis process is very small, generally no more than 750 μL, and the detection sensitivity is high. The kinetic constants of the reaction process, including association constants, dissociation constants, and other data, can be directly recorded. This technology has been widely used to study the interaction of biomolecules such as proteins, nucleic acids, peptides, and small molecular compounds.

### 4.5.2 FORTEBIO/SARTORIUS OCTET®

In 2006, the American company ForteBio launched a BLI molecular interaction analysis system based on fiber optics. In 2020, the brand was acquired by Sartorius of Germany and changed to Octet®. For the first time, this instrument replaces the sensing chip with a sensing "fiber optic" probe, replaces the microfluidic chip with a 96- or 384-microwell plate, and directly inserts the probe into the well to read molecular interaction signals. The instrument uses BLI to provide high-throughput biomolecular interaction information in real time. The instrument emits white light onto the sensor surface and collects the reflected light. The reflected light spectrum of different frequencies is affected by the thickness of the optical film layer of the biosensor. Some frequencies of reflected light form constructive interference (blue), while others are subject to destructive interference (red). These interferences are detected using a spectrometer and form an interference spectrum, which is displayed by the phase shift intensity (nm) of the interference spectrum. Therefore, once the number of molecules bound to the sensor surface increases or decreases, the spectrometer will detect the shift of the interference spectrum in real time, and this shift directly reflects the thickness of the biofilm on the sensor surface, enabling high-throughput, rapid, and real-time detection of the interaction between biomolecules and compounds. It can detect the interaction of biomolecules, including the affinity and kinetics determination of antibodies and small-molecule drugs, the titer identification of a certain trace protein in vaccines, protein–protein interaction, and protein–DNA interaction. It is label-free, highly sensitive, and has a wide application range.

Both of the above commercial instruments can monitor and analyze biomolecular interactions in real time without labeling and quantitatively analyze their binding strengths and dynamics. But each has its advantages and disadvantages. Biacore instruments have a high degree of automation, high detection sensitivity, high throughput, a wide range of applications, and relatively high accuracy and repeatability. The Biacore system can in principle be used to study interactions between any kind of molecules, from candidate organic drug molecules to proteins, nucleic acids, sugars, and even viruses and whole cells. A lot of information is provided, including kinetics, affinity, specificity, thermodynamics, and concentration. However, the Biacore instrument is more demanding on external conditions, including solutions and sample processing before experimental design, and the operation is more complicated. Biacore instruments are expensive and require special detection chips with high consumable costs. The BLI molecular interaction analysis system is a dip-and-read detection method with high throughput, low cost, and relatively simple operation. The experimental equipment has no complex flow path system, no pump, and no sampling system to reduce the loss of samples, and the remaining samples can be recovered after the experiment. But the detection sensitivity and repeatability are relatively poor. Of course, these two commercial instruments are large in size and mainly used for laboratory analysis. They are not portable for on-site testing. Commercial instruments for biomolecular interaction analysis can be further integrated with sample collection and preprocessing systems to make them more intelligent and portable, enabling rapid on-site analysis.

## 4.6 CONCLUSIONS AND PERSPECTIVES

This chapter briefly expounds upon the principles and traditional analysis methods of biomolecular interactions. The detection principle and classification of the microfluidic detection system, as well as common nanomaterials, are described in detail. At present, microfluidic nanodetection systems have been widely used in single-cell detection, protein–protein interaction analyses, protein–DNA interaction analyses, and other biomolecular interaction analyses. In addition, various commercial instruments have been developed for biomolecular interaction analysis.

The development of microfluidic technology has brought great convenience to the detection of biomolecules. The microfluidic systems integrate and miniaturize a variety of large analytical chemical equipment to complete sample pretreatment, mixing, reaction, separation, detection, and other operating steps on a chip of a few square centimeters or even smaller, as well as the basic operating units of a cell or other biological cultures, separation, and screening, enabling the portability and automation of biological detection. Compared to traditional analysis methods, the microfluidic detection system can reduce the consumption of samples and reagents, shorten detection times, and significantly improve the detection sensitivity when combined with nanomaterials. Due to their unique properties, nanomaterials are very sensitive to the binding of biomolecules on their surfaces, but they are also easily disturbed by the external environment. Therefore, nanomaterials need to be further modified or augmented with other materials to develop them in the direction of high sensitivity, high stability, and low cost. They can also be coupled with highly specific biomolecules to prepare biosensing materials with high specificity.

At present, commercial biomolecular interaction analyzers have been highly integrated, but these instruments are still large in size, basically used in laboratory conditions, and their cost is high. Therefore, the biomolecular interaction analysis equipment based on the microfluidic detection systems still needs to be miniaturized, portable, require low energy consumption, and have a low cost to be able to achieve real-time online analysis and to be applicable to a large number of users. In addition, with the development of artificial intelligence and Internet of Things technology, microfluidic nanodetection systems can also integrate functions such as data processing, data analysis, self-diagnosis, image output, and real-time information sharing. Collaborate with multiple disciplines, especially artificial intelligence, machine learning, and deep learning, to streamline equipment operation, automate high-throughput reagent processing, and direct result readout. Microfluidic nanodetection platforms can use robots to handle liquids, minimizing error rates and thus improving reproducibility. The continuous development and improvement of microfluidic nanodetection systems will become an indispensable analytical tool in the fields of life sciences, food safety, medical diagnosis, drug development, drug delivery, and microbial detection.

## REFERENCES

Alhalaili, B., Popescu, I. N., Rusanescu, C. O., and Vidu, R. 2022. Microfluidic devices and microfluidics-integrated electrochemical and optical (bio) sensors for pollution analysis: A review. *Sustainability* 14(19): 12844. https://doi.org/10.3390/su141912844

Alsabbagh, K., Hornung, T., Voigt, A., Sadir, S., Rajabi, T., and Länge, K. 2021. Microfluidic impedance biosensor chips using sensing layers based on DNA-based self-assembled monolayers for label-free detection of proteins. *Biosensors* 11(3): 80. https://doi.org/10.3390/bios11030080

Altemose, N., Maslan, A., Rios-Martinez, C., Lai, A., White, J. A., and Streets, A. 2020. μDamID: A microfluidic approach for joint imaging and sequencing of protein-DNA interactions in single cells. *Cell Systems* 11(4): 354–366. https://doi.org/10.1016/j.cels.2020.08.015

Amir, S., Arathi, A., Reshma, S., and Mohanan, P. V. 2023. Microfluidic devices for the detection of disease-specific proteins and other macromolecules, disease modelling and drug development: A review. *International Journal of Biological Macromolecules* 123784. https://doi.org/10.1016/j.ijbiomac.2023.123784

An, X., Zuo, P., and Ye, B. C. 2020. A single cell droplet microfluidic system for quantitative determination of food-borne pathogens. *Talanta* 209: 120571. https://doi.org/10.1016/j.talanta.2019.120571

Arter, W. E., Levin, A., Krainer, G., and Knowles, T. P. 2020. Microfluidic approaches for the analysis of protein–protein interactions in solution. *Biophysical Reviews* 12: 575–585. https://doi.org/10.1007/s12551-020-00679-4

Asgari, S., Dhital, R., Aghvami, S. A., Mustapha, A., Zhang, Y., and Lin, M. 2022. Separation and detection of *E. coli* O157: H7 using a SERS-based microfluidic immunosensor. *Microchimica Acta* 189(3): 111. https://doi.org/10.1007/s00604-022-05187-8

Asmari, M., Ratih, R., Alhazmi, H. A., and El Deeb, S. 2018. Thermophoresis for characterizing biomolecular interaction. *Methods* 146: 107–119. https://doi.org/10.1016/j.ymeth.2018.02.003

Bodin-Thomazo, N., Malloggi, F., Pantoustier, N., Perrin, P., Guenoun, P., and Rosilio, V. 2022. Formation and stabilization of multiple w/o/w emulsions encapsulating catechin, by mechanical and microfluidic methods using a single pH-sensitive copolymer: Effect of copolymer/drug interaction. *International Journal of Pharmaceutics* 622: 121871. https://doi.org/10.1016/j.ijpharm.2022.121871

Chen, P., Chen, D., Li, S., Ou, X., and Liu, B. F. 2019. Microfluidics towards single cell resolution protein analysis. *TrAC Trends in Analytical Chemistry* 117: 2–12. https://doi.org/10.1016/j.trac.2019.06.022

Chen, T. Y., Yang, T. H., Wu, N. T., Chen, Y. T., and Huang, J. J. 2017. Transient analysis of streptavidin-biotin complex detection using an IGZO thin film transistor-based biosensor integrated with a microfluidic channel. *Sensors and Actuators B: Chemical* 244: 642–648. https://doi.org/10.1016/j.snb.2017.01.050

Chiesa, M., Cardenas, P. P., Otón, F., Martinez, J., Mas-Torrent, M., Garcia, F., Alonso, J. C., Rovira, C., and Garcia, R. 2012. Detection of the early stage of recombinational DNA repair by silicon nanowire transistors. *Nano Letters* 12(3): 1275–1281. https://doi.org/10.1021/nl2037547

Ebrahimi, G., Samadi Pakchin, P., Shamloo, A., Mota, A., de la Guardia, M., Omidian, H., and Omidi, Y. 2022. Label-free electrochemical microfluidic biosensors: Futuristic point-of-care analytical devices for monitoring diseases. *Microchimica Acta* 189(7): 252. https://doi.org/10.1007/s00604-022-05316-3

Fattahi, Z., and Hasanzadeh, M. 2022. Nanotechnology-assisted microfluidic systems for chemical sensing, biosensing, and bioanalysis. *TrAC Trends in Analytical Chemistry* 152: 116637. https://doi.org/10.1016/j.trac.2022.1 16637

Fong Lei, K. 2013. Recent developments and patents on biological sensing using nanoparticles in microfluidic systems. *Recent Patents on Nanotechnology* 7(1): 81–90. https://doi.org/10.2174/18722101380448481 8

Freitag, R. 1999. Utilization of enzyme–substrate interactions in analytical chemistry. *Journal of Chromatography B: Biomedical Sciences and Applications* 722(1–2): 279–301. https://doi.org/10.1016/S0378-4347(98)00507-6

Gao, J., Wang, C., Chu, Y., Han, Y., Gao, Y., Wang, Y., Wang, C., Liu, H., Han, L., and Zhang, Y. 2022. Graphene oxide-graphene Van der Waals heterostructure transistor biosensor for SARS-CoV-2 protein detection. *Talanta* 240: 123197. https://doi.org/10.1016/j.talanta.2021.123197

Goral, V., Wu, Q., Sun, H., and Fang, Y. 2011. Label-free optical biosensor with microfluidics for sensing ligand-directed functional selectivity on trafficking of thrombin receptor. *FEBS Letters* 585(7): 1054–1060. https://doi.org/10.1016/j.febslet.2011.03.003

Hossain, M. M., Kim, K. B., Jannath, K. A., Park, D. S., and Shim, Y. B. 2023. Separation detection of saccharides in whole blood using an electrodynamic microfluidic channel sensor with AuCo dendrite-anchored conductive polymer. *Sensors and Actuators B: Chemical* 389: 133843. https://doi.org/10.1016/j.snb.2023.133843

Hung, L. Y., Wu, H. W., Hsieh, K., and Lee, G. B. 2014. Microfluidic platforms for discovery and detection of molecular biomarkers. *Microfluidics and Nanofluidics* 16: 941–963. https://doi.org/10.1007/s10404-014-1354-6

Jammes, F. C., and Maerkl, S. J. 2020. How single-cell immunology is benefiting from microfluidic technologies. *Microsystems & Nanoengineering* 6(1): 45. https://doi.org/10.1038/s41378-020-0140-8

Javanmard, M., Talasaz, A. H., Nemat-Gorgani, M., Huber, D. E., Pease, F., Ronaghi, M., and Davis, R. W. 2009. A microfluidic platform for characterization of protein–protein interactions. *IEEE Sensors Journal* 9(8): 883–891. https://doi.org/10.1109/JSEN.2009.2022558

Jeong, K. B., Kim, J. S., Dhanasekar, N. N., Lee, M. K., and Chi, S. W. 2022. Application of nanopore sensors for biomolecular interactions and drug discovery. *Chemistry – An Asian Journal* 17(19): e202200679. https://doi.org/10.1002/asia.202200679

Jiang, Y., Qiu, Z., Le, T., Zou, S., and Cao, X. 2020. Developing a dual-RCA microfluidic platform for sensitive *E. coli* O157: H7 whole-cell detections. *Analytica Chimica Acta* 1127: 79–88. https://doi.org/10.1016/j.aca.2020.06.046

Jimenez, R. M., Creton, B., Marliere, C., Teule-Gay, L., Nguyen, O., and Marre, S. 2023. A microfluidic strategy for accessing the thermal conductivity of liquids at different temperatures. *Microchemical Journal* 193: 109030. https://doi.org/10.1016/j.microc.2023.109030

Khatri, K., Klein, J. A., Haserick, J. R., Leon, D. R., Costello, C. E., McComb, M. E., and Zaia, J. 2017. Microfluidic capillary electrophoresis–mass spectrometry for analysis of monosaccharides, oligosaccharides, and glycopeptides. *Analytical Chemistry* 89(12): 6645–6655. https://doi.org/10.1021/acs.analchem.7b00875

Kovalchuk, N. M., and Simmons, M. J. 2023. Review of the role of surfactant dynamics in drop microfluidics. *Advances in Colloid and Interface Science* 102844. https://doi.org/10.1016/j.cis.2023.102844

Lee, J. O., Choi, N., Lee, J. W., Song, S., and Kim, Y. P. 2021. Rapid electrokinetic detection of low-molecular-weight thiols by redox regulatory protein-DNA interaction in microfluidics. *Sensors and Actuators B: Chemical* 336: 129735. https://doi.org/10.1016/j.snb.2021.129735

Li, T., Shang, D., Gao, S., Wang, B., Kong, H., Yang, G., Shu, W., Xu, P., and Wei, G. 2022. Two-dimensional material-based electrochemical sensors/biosensors for food safety and biomolecular detection. *Biosensors* 12(5): 314. https://doi.org/10.3390/bios12050314

Li, X., Soler, M., Szydzik, C., Khoshmanesh, K., Schmidt, J., Coukos, G., Mitchell, A., and Altug, H. 2018. Label-free optofluidic nanobiosensor enables real-time analysis of single-cell cytokine secretion. *Small* 14(26): 1800698. https://doi.org/10.1002/smll.201800698

Liang, Z. Y., Deng, Y. Q., and Tao, Z. Z. 2020. A quantum dot-based lateral flow immunoassay for the rapid, quantitative, and sensitive detection of specific IgE for mite allergens in sera from patients with allergic rhinitis. *Analytical and Bioanalytical Chemistry* 412: 1785–1794. https://doi.org/10.1007/s00216-020-02422-0

Lin, W., Wei, W., Wu, J., Cao, Q., Bi, H., Zhang, J., Mei, Z., Jin, J., and Wang, X. 2023. Development of microfluidic chip flowmeter-based constant pressure system for analysing the hydrogen adsorption performance of non-evaporable getters. *Analytica Chimica Acta* 1278: 341690. https://doi.org/10.1016/j.aca.2023.341690

Madeira, A., Öhman, E., Nilsson, A., Sjögren, B., Andrén, P. E., and Svenningsson, P. 2009. Coupling surface plasmon resonance to mass spectrometry to discover novel protein–protein interactions. *Nature Protocols* 4(7): 1023–1037. https://doi.org/10.1038/nprot.2009.84

Mi, F., Hu, C., Wang, Y., Wang, L., Peng, F., Geng, P., and Guan, M. 2022. Recent advancements in microfluidic chip biosensor detection of foodborne pathogenic bacteria: A review. *Analytical and Bioanalytical Chemistry* 414(9): 2883–2902. https://doi.org/10.1007/s00216-021-03872-w

Miernicki, M., Hofmann, T., Eisenberger, I., von der Kammer, F., and Praetorius, A. 2019. Legal and practical challenges in classifying nanomaterials according to regulatory definitions. *Nature Nanotechnology* 14(3): 208–216. https://doi.org/10.1038/s41565-019-0396-z

Ordinario, D. D., Burke, A. M., Phan, L., Jocson, J. M., Wang, H., Dickson, M. N., and Gorodetsky, A. A. 2014. Sequence specific detection of restriction enzymes at DNA-modified carbon nanotube field effect transistors. *Analytical Chemistry* 86(17): 8628–8633. https://doi.org/10.1021/ac501441d

Parvizi, F., Parvareh, A., and Heydari, R. 2023. Fabrication of a hydrophobic surface as a new supported liquid membrane for microfluidic based liquid phase microextraction device using modified boehmite nanoparticles (AlOO-NSPO). *Microchemical Journal* 189: 108514. https://doi.org/10.1016/j.microc.2023.108514

Qi, W., Zheng, L., Wang, S., Huang, F., Liu, Y., Jiang, H., and Lin, J. 2021. A microfluidic biosensor for rapid and automatic detection of *Salmonella* using metal-organic framework and Raspberry Pi. *Biosensors and Bioelectronics* 178: 113020. https://doi.org/10.1016/j.bios.2021.113020

Resch-Genger, U., Grabolle, M., Cavaliere-Jaricot, S., Nitschke, R., and Nann, T. 2008. Quantum dots versus organic dyes as fluorescent labels. *Nature Methods* 5(9): 763–775. https://doi.org/10.1038/nmeth.1248

Schneider, M. M., Scheidt, T., Priddey, A. J., Xu, C. K., Hu, M., Meisl, G., Devenish, S. R., Dobson, C. M., Kosmoliaptsis, V., and Knowles, T. P. (2023). Microfluidic antibody affinity profiling of alloantibody-HLA interactions in human serum. *Biosensors and Bioelectronics* 228: 115196. https://doi.org/10.1016/j.bios.2023.115196

Srikanth, S., Jayapiriya, U. S., Dubey, S. K., Javed, A., and Goel, S. 2022. A lab-on-chip platform for simultaneous culture and electrochemical detection of bacteria. *Iscience* 25(11): 105388. https://doi.org/10.1016/j.isci.2022.105388

Sun, J., Gao, L., Wang, L., and Sun, X. 2021. Recent advances in single-cell analysis: Encapsulation materials, analysis methods and integrative platform for microfluidic technology. *Talanta* 234: 122671. https://doi.org/10.1016/j.talanta.2021.122671

Szittner, Z., Péter, B., Kurunczi, S., Székács, I., and Horváth, R. 2022. Functional blood cell analysis by label-free biosensors and single-cell technologies. *Advances in Colloid and Interface Science* 308: 102727. https://doi.org/10.1016/j.cis.2022.102727

Tavakoli, H., Mohammadi, S., Li, X., Fu, G., and Li, X. 2022. Microfluidic platforms integrated with nanosensors for point-of-care bioanalysis. *TrAC Trends in Analytical Chemistry* 157: 116806. https://doi.org/10.1016/j.trac.2022.1 16806

Tokel, O., Yildiz, U. H., Inci, F., Durmus, N. G., Ekiz, O. O., Turker, B., Cetin, C., Rao, S., Sridhar, K., Natarajan, N., Shafiee, H., Aykutlu Dana, A., and Demirci, U. 2015. Portable microfluidic integrated plasmonic platform for pathogen detection. *Scientific Reports* 5(1): 9152. https://doi.org/10.1038/srep09152

Vloemans, D., Van Hileghem, L., Verbist, W., Thomas, D., Dal Dosso, F., and Lammertyn, J. 2021. Precise sample metering method by coordinated burst action of hydrophobic burst valves applied to dried blood spot collection. *Lab on a Chip* 21(22): 4445–4454. https://doi.org/10.1039/D1LC00422K

Walgama, C., Al Mubarak, Z. H., Zhang, B., Akinwale, M., Pathiranage, A., Deng, J., Berlin, K. D., Benbrook, D. M., and Krishnan, S. 2016. Label-free real-time microarray imaging of cancer protein–protein interactions and their inhibition by small molecules. *Analytical Chemistry* 88(6): 3130–3135. https://doi.org/10.1021/acs.analchem.5b04234

Wang, C., Madiyar, F., Yu, C., and Li, J. 2017. Detection of extremely low concentration waterborne pathogen using a multiplexing self-referencing SERS microfluidic biosensor. *Journal of Biological Engineering* 11(1): 1–11. https://doi.org/10.1186/s13036-017-0051-x

Wang, X., Liu, A., Xing, Y., Duan, H., Xu, W., Zhou, Q., Wu, H., Chen, C., and Chen, B. 2018. Three-dimensional graphene biointerface with extremely high sensitivity to single cancer cell monitoring. *Biosensors and Bioelectronics* 105: 22–28. https://doi.org/10.1016/j.bios.2018.01.012

Wang, Z., Liu, W., Wang, J., Huang, L., Cui, S., and He, X. 2023. Structure-controllable Ag aerogel optimized SERS-digital microfluidic platform for ultrasensitive and high-throughput detection of harmful substances. *Sensors and Actuators B: Chemical* 134934. https://doi.org/10.1016/j.snb.2023.134934

Welch, E. C., Powell, J. M., Clevinger, T. B., Fairman, A. E., and Shukla, A. 2021. Advances in biosensors and diagnostic technologies using nanostructures and nanomaterials. *Advanced Functional Materials* 31(44): 2104126. https://doi.org/10.1002/adfm

Weng, X., and Neethirajan, S. 2018. Paper-based microfluidic aptasensor for food safety. *Journal of Food Safety* 38(1): e12412. https://doi.org/10.1111/jfs.12412

Wu, J., Fang, H., Zhang, J., and Yan, S. 2023. Modular microfluidics for life sciences. *Journal of Nanobiotechnology* 21(1): 1–30. https://doi.org/10.1186/s12951-023-01846-x

Xie, X., Ma, J., Wang, H., Cheng, Z., Li, T., Chen, S., Du, Y., Wu, J., Wang, C., and Xu, X. 2022. A self-contained and integrated microfluidic nano-detection system for the biosensing and analysis of molecular interactions. *Lab on a Chip* 22(9): 1702–1713 https://doi.org/10.1039/D1LC01056E

Yamaguchi, H., and Miyazaki, M. 2022. Enzyme-immobilized microfluidic devices for biomolecule detection. *TrAC Trends in Analytical Chemistry* 116908. https://doi.org/10.1016/j.trac.2022.116908

Yan, X., Zhang, Z., Zhang, R., Yang, T., Hao, G., Yuan, L., and Yang, X. 2021. Rapid detection of dimethoate in soybean samples by microfluidic paper chips based on oil-soluble CdSe quantum dots. *Foods* 10(11): 2810. https://doi.org/10.3390/foods10112810

Yang, H., and Gijs, M. A. 2018. Micro-optics for microfluidic analytical applications. *Chemical Society Reviews* 47(4): 1391–1458. https://doi.org/10.1039/c5cs00649j

Yang, X., Kanter, J., Piety, N. Z., Benton, M. S., Vignes, S. M., and Shevkoplyas, S. S. 2013. A simple, rapid, low-cost diagnostic test for sickle cell disease. *Lab on a Chip* 13(8): 1464–1467. https://doi.org/10.1039/C3LC41302K

Yang, Y., Chen, Y., Tang, H., Zong, N., and Jiang, X. 2020. Microfluidics for biomedical analysis. *Small Methods* 4(4): 1900451. https://doi.org/10.1002/smtd.201900451

Zhou, W., Le, J., Chen, Y., Cai, Y., Hong, Z., and Chai, Y. 2019. Recent advances in microfluidic devices for bacteria and fungus research. *TrAC Trends in Analytical Chemistry* 112: 175–195. https://doi.org/10.1016/j.trac.2018.12.024

Zhou, X., Li, Z., Zhang, Z., Zhu, L., and Liu, Q. 2022. A rapid and label-free platform for virus enrichment based on electrostatic microfluidics. *Talanta* 242: 122989. https://doi.org/10.1016/j.talanta.2021.122989

Zijlstra, N., Dingfelder, F., Wunderlich, B., Zosel, F., Benke, S., Nettels, D., and Schuler, B. 2017. Rapid microfluidic dilution for single-molecule spectroscopy of low-affinity biomolecular complexes. *Angewandte Chemie International Edition* 56(25): 7126–7129. https://doi.org/10.1002/anie.201702439

# 5 Microfluidic Devices for Synthesizing Nanomaterials

*Khairunnisa Amreen, Ramya K., and Sanket Goel*

## 5.1 INTRODUCTION

Nanomaterials (NMs) have been proven to be revolutionary in wide-ranging applications pertaining to biomedical, pharmaceutical, environmental, industrial, electronics, etc. Their assistance in drug delivery, electrochemical/biosensing, energy storage and harvesting, imaging, diagnosis, therapeutics, etc. is advancing swiftly (Lines, 2008). Hence, the competence for designing morphologically and chemically structured NMs via various nanotechnology approaches is continually growing (Abid et al., 2022). Custom-engineered properties and structure of NMs have led to the development of various fabrication methods (Mtibe et al., 2018). Lesser cost, high throughput, and sustainable and reproducible methods of synthesis are being explored. The shape, chemical composition, structure, size, porosity, crystallinity, etc. (Asha and Narain, 2020), are some of the focus areas while synthesizing NMs. However, it is often challenging to develop a strategy that gives homogenous, reproducible, and uniform nanoparticles for each batch, especially while adapting wet-chemical synthesis process. Therefore, microfluidic-based devices are an excellent alternative to conventional methods as they are fully integrated and, automated so all the optimal conditions are repeatable while synthesizing (Niculescu et al., 2021). Further, salient features like controlled size, portability, economic efficiency, low reagent usage, rapid synthesis, handling, and storage are added advantages. The micro-devices proffer precise reagent flow and mixing, adequate temperature control and heat transfer, lesser reaction time, and specified geometrical microchannels enabling uniform size and shape of nanoparticles (L. J. Pan et al., 2018b; Song et al., 2008). The wet-chemical synthesis of NMs in bulk has two broader methods: (1) top-down method and (2) bottom-up method. Figure 5.1 gives the schematic of these two methods in general.

In top-down method, bulk starting material is broken down gradually into fragments and then into particles of <100 nm size. Often, milling and attiring of the bulk material are done via micromachining, lithography, electron beam, etching, mechanical and thermal energy, etc. Although this approach gives reproducible nanoparticles, it is unfit for industrial and large-scale production. Hence, bottom-up approach is a substitute to this. In this, NM is synthesized by the addition of atom-by-atom to form clusters and finally the ideal morphological size is achieved. Herein, chemical growth and assembling of atoms to form NMs is done via reagents and physical parameters like temperature, pressure, and pH. This method is useful for industrial-scale production; however, it is prone to limitations like a large volume of reagents making it expensive (Iqbal et al., 2012; Arole and Munde, 2014). In order to resolve this, bottom-up approach method can be implemented in a fully integrated microfluidic device. These devices enable to resolve the limitations of the bulk synthesis approach owing to reduced sizes of the reaction zones, yielding a high surface-to-volume ratio.

## 5.2 MICROFLUIDIC DEVICES FUNDAMENTALS

The principle behind these microdevices is the fluid flow control in the microchannels of the specified geometry, mixing of the reagents, and the separation of the product obtained. The fluid flow is generally of two types: (i) single phase or continuous flow and (ii) multiphase or microdroplet formation.

DOI: 10.1201/9781032632599-5

# Microfluidic Devices for Synthesizing Nanomaterials

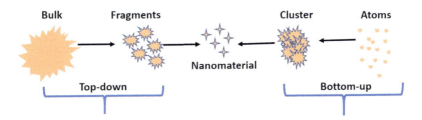

**FIGURE 5.1** Schematic representation of the conventional NM synthesis: top-down and bottom-up approaches.

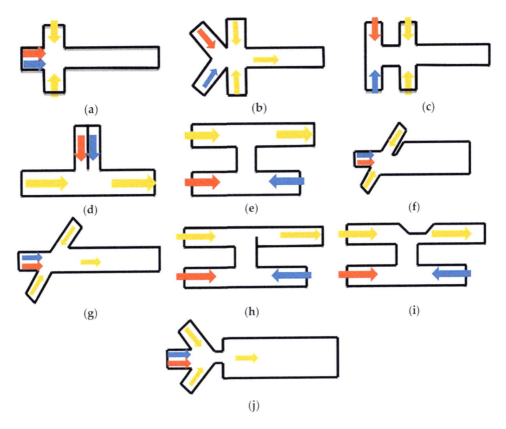

**FIGURE 5.2** Reprint of the schematic diagrams of various designs of microchannels for mixing. (MDPI (Open access) (Chen et al., 2018).)

1. *Single Phase or Continuous Flow*: Mostly adapted for NM synthesis. Herein, a continuous flow of reagents is used. The mixing of reagents at different stages allows the chemical reaction to give NMs. Mixing is a vital step and often carried out via laminar flow. A low Reynolds number (preferably <10) helps in smooth flow. Therefore, various geometrical designs are implemented to get a turbulent mixing in these microdevices (Lee et al., 2011; Ward and Fan, 2015). Figure 5.2 gives the various designs of the microchannels implemented most commonly. Figure 5.2a–c shows simple microchannels wherein a "T" or "Y" junction inlet is designed. Figure 5.2d–g is an asymmetric model of microchannel design wherein the mixing of various reagents at different times can be carried out allowing the reactions to take place accordingly. Figure 5.2h–j shows the schematic of convergent–divergent models of microchannels wherein the flow speed can be increased and swirling of the reagent fluid can be done (Chen et al., 2018).

2. *Multiphase or Microdroplet Formation*: Herein, microdroplets are generated within the microchannels using different immiscible fluids like water-in-oil and vice versa. This approach often gives a high throughput owing to appropriate mixing of the reagents.

The two immiscible liquids in the microchannels, one is in continuous phase and the other is dispersed, with droplets forming at the junction. These droplets are controlled by monitoring the flow rate, reagent viscosity, surface tension, and pressure (Wehking et al., 2014). Microdroplet formation can take place via two modes: passive and active generation. Passive approach employs dripping, squeezing, jetting, etc., whereas active approach uses external energy sources like heat, force, pressure, and electrical or magnetic field for the formation of droplets (Zhu and Wang, 2017). Figure 5.3 shows the schematic representation of these methods.

Monodispersed droplets are formed using these methods. To calculate the degree of monodispersity, coefficient of variation is determined. A set of droplets are taken, and their standard deviation of size is measured followed by dividing it with an arithmetic average. Often, <5% coefficient of variation is considered as optimal (Braunger et al., 2020).

## 5.2.1 Flow Rate Control

The manipulation of reagent flow is carried out via flow monitoring systems. Effectively, there are three variants of these based on the pressure controller used: (1) compressed air controller, (2) mechanical energy-dependent controller, and (3) non-mechanical energy controller (Li et al., 2012).

(i). *Compressed Air*: This type of controller operates based on using a compressed air source for pressurizing the reservoir of the reagents. This allows the fluid to flow through the interconnected microchannel based on pressure difference.

(ii). *Mechanical Energy-Dependent Controller*: External mechanical pumps like syringe pumps and peristaltic pumps are employed at the reagent reservoir to allow the fluid to flow through the microchannels. Electrical stepper motor-integrated syringe pumps allow a precise fluid control. The size of syringe, nozzle, and speed of motor can control the rate. Peristaltic pumps have rotors to allow relaxation and compression that make the reagent flow.

(iii). *Non-mechanical Energy Controller*: The flow rate is manipulated with forces like osmosis, capillary, and hydrostatic.

Irrespective of the type of controllers used, the flow rate is examined via flow sensors like thermal/heat dependent and mass dependent.

## 5.2.2 Temperature Controller

NM synthesis requires specified temperatures to be maintained while the reaction occurs. Hence, a miniaturized thermal management system is used to maintain the optimum temperature throughout. Of late, Srikanth et al. reported a laser-induced graphene (LIG)-based miniaturized heating system integrated with a lab-on-chip platform for gold nanoparticle synthesis. LIG heaters are flexible, thin, and cost-effective. These can be fabricated in various geometric shapes and sizes using variable speed and power of the laser. With a combination of laser power of 15% and speed of 5.5%, LIG heaters are obtained with a maximum temperature of 589 °C. These heaters are coupled with a droplet microfluidic device for the synthesis of gold nanoparticles. Figure 5.4a is the reprint of their device and the thermal camera image of the heater (Srikanth et al., 2021). Similarly, Madhusudan et al. developed a portable thermal monitoring system with

# Microfluidic Devices for Synthesizing Nanomaterials

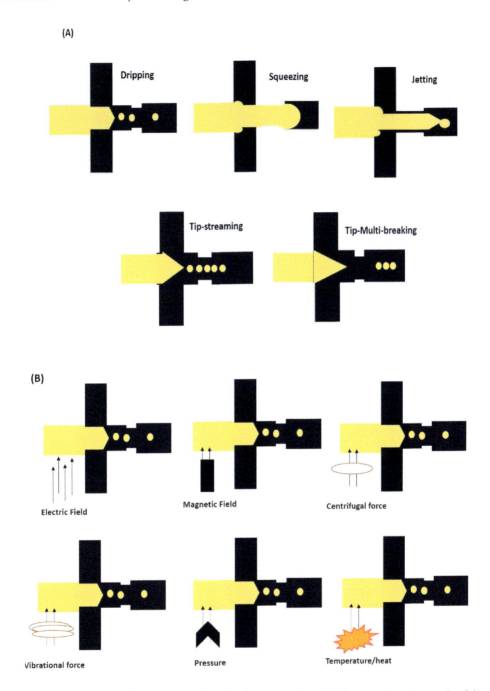

**FIGURE 5.3** (a) Schematic for passive mode of droplet generation. (b) Schematic for active mode of droplet generation.

Internet of Things (IoT) integration. Herein, a microcontroller with cartridge heaters, DC–DC converter, and feedback thermocouple are assembled. This set-up monitored temperature by a proportional-derivative-integral (PID) approach. Figure 5.4b is the real image reprinted of their microfluidic reactor device. They demonstrated the synthesis of $MnO_2$ nanoparticles in a poly(methyl methacrylate) (PMMA) device (Kulkarni et al., 2020).

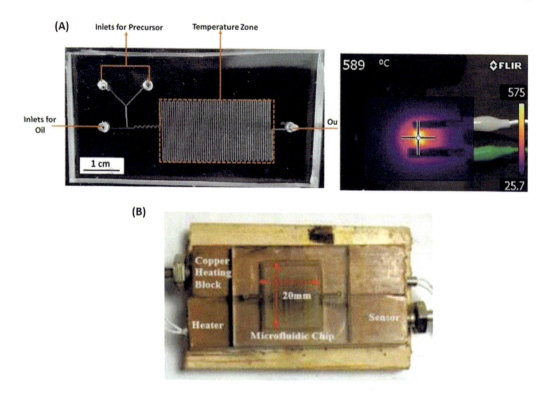

**FIGURE 5.4** (a) Reprint of the microfluidic device and real thermal camera image of LIG heater. (b) Reprint of the real image of thermal management system. ((a) (Srikanth et al., 2021) (open access). (b) Copyright@2020, IOP from (Kulkarni et al., 2020) (open access).)

## 5.3 FABRICATION OF MICROFLUIDIC DEVICES

The state-of-the-art approaches for fabrication of microdevices have evolved significantly. Earlier, the most effective method was soft lithography. Several other methods are being utilized now, for instance, photolithography, three-dimensional (3D) printing, laser-assisted, moulding, and laminating. Figure 5.5 is the schematic of various processes which are used for fabrication of these devices.

1. *Photolithography Method*: One of the most preferred approaches wherein an optical beam of photons is used to design the microchannels thus also known as optical lithography. Substrates like glass, polymers, and flexible films can be used. A photoresistive mask is made with a desired pattern and placed over the substrate. Optical beams like UV, e-beam, and ionic beam are incident over the mask drawing the patterns over the substrate (Goel and Amreen, 2022).
2. *Softlithography Method*: Polymer-based substrates like polyimides, polydimethylsiloxane (PDMS), and polyurethanes are used here. Micromachined casting mould or master mould is made with microchannel patterns. Over this, liquid state polymer proportionately mixed with curing agent is poured. Elevating the temperature solidifies this liquid polymer giving it the pattern and shape of the mould (Goel and Amreen, 2022).
3. *Embossing Method*: Also known as hot embossing, it is a method of using a heat-tolerable material like silicon to make the mould of specified geometry. Solid substrate of polymer is placed between the mould and heated to its melting point. The molten polymer now takes the shape and pattern of the silicon mould. The temperature is cooled to solidify the heated polymer and desired patterned microdevice is formed (Goel and Amreen, 2022).

# Microfluidic Devices for Synthesizing Nanomaterials

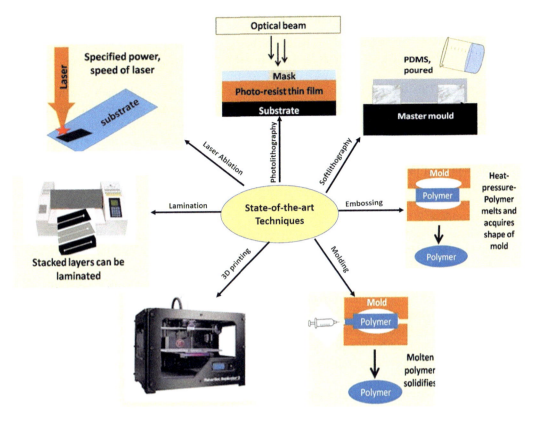

**FIGURE 5.5** The schematic representation of various state-of-the-art approaches of microfluidic device fabrication.

4. *Moulding Method*: Silicon-based master mould is made with patterns and microchannels. Liquid polymer is poured into the mould. Curing is carried out to give a solid structure. Hence, it is also known as inject moulding.
5. *3D Printing Method*: It is the approach of layer-by-layer deposition of conductive materials like polylactic acid (PLA), polyethylene terephthalate (PET), acrylonitrile butadiene styrene (ABS), and polyvinyl alcohol (PVA). Hence, it is also known as additive manufacturing. The desired patterns of the microchannels are made in the computer-aided design (CAD) software. It is then converted to compatible file format and fed to a 3D printer that deposits the material in layers to give a firm 3D structure of optimized geometry. Several types of 3D printers are available in the market for these, for example, stereolithography (SLA), fused deposition moulding (FDM), extrusion, and photopolymerization (Goel and Amreen, 2022).
6. *Lamination Method*: This method includes stacking layers of substrates like glass, PET, PMMA, etc. Usually, the design has three layers, bottom layer as a support, intermediate layer where microchannels are made, and top layer that makes a close unit. Stacking and thermal bonding of these layers via laminator are done to get solid device (Goel and Amreen, 2022).
7. *Laser-Assisted Method*: This method uses high-speed, power lasers like $CO_2$, pulsed, UV, etc. to draw microchannels over the substrates. Often, paper, glass, polymer sheets, silicon, etc. are used here to cut and draw the patterns (Goel and Amreen, 2022).

## 5.4 CATEGORIES OF MICRODROPLET REACTORS

Microdroplet reactors can be categorized based on their chemical and physical nature into different forms like capsules, emulsions, vesicles, and micelles (Shchukin and Sukhorukov, 2004).

1. *Capsules*: This technique was introduced in the early 1990s by Donath and his co-workers. Herein, polyelectrolyte-based capsules, that is, polymer capsules such as poly-allylamine and polystyrene, are prepared. Colloid suspensions when immersed in the ionic solution of polyelectrolytes form a self-assembled multilayer. These give hollow centres to capsule structures. Owing to the permeability of polyelectrolytes, the chemicals can enter the capsules where the reaction can take place. Capsules help in regulating physical parameters like pH and concentration but have a lacuna relating to permeability. Hence, this is a lesser-used approach of microfluidic synthesis (Shchukin and Sukhorukov, 2004).
2. *Emulsions*: Suspensions of water-in-oil give microdroplets that act as microreactors. Herein, the chemical reagent is present in the emulsion droplet. For this, the reagents are dissolved in a dispersion medium and injected into the microfluidic devices which give microdroplet formation. Two or more reagents are fed together as emulsions which undergo reaction under optimal conditions to give NMs (Katsura et al., 1999).
3. *Micelles*: In these, the reagents are dissolved in a medium and encapsulated with surfactants. These reagent molecules can aggregate and form micelles with hydrophilic heads and hydrophobic tails. The reaction occurs in the centre of a micelle. Several metallic nanoparticles have been reportedly synthesized using this method (Shchukin and Sukhorukov, 2004).
4. *Capillary Tubular*: Micro-sized capillary tubes with a compromised flow cross-sectional area are used here. Tubes are often either glass, silica, or steel to make them inert to the chemical reaction taking place inside. Owing to the smaller dimensions of these tubes, thermal management is easy giving uniform-sized particles. Literature reports certain metallic nanoparticles with this method (Zhao et al., 2011).

## 5.5 MICROFLUIDIC-ASSISTED SYNTHESIZED NANOMATERIALS

Of late, a variety of NMs like carbon-based, metallic, and polymer-based are being synthesized using these devices.

### 5.5.1 ORGANIC NANOMATERIALS

Several organic polymers and pharmaceutical molecule-based nanoparticles have reportedly been synthesized using microfluidic devices. These are often useful in drug delivery and therapeutic applications. In this context, liposomes are one such NM which are of significance. Liposome nanoparticles can entrap the drug molecule. The advantage of these being biocompatible makes it an excellent choice for drug delivery application. Further, timely and sustained release of drug at the site of disease is also possible, especially in the treatment of inflammation, cancer, and infections. Droplet-based microfluidic approach for the synthesis of liposomes nanoparticles is substantially attempted. Likewise, organic polymer-based NMs are also used for drug delivery. For instance, poly-(lactic-co-glycolic acid), poly-(ethylene glycol), polycaprolactone, chitosan, heparin, hyaluronic acid nanoparticles, etc., are synthesized in microfluidic devices and used for drug delivery, release, and administration. In addition, microfluidic synthesis of pharmaceutical molecules like lactose, aspirin, nitroglycerin, and ibuprofen has also been attempted. Niculescu et al. gave a detailed summarized table of some of the recent updates in the microfluidic synthesis of these organic NMs. Table 5.1 is adapted from Niculescu et al. (2021).

# TABLE 5.1
## Summary of Organic Polymer and Pharmaceutical Molecule Microfluidic Synthesis

| S. No. | Synthesized Nanomaterial | Type of Microreactor | Reference |
|---|---|---|---|
| 1 | Liposomes | Vertical flow-focusing, thermoplastic material device | Hood and Devoe (2015) |
| 2 | Liposomes | V-shaped mixer device connected with a Teflon tubing | Kawamura et al. (2020) |
| 3 | Liposomes | Ultrasound-assisted microfluidic device | Huang et al. (2010) |
| 4 | Poly-(lactic-co-glycolic acid) | Plus-shaped, Teflon microfluidic chip with flow focus | Shokoohinia et al. (2019) |
| 5 | Poly-(ethylene glycol) | Polyimide film microreactor with 3D flow-focus geometry | Min et al. (2014) |
| 6 | Polycaprolactone | Glass-based microfluidic device with microchannels | Heshmatnezhad and Solaimany Nazar (2020) |
| 7 | Heparin | Glass-based cross junctional | Bicudo and Santana (2012) |
| 8 | Nitroglycerin | Acrylic microfluidic chip | Susilo and Gumono (2020) |
| 9 | Telmisartan | Glass device with silicone tubing | Shrimal et al. (2021) |
| 10 | Hydrocortisone | "Y"-shaped microchannel device | Ali et al. (2009) |
| 11 | Indomethacin nanocrystals | PDMS microdroplet-based chip | Su et al. (2020) |
| 12 | Danazol | "Y"-shaped microchannel device | Zhao et al. (2007) |
| 13 | Cefuroxime | "Y"-shaped microchannel device | Wang et al. (2010) |
| 14 | Piroxicam | PDMS microfluidic platform with 72 wells | Horstman et al. (2015) |
| 15 | Piracetam | PDMS microfluidic platform with 72 wells | Horstman et al. (2015) |
| 16 | Carbamazepine | PDMS microfluidic platform with 72 wells | Horstman et al. (2015) |

## 5.5.2 Inorganic or Metallic Nanomaterials

Synthesis of metallic or inorganic NMs is quite substantially reported in the literature. Different metallic NMs like silver, gold, copper, and zinc are synthesized. Herein, a simple metal salt solution and reducing agent are fed as reagents into the microreactors. Amreen et al. gave a critical review about recent advances in the inorganic NM synthesis (Amreen and Goel, 2021). Some of the significant works from the review are discussed here, for example, gold (Au) nanoparticle synthesis from chloroauric acid precursor. A microfluidic pyrex glass-based device with silicon tubing is designed with three inlets, two mixing zones, and single outlet. Syringe pumps are used to inject the reagents. Figure 5.6a is schematic of their device reprinted from Amreen and Goel (2021). Au nanoparticles are prepared in a PDMS microfluidic device using ionic solutions. Device showcases four inters and single outlet with a total volume of 7.6 μL. Serpentine channel is made for appropriate mixing. Figure 5.6b shows the schematic diagram of the device reprinted from Amreen and Goel (2021) of flow focusing, micromixer, micro thermo-controller, micro-valve, and micro-pumps integrated device for Au nanoparticle synthesis using chloroauric acid precursor. Figure 5.6c is the schematic representation of the device reprinted from Amreen and Goel (2021) of glass-based, Y-shaped microchannel device for Au nanoparticles synthesis using tetracholoauric acid solution. The device is integrated with syringe pumps. Figure 5.6d is the diagrammatic representation of the microfluidic device reprinted from Amreen and Goel (2021) of silver (Ag) nanoparticles preparation in a simple Y-shaped microfluidic channel. Figure 5.6e is the diagrammatic representation reprinted from Amreen and Goel (2021).

A unique impinging jet reactor is used for Ag nanoparticle synthesis from silver nitrate salt solution. Figure 5.6f is the reprint of the device from (Amreen and Goel, 2021): PDMS-based microfluidic chip with three inlets and T-shaped microchannel is designed via soft lithography. Optimum room temperature is maintained, and syringe pumps are integrated for monitoring fluid flow. Figure 5.6g is the schematic representation of the device reprinted from Amreen and Goel (2021).

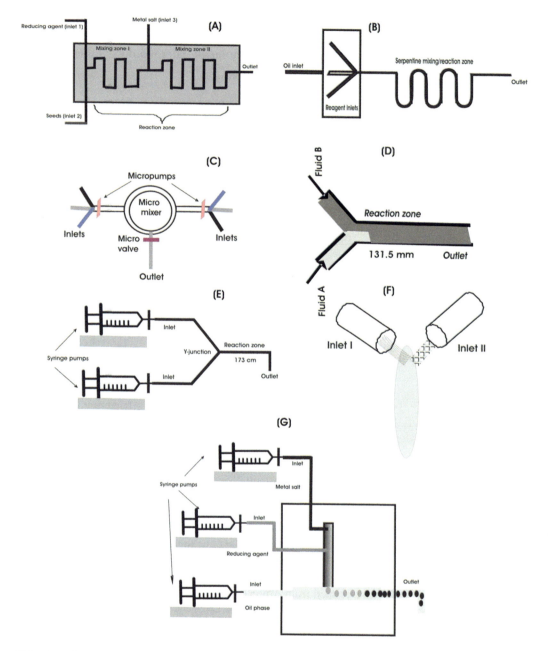

**FIGURE 5.6** Schematic of various microfluidic devices represented in the literature. (Reprinted from Amreen and Goel (2021), IOP (open access).)

Palladium nanoparticle is prepared using a microreactor with T-shaped geometry. Palladium chloride salt is used as precursor. Syringe pumps control the fluid flow. Coiled tubing is integrated for mixing. The temperature is maintained at 60 °C using a water bath. Figure 5.7a is the schematic representation of the device reprinted from Amreen and Goel (2021). Similar T-shaped microchannel device with dual inlets and one outlet over a silicon base material is reported for zinc oxide nanoparticle synthesis. The device uses syringe pumps for flow monitoring. Figure 5.7b is the reprint of the device schematic from Amreen and Goel (2021).

# Microfluidic Devices for Synthesizing Nanomaterials    115

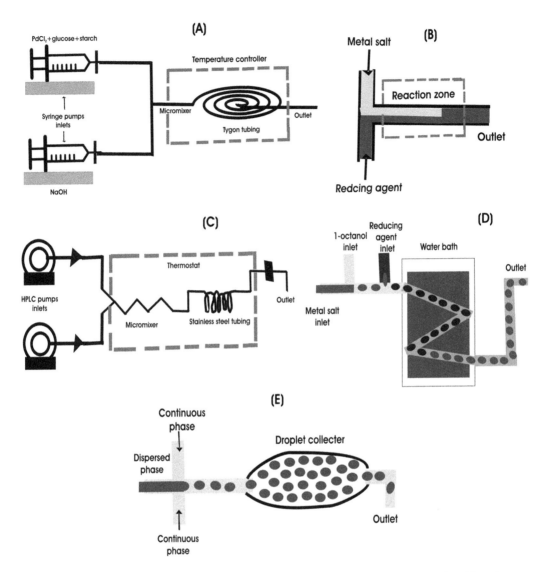

**FIGURE 5.7** Schematic of various microfluidic devices represented in the literature. (Reprinted from Amreen and Goel (2021), IOP (open access).)

Zinc oxide nanoparticles in an emulsion-based microreactor are also reported. Dual inlet device integrated with HPLC pressure pumps has been developed. The device has a micromixer with steel tubing and a temperature controller. Figure 5.7c is the reprint of the schematic of device from Amreen and Goel (2021). PDMS base and T microchannel-integrated device with multiple inlets and one outlet is reported for zinc oxide nanoparticle synthesis. Tubular channels of the device are kept in water bath for maintaining temperature. Figure 5.7d is the schematic diagram of the device reprinted from Amreen and Goel (2021). Likewise, PDMS device for zinc oxide synthesis using syringe pumps and T-shaped microchannel is also reported. Figure 5.7e is the reprint of the device schematic from Amreen and Goel (2021).

Interesting studies pertaining to superparamagnetic nanoparticles are also done, wherein iron oxide nanoparticles are synthesized in a capillary glass tube device. Syringe pumps are used to let the reagents through tubes. Figure 5.8a is the diagrammatic representation of the microdevice reprinted from Amreen and Goel (2021). Another capillary tube microreactor for iron oxide and

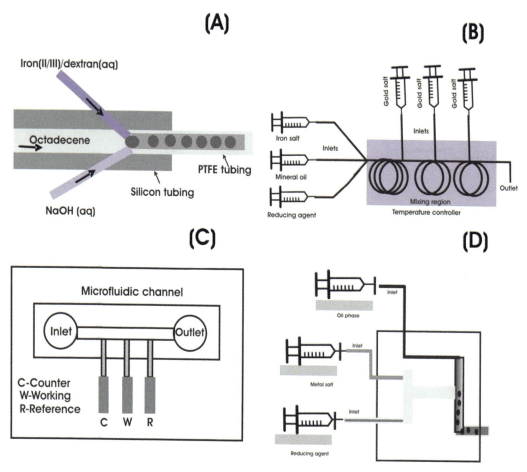

**FIGURE 5.8** Schematic of various microfluidic devices represented in the literature. (Reprinted from Amreen and Goel (2021), IOP (open access).)

gold core nanoparticles is reported using Tygon tubing. Multiple inlets allow injection of reagents in the device at different times and allow appropriate mixing. Figure 5.8b is the schematic representation of the device reprinted from Amreen and Goel (2021). Electrodeposition microfluidic device with three electrodes over a PDMS substrate is made via soft lithography for copper nanoparticle synthesis. Figure 5.8c represents the schematic diagram of the device reprinted from Amreen and Goel (2021). A T-shaped microchannel device with multiple inlets and syringe pumps for copper nanoparticle synthesis is discussed in the review. Figure 5.8d is a reprint of the schematic diagram of device from Amreen and Goel (2021).

### 5.5.3 Composite or Hybrid Nanomaterials

Microfluidic devices can be used for structurally designing the hybrid or composite NMs, wherein an organic polymer encapsulating a metallic nanoparticle or more than two or more metallic nanoparticles or organic nanoparticles can be formed. Such association changes the overall chemical composition and hence the properties. Often, hybrid NMs have shown significant applications in real time. For instance, nucleic acid (messenger RNA) molecules encapsulated in lipid nanoparticles are synthesized in a droplet microreactor for preparing COVID-19 vaccine (Aldosari et al., 2021). Similarly, lipid–polymer–drug composite nanoparticles are used in drug delivery. Microfluidic devices have also proven useful in effective synthesis of lipoproteins. Niculescu et al. in a detailed

**TABLE 5.2**

**Summary of Different Hybrid/Composite Nanomaterials Synthesized**

| S. No. | Synthesized Nanocomposite | Type of Microreactor | Reference |
|---|---|---|---|
| 1 | ZnS–CdSe | Continuous flow, multistep microfluidic device | Wang et al. (2004) |
| 2 | Pt–Sn | Segmented microreactor with water bath and hot plate | Zhang et al. (2020) |
| 3 | Polystyrene–iron oxide | Continuous flow microfluidic device | Taddei et al. (2019) |
| 4 | Chitosan–silver | Cross-junction microchannel device over PDMS substrate | Yang et al. (2016) |
| 5 | Gold–liposome | Fully automated microfluidic device | Al-Ahmady et al. (2019) |
| 6 | Liposome–hydrogel | Glass–silicon device | Hong et al. (2010) |
| 7 | Pegylated-hyaluronic acid | X-shaped microchannel | Tammaro et al. (2020) |
| 8 | Poly (d,l-lactic acid-co-caprolactone)-pegylated | Integrated two microfluidic chips: X- and Y-shaped cross-flow microchannels with micromixer and laminar flow | Lallana et al. (2018) |
| 9 | Ribavirin–poly-(lactic-co-glycolic acid) | Continuous flow microreactor | Bramosanti et al. (2017) |
| 10 | Ketoprofen-PMMA | Integrated three chips with micromixer | Ding et al. (2019) |
| 11 | Paclitaxel–chitosan | PDMS microchip | Majedi et al. (2013)) |
| 12 | (Pluronic F68)-poly-(lactic-co-glycolic acid) | Cross-linking microchannel device | Lababidi et al. (2019) |
| 13 | Efavirenz-poly-(lactic-co-glycolic acid) | Borosilicate glass slide with capillaries | Martins et al. (2019) |

*Source*: Niculescu et al. (2021).

review have given a summarized table of the various hybrid NMs synthesized in microdevices. Table 5.2 is adapted from Niculescu et al. (2021).

### 5.5.4 QUANTUM DOTS

These are nanocrystalline structures that can enhance electron transfer behaviour. Microfluidic synthesis of quantum dots has been successfully explored. For example, microdevices for cadmium-selenium (CdSe) are the most commonly reported and substantially explored quantum dots (Hung et al., 2006). A silicon substrate, segmented flow-based microfluidic device for CdSe quantum dots is reported. A thermal management system was integrated to supply high temperature. A serpentine channel is used as heating and reaction zone followed by a lower temperature zone where the prepared particles are cooled (Yen et al., 2005). Similarly, a Y-shaped microchannel device with dual inlets and single outlet and tubular coiled reaction zone is reported for CdSe quantum dots. The reaction completed in about 40 seconds at 270 °C (Tian et al., 2016). The same group also reported a dual-inlet, PMMA microfluidic chip with a Y junction and serpentine microchannel as reaction zone. Syringe pumps are used for feeding the reagents in the inlets. A polytetrafluoroethylene and steel tubular channel device is also reported which gave controlled and uniform-shaped CdSe quantum dots (Wang et al., 2017). A capillary tubular microreactor with a continuous flow and syringe pumps is also reported (Kwak et al., 2018). Triple inlet, one outlet-based microfluidic device with spiral coiled tubular reaction zone is reported for CdSe. Herein, syringe pumps are used to let the reagents in the reaction zone (Wang et al., 2017). Preparation of other metallic quantum dots is also attempted like $Ag_2S$. A serpentine microchannel, PDMS substrate device is made with flow-focused geometry (Shu et al., 2015). Zinc metal-based quantum dots are made in a Y-shaped junction device with coiled tubing as the reaction zone (Schejn et al., 2014) (Kwon et al., 2012). Carbon quantum dots are also reportedly synthesized using a glucose precursor in a simple capillary tubular microreactor (Lu et al., 2014).

## 5.5.5 Nanosized Molecularly Imprinted Polymers (NanoMIPs)

Molecularly imprinted polymers (MIPs) are synthetic receptors designed to recognize their targets with high specificity (Chen et al., 2016). MIPs' polymeric network contains an adjustable polymer that acts like a receptor, enabling the detection of chemicals (J. Pan et al., 2018a). They have shown great promise as replacements for biological receptors (such as antibodies and enzymes) in many different fields, including purification and separation as well as multiple biomedical uses, due to their significant binding affinity and selectivity as well as outstanding thermal and mechanical strength, simple preparation, and low cost. With their high surface-to-volume ratio, ease of template removal, superior dispersion, simple activation and surface modification, and high compatibility with a wide range of nanosized devices and in vivo biomedical applications, sphere-shaped MIP nanoparticles (nanoMIPs) within diameters typically below 200 nm have attracted considerable interest. MIPs were prepared by copolymerizing a functional monomer in a suitable pore with the compound targeted, which has been preassembled with a functional monomer or several monomers employing a combination of driving factors (such as covalent binding, hydrogen bonding, van der Waals forces, metallic and ionized interactions, or a hydrophilic effect). Selective template binding MIPs were developed by first cross-linking a template to generate a polymer network. Subsequent to the removal of the template, the resultant structure comprised a cross-linked polymer network and MIPs possessing binding sites that closely resembled the template in terms of its dimensions, morphology, and functional group configuration. The three molecular imprinting approaches were developed by utilizing covalent, noncovalent, and semicovalent interactions between monomers and templates inside their binding sites over the imprinting and template rebinding phases. Noncovalent imprinting is a widely used technique due to its compatibility with various commercially available templates and functional monomers, as well as its lack of requirement for intricate organic synthesis. Figures 5.9 and 5.10 depict various monomer structures and the prevalent cross-linker structure cited in the scientific literature.

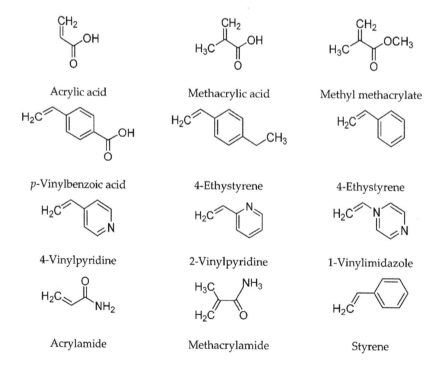

**FIGURE 5.9** Diverse kinds of commonly employed monomers and their structures. (Reprinted from Lusina and Cegłowski (2022), MDPI (open access).)

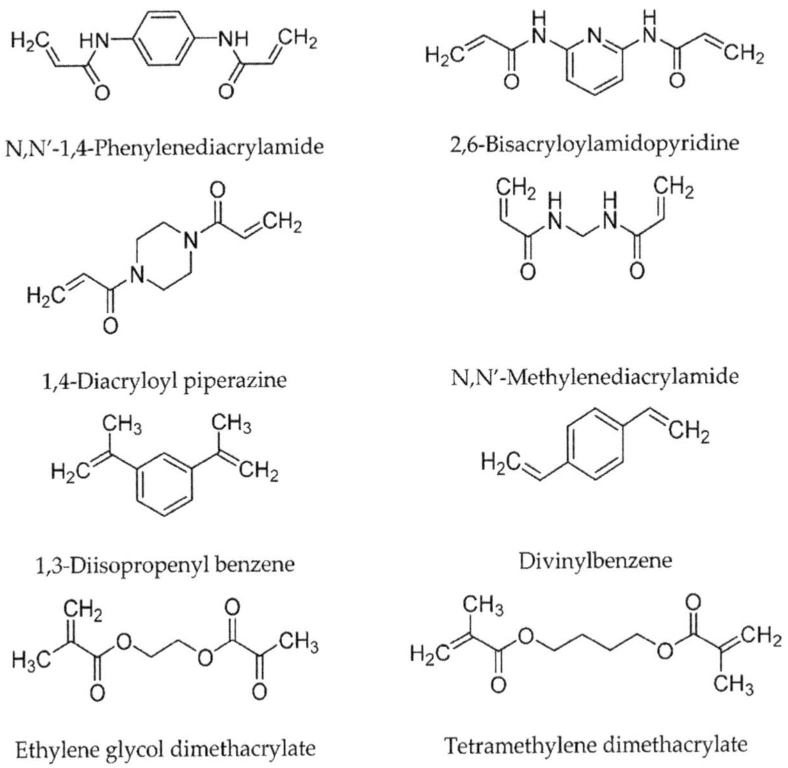

N,N'-1,4-Phenylenediacrylamide

2,6-Bisacryloylamidopyridine

1,4-Diacryloyl piperazine

N,N'-Methylenediacrylamide

1,3-Diisopropenyl benzene

Divinylbenzene

Ethylene glycol dimethacrylate

Tetramethylene dimethacrylate

**FIGURE 5.10** Structure of numerous cross-linker types. (Reprinted from Lusina and Cegłowski (2022), MDPI (open access).)

### 5.5.5.1 Techniques for the Synthesis of NanoMIPs

According to the field of nanotechnology, "nanoparticles" are any particles with a size between 1 and 100 nm. Yet, in the context of molecular imprinting, nanoMIPs often refer to MIPs having diameters of many hundreds of nanometres (Chianella et al., 2013; Gagliardi et al., 2017). In view of this accord, this definition of nanoMIPs was used throughout this section. Furthermore, nanosized MIPs of different shapes, including MIP nanorods, MIP nanofibres, thin, and planar MIP nanofilms, were not given as much attention as spherical nanoMIPs (Ye and Mosbach, 2008; Beyazit et al., 2016). Precipitation polymerization, solid-phase imprinting, emulsion polymerization, and surface imprinting (on nanoparticles) were just some of the effective synthetic techniques explored so far for nanoMIPs. Nanoprecipitation imprinting, self-assemblage imprinting, as well as monomolecular imprinting were other less-used methods for creating nanoMIPs. Several research groups has various strategies for exploring nanoMIPs synthesis for instance, Ding and Heiden, 2014; Gao et al., 2007; Li et al., 2006; Liu et al., 2017; Pan et al., 2018a; Pan et al., 2018b; Pé Rez et al., 2000; Poma et al., 2013; Wan et al., 2017; Ye et al., 1999; Zhang et al., 2016; Zimmerman et al., 2002

### 5.5.5.2 Microfluidic Methods for NanoMIP Synthesis

For the synthesis of extremely high-affinity sites for binding, a microfluidic method has been utilized, which has produced micro- or nanosized MIP particles.

The rapid enantioseparation achieved by Qu, Ping, and colleagues through capillary electrochromatography has led to the creation of a microfluidic apparatus that integrates nanoMIPs as the stationary phase. The nanoparticles that were produced by the co-polymerization of methacrylic acid and ethylene glycol dimethacrylate on 3-(methacryloyloxy)propyltrimethoxysilane-functionalized magnetic nanoparticles (with a diameter of 25 nm) in the presence of a template molecule were characterized using infrared spectroscopy, thermal gravimetric analysis, and transmission electron microscopy. The nanoparticles, which have a diameter of 200 nm, were immobilized as a stationary phase within a microchannel of a microfluidic device through the application of an external magnetic field. The packing length was controllable in this process. The separation of enantiomers of ofloxacin was accomplished within a baseline timeframe of 195 seconds by utilizing S-ofloxacin as the template molecule. This was achieved through additional refinement of the synthesis of imprinted nanoparticles, optimization of the mobile phase composition and pH, and the separation voltage. Compared to traditional packed capillary electrochromatography, changing the packing length of the nanoparticles' zone is a simple way to boost analytical performance. When utilizing a carbon fibre microdisc electrode at +1.0 V (vs. Ag/AgCl), the linear ranges were 1.0–500 M and 5.0–500 M, with detection limits of 0.4 and 2.0 M, respectively (Qu et al., 2010).

Cáceres, César et al. proposed an automated synthesizer that was used to create nanoMIPs. The immobilized template was housed within glass beads in a temperature-controlled column reactor. The inlets including a stirring system were integrated into the column's sliding lid mechanism. An in-built thermocouple in the reactor's cover allowed for precise regulation of the working temperature. At the output, a portion of the collectors segregated waste products into high-affinity product fractions after a set of pumps provided the monomer mixture, the initiator, extra solutions to be used for post-differentiation, as well as washing and elution solvents. NanoMIPs can be post-derivatized with a variety of chemicals in solution. The reactor was flushed with nitrogen before polymerization, and the machine used positive pressure to expel the fluid and empty the reactor. The computer and appropriate software (WinISO) allowed the operator to pre-program all of the reactor's parameters and components as shown in Figure 5.11 (Cáceres et al., 2021).

Li, Qianjin et al. designed fluorescent MIP nanoparticles for fluorescence resonance energy transfer (FRET)-based turn-on fluorescence assay using a one-pot synthetic process and a preliminary microfluidic methodology. Propranolol was chosen as a medicinal medication and acted as both the molecular model and the source of fluorescence. N-allyl-5-(dimethylamino) naphthalene-1-sulphonamide (ADS) was added to the pre-polymerization mixture as a polymerizable fluorescent monomer as the fluorescence acceptor to enable one-pot synthesis. Spectrum overlap between the fluorescence acceptor ADS and the fluorescence donor was a necessary condition for the FRET system to function. The one-pot polymerization strategy established in this technique requires simply the addition of an appropriate fluorescent monomer into the pre-polymerization mixture, making it easier and faster to implement than post-imprinting modification methods. This allows for more flexible control over how much fluorescence acceptor was integrated into MIP nanoparticles. The initial continuous synthesis of fluorescent MIP nanoparticles was carried out using a microfluidic reactor and an optimized pre-polymerization mixture. A separation-free method for the monitoring of the target analyte was created by dispersing the synthesized and purified MIP nanoparticles in an optimized solvent to form a stable suspension. An increase in fluorescence at the acceptor's emission band was caused by FRET between the MIP nanoparticles and a target analyte that was either a fluorescence donor or labelled with a fluorescence donor (Li et al., 2018).

Jin et al. developed glycoprotein-imprinted nanospheres. Microfluidics reactors were made from commercial PTFE capillary columns. The structure of the template, active monomer, PTFE capillary column inner diameters, initiator concentrations, and flow rates affected nanosphere morphology, size, and selectivity. Figure 5.12 shows three solutions—Solutions A, B, and C—used to make MIPs. Solution A had template molecule ovalbumin (OVA) and functional monomer phenylboronic acid, while Solution B had cross-linker tetraethyl orthosilicate (TEOS). Solution C employed $NH_3$–$H_2O$ as a basic initiator. Nanospheres formed when Solutions A and B were mixed with Solution C at the capillary microreactor intersection. SEM and DLS characterized the nanospheres as 193–653 nm. Adjusting synthetic parameters yielded a library of nanoparticles to study how each factor affects nanosphere form and size. Optimized conditions yielded imprinted

# Microfluidic Devices for Synthesizing Nanomaterials 121

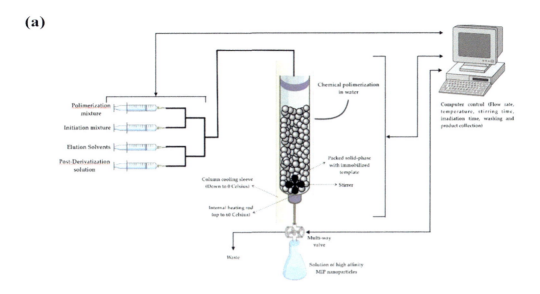

**FIGURE 5.11** Schematic and photographic depiction of an autonomous chemical reactor. (Reprinted from Cáceres et al. (2021), MDPI (open access).)

nanoparticles in under 2 h. Even with competing molecules like horseradish peroxidase (HRP), β-lactoglobulin (BLG), and bovine serum albumin (BSA), they extracted ovalbumin. MIP particles could be reused five times without losing binding capacity. This study shows that imprinted nanospheres with limited dispersity and good selectivity were produced quickly using a microfluidics-based design (Jin et al., 2022).

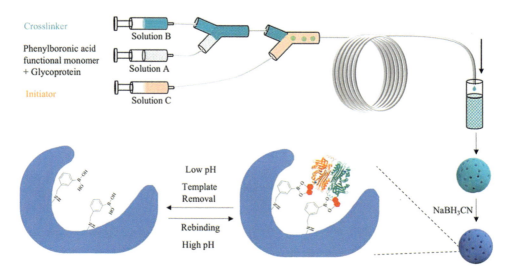

**FIGURE 5.12** Schematic of the microfluidics apparatus used to create droplets of molecularly imprinted glycoprotein. (Reprinted from Orbay and Sanyal (2023), MDPI (open access).)

## 5.6 CONCLUSIONS AND PERSPECTIVES

The multifaceted nature of microfluidic devices, encompassing their construction and manufacturing capabilities, renders them an appealing choice for nanoparticle synthesis. The wide range of varied geometries that currently exist or are being developed provides a compelling rationale for choosing microfluidics methodologies over traditional synthesis techniques. This is because microfluidics enables the production of high-quality NMs in a shorter period and at lower production costs. Furthermore, the utilization of 3D printing technology presents significant potential in the production of microfluidic devices that possess hybrid structures, high-resolution features, and customizable configurations tailored to specific requirements.

Moreover, the examination of chemical synthesis within diverse microreactor configurations may establish a fundamental basis for the development of microchips tailored to the specific needs of researchers engaged in the production of innovative biomedical microfluidic devices.

Despite the numerous advantages offered by microfluidic devices over conventional batch reactors, they are also constrained by inherent limitations. The production of most microfluidic devices typically involves the application of lithography techniques, which require access to a cleanroom facility. The commonly employed photoresist moulds, such as SU-8, are often accompanied by a substantial expense. The utilization of conventional batch techniques for synthesizing NMs obviates the need for clean room facilities or costly photoresists. The following issue relates to the constrained production capacity that arises from a decrease in operational volume. One of the main advantages of utilizing microfluidics in the synthesis of NMs is its ability to minimize reagent consumption, leading to cost and waste reduction. However, it should be noted that the production rate is limited to grams per hour, even with an increase in the flow rate. The current body of literature on the amplification of NMs using identical microreactors or combinatorial reactors is limited, with only a few sources reporting on this topic. Nanoparticles synthesized through microfluidics exhibit the presence of supplementary reagents, such as oil and toluene, owing to the multiphase flow, which distinguishes them from conventional batch reactors employed for NM synthesis. Additional effort is required in order to completely remove these reagents before clinical application. Frequently employed materials for various devices, such as PDMS and PMMA, demonstrate restricted compatibility with solvents and inadequate resistance to high temperatures. Another issue that arises is the potential obstruction of microchannels in a PDMS material, which has the potential to adversely affect the merging of reagents. The current state of the field of biomedical NMs is characterized by limited availability of in-line control for real-time analysis.

Biomimetic nanoparticles represent a distinct class of NMs which can be conveniently synthesized using microfluidic techniques, which may pose a significant challenge in the clinical market in the future. A limited number of studies have employed the characteristics of microfluidics and exogenous energy sources to produce these NMs. Likewise, it is feasible to replicate biological environments using NMs, since microchannels can promptly conform for physical flow rates and chemical gradients. One potential approach involves converting biological mechanisms that generate anticancer activity in the human body into microchannels. The preparation of such specific materials can facilitate the development of a novel therapeutic approach. Currently, most microfluidic device materials encounter fouling concerns, which are anticipated to be resolved through the utilization of bioinspired intelligent material design. The phenomenon results in the emergence of novel biofabrication methodologies, including 4D bioprinting, which introduces an additional dimension to the conventional 3D bioprinting technique, enabling the manipulation of shape in response to external stimuli.

Anticipated advancements in the foreseeable future encompass innovative material designs and modifications to existing materials, alongside progress in fabrication technologies. However, the achievement of a comprehensive automated procedure for synthesizing, screening, and clinically assessing remains a significant challenge to be realized in the near future. The screening process is performed on the crude reaction product, irrespective of the complexity of the optimization algorithm or screening parameters employed to improve the process. The purification of products is a crucial stage in the synthesis of NMs. The successful incorporation of various constituents for this objective would represent a significant achievement in the fabrication of NMs using microfluidic techniques. In the field of microfluidics, it is possible to utilize an external energy-based centrifugation technique for the purpose of purifying the solution of nanoparticles following their synthesis. Rotations can be induced in NMs solution through the application of external electric fields with opposite polarities. Adequate voltage has the potential to induce substantial centrifugal forces.

## REFERENCES

Abid, N., Khan, A.M., Shujait, S., Chaudhary, K., Ikram, M., Imran, M., Haider, J., Khan, M., Khan, Q., Maqbool, M., 2022. Synthesis of nanomaterials using various top-down and bottom-up approaches, influencing factors, advantages, and disadvantages: A review. Adv Colloid Interf Sci. https://doi.org/10.1016/j.cis.2021.102597

Al-Ahmady, Z.S., Donno, R., Gennari, A., Prestat, E., Marotta, R., Mironov, A., Newman, L., Lawrence, M.J., Tirelli, N., Ashford, M., Kostarelos, K., 2019. Enhanced intraliposomal metallic nanoparticle payload capacity using microfluidic-assisted self-assembly. *Langmuir* 35, 13318–13331. https://doi.org/10.1021/acs.langmuir.9b00579

Aldosari, B.N., Alfagih, I.M., Almurshedi, A.S., 2021. Lipid nanoparticles as delivery systems for RNA-based vaccines. *Pharmaceutics*. https://doi.org/10.3390/pharmaceutics13020206

Ali, H.S.M., York, P., Blagden, N., 2009. Preparation of hydrocortisone nanosuspension through a bottom-up nanoprecipitation technique using microfluidic reactors. *Int J Pharm* 375, 107–113. https://doi.org/10.1016/j.ijpharm.2009.03.029

Amreen, K., Goel, S., 2021. Review – Miniaturized and microfluidic devices for automated nanoparticle synthesis. *ECS J Solid State Sci Technol* 10, 017002. https://doi.org/10.1149/2162-8777/abdb19

Arole, V.M., Munde, S.V., 2014. Fabrication of nanomaterials by top-down and bottom-up approaches-an overview. *JAAST: Mat Sci* 1 (2), 89–93.

Asha, A.B., Narain, R., 2020. Nanomaterials properties, in: *Polymer Science and Nanotechnology: Fundamentals and Applications*. Elsevier, pp. 343–359. https://doi.org/10.1016/B978-0-12-816806-6.00015-7

Beyazit, S., Tse Sum Bui, B., Haupt, K., Gonzato, C., 2016. Molecularly imprinted polymer nanomaterials and nanocomposites by controlled/living radical polymerization. *Prog Polym Sci* 62, 1–21. https://doi.org/10.1016/J.PROGPOLYMSCI.2016.04.001

Bicudo, R.C.S., Santana, M.H.A., 2012. Production of hyaluronic acid (HA) nanoparticles by a continuous process inside microchannels: Effects of non-solvents, organic phase flow rate, and HA concentration. *Chem Eng Sci* 84, 134–141. https://doi.org/10.1016/j.ces.2012.08.010

Bramosanti, M., Chronopoulou, L., Grillo, F., Valletta, A., Palocci, C., 2017. Microfluidic-assisted nanoprecipitation of antiviral-loaded polymeric nanoparticles. *Colloids Surf A Physicochem Eng Asp* 532, 369–376. https://doi.org/10.1016/j.colsurfa.2017.04.062

Braunger, M.L., Fier, I., Rodrigues, V., Arratia, P.E., Riul, A., 2020. Microfluidic mixer with automated electrode switching for sensing applications. *Chemosensors* 8. https://doi.org/10.3390/chemosensors8010013

Cáceres, C., Moczko, E., Basozabal, I., Guerreiro, A., Piletsky, S., 2021. Molecularly imprinted nanoparticles (nanoMIPs) selective for proteins: Optimization of a protocol for solid-phase synthesis using automatic chemical reactor. *Polymers* 13(3), 314. https://doi.org/10.3390/POLYM13030314

Chen, C., Zhao, Y., Wang, J., Zhu, P., Tian, Y., Xu, M., Wang, L., Huang, X., 2018. Passive mixing inside microdroplets. *Micromachines (Basel)* https://doi.org/10.3390/mi9040160

Chen, L., Wang, X., Lu, W., Wu, X., Li, J., 2016. Molecular imprinting: Perspectives and applications. *Chem Soc Rev* 45, 2137–2211. https://doi.org/10.1039/C6CS00061D

Chianella, I., Guerreiro, A., Moczko, E., Caygill, J.S., Piletska, E.V., De Vargas Sansalvador, I.M.P., Whitcombe, M.J., Piletsky, S.A., 2013. Direct replacement of antibodies with molecularly imprinted polymer nanoparticles in ELISA – Development of a novel assay for vancomycin. *Anal Chem* 85, 8462–8468. https://doi.org/10.1021/AC402102J/SUPPL_FILE/AC402102J_SI_002.PDF

Ding, S., Serra, C.A., Anton, N., Yu, W., Vandamme, T.F., 2019. Production of dry-state ketoprofen-encapsulated PMMA NPs by coupling micromixer-assisted nanoprecipitation and spray drying. *Int J Pharm* 558, 1–8. https://doi.org/10.1016/j.ijpharm.2018.12.031

Ding, X., Heiden, P.A., 2014. Recent developments in molecularly imprinted nanoparticles by surface imprinting techniques. *Macromol Mater Eng* 299, 268–282. https://doi.org/10.1002/MAME.201300160

Gagliardi, M., Bertero, A., Bifone, A., 2017. Molecularly imprinted biodegradable nanoparticles. *Sci Rep* 7(1), 1–9. https://doi.org/10.1038/srep40046

Gao, D., Zhang, Z., Wu, M., Xie, C., Guan, G., Wang, D., 2007. A surface functional monomer-directing strategy for highly dense imprinting of TNT at surface of silica nanoparticles. *J Am Chem Soc* 129, 7859–7866. https://doi.org/10.1021/JA070975K/ASSET/IMAGES/MEDIUM/JA070975KE00002.GIF

Goel, S., Amreen, K., 2022. Laser induced graphanized microfluidic devices. *Biomicrofluidics*. https://doi.org/10.1063/5.0111867

Heshmatnezhad, F., Solaimany Nazar, A.R., 2020. On-chip controlled synthesis of polycaprolactone nanoparticles using continuous-flow microfluidic devices. *J Flow Chem* 10, 533–543. https://doi.org/10.1007/s41981-020-00092-8

Hong, J.S., Stavis, S.M., Depaoli Lacerda, S.H., Locascio, L.E., Raghavan, S.R., Gaitan, M., 2010. Microfluidic directed self-assembly of liposome-hydrogel hybrid nanoparticles. *Langmuir* 26, 11581–11588. https://doi.org/10.1021/la100879p

Hood, R.R., Devoe, D.L., 2015. High-throughput continuous flow production of nanoscale liposomes by microfluidic vertical flow focusing. *Small* 11, 5790–5799. https://doi.org/10.1002/smll.201501345

Horstman, E.M., Goyal, S., Pawate, A., Lee, G., Zhang, G.G.Z., Gong, Y., Kenis, P.J.A., 2015. Crystallization optimization of pharmaceutical solid forms with X-ray compatible microfluidic platforms. *Cryst Growth Des* 15, 1201–1209. https://doi.org/10.1021/cg5016065

Huang, Z., Mayr, N.A., Yuh, W.T.C., Lo, S.S., Montebello, J.F., Grecula, J.C., Lu, L., Li, K., Zhang, H., Gupta, N., Wang, J.Z., 2010. Predicting outcomes in cervical cancer: A kinetic model of tumor regression during radiation therapy. *Cancer Res* 70, 463–470. https://doi.org/10.1158/0008-5472.CAN-09-2501

Hung, L.H., Choi, K.M., Tseng, W.Y., Tan, Y.C., Shea, K.J., Lee, A.P., 2006. Alternating droplet generation and controlled dynamic droplet fusion in microfluidic device for CdS nanoparticle synthesis. *Lab Chip* 6, 174–178. https://doi.org/10.1039/b513908b

Iqbal, P., Preece, J.A., Mendes, P.M., 2012. Nanotechnology: The "top-down" and "bottom-up" approaches, in: *Supramolecular Chemistry*. John Wiley & Sons, Ltd. https://doi.org/10.1002/9780470661345.smc195

Jin, Y., Wang, T., Li, Q., Wang, F., Li, J., 2022. A microfluidic approach for rapid and continuous synthesis of glycoprotein-imprinted nanospheres. *Talanta* 239, 123084. https://doi.org/10.1016/J.TALANTA.2021.123084

Katsura, S., Yamaguchi, A., Harada, N., Hirano, K., Mizuno, A., 1999 "Micro-reactors based on water-in-oil emulsion," *Conference Record of the 1999 IEEE Industry Applications Conference. Thirty-Forth IAS Annual Meeting (Cat. No.99CH36370), Phoenix, AZ, USA*, pp. 1124–1129 vol. 2, https://doi.org/10.1109/IAS.1999.801645

Kawamura, J., Kitamura, H., Otake, Y., Fuse, S., Nakamura, H., 2020. Size-controllable and scalable production of liposomes using a V-shaped mixer micro-flow reactor. *Org Process Res Dev* 24, 2122–2127. https://doi.org/10.1021/acs.oprd.0c00174

Kulkarni, M.B., Enaganti, P.K., Amreen, K., Goel, S., 2020. Internet of Things enabled portable thermal management system with microfluidic platform to synthesize $MnO_2$ nanoparticles for electrochemical sensing. *Nanotechnology* 31. https://doi.org/10.1088/1361-6528/ab9ed8

Kwak, C.H., Kang, S.M., Jung, E., Haldorai, Y., Han, Y.K., Kim, W.S., Yu, T., Huh, Y.S., 2018. Customized microfluidic reactor based on droplet formation for the synthesis of monodispersed silver nanoparticles. *J Ind Eng Chem* 63, 405–410. https://doi.org/10.1016/j.jiec.2018.02.040

Kwon, B.H., Lee, K.G., Park, T.J., Kim, H., Lee, T.J., Lee, S.J., Jeon, D.Y., 2012. Continuous in situ synthesis of ZnSe/ZnS core/shell quantum dots in a microfluidic reaction system and its application for light-emitting diodes. *Small* 8, 3257–3262. https://doi.org/10.1002/smll.201200773

Lababidi, N., Sigal, V., Koenneke, A., Schwarzkopf, K., Manz, A., Schneider, M., 2019. Microfluidics as tool to prepare size-tunable PLGA nanoparticles with high curcumin encapsulation for efficient mucus penetration. *Beilst J Nanotechnol* 10. https://doi.org/10.3762/bjnano.10.220

Lallana, E., Donno, R., Magrì, D., Barker, K., Nazir, Z., Treacher, K., Lawrence, M.J., Ashford, M., Tirelli, N., 2018. Microfluidic-assisted nanoprecipitation of (PEGylated) poly (D,L-lactic acid-co-caprolactone): Effect of macromolecular and microfluidic parameters on particle size and paclitaxel encapsulation. *Int J Pharm* 548, 530–539. https://doi.org/10.1016/j.ijpharm.2018.07.031

Lee, C.Y., Chang, C.L., Wang, Y.N., Fu, L.M., 2011. Microfluidic mixing: A review. *Int J Mol Sci* 12, 3263–3287. https://doi.org/10.3390/ijms12053263

Li, Q., Jiang, L., Kamra, T., Ye, L., 2018. Synthesis of fluorescent molecularly imprinted nanoparticles for turn-on fluorescence assay using one-pot synthetic method and a preliminary microfluidic approach. *Polymer (Guildf)* 138, 352–358. https://doi.org/10.1016/J.POLYMER.2018.01.086

Li, X., Ballerini, D.R., Shen, W., 2012. A perspective on paper-based microfluidics: Current status and future trends. *Biomicrofluidics*. https://doi.org/10.1063/1.3687398

Li, Z., Ding, J., Day, M., Tao, Y., 2006. Molecularly imprinted polymeric nanospheres by diblock copolymer self-assembly. *Macromolecules* 39, 2629–2636. https://doi.org/10.1021/MA0526793/ASSET/IMAGES/LARGE/MA0526793F00009.JPEG

Lines, M.G., 2008. Nanomaterials for practical functional uses. *J Alloys Compd* 449, 242–245. https://doi.org/10.1016/j.jallcom.2006.02.082

Liu, R., Cui, Q., Wang, C., Wang, X., Yang, Y., Li, L., 2017. Preparation of sialic acid-imprinted fluorescent conjugated nanoparticles and their application for targeted cancer cell imaging. *ACS Appl Mater Interfaces* 9, 3006–3015. https://doi.org/10.1021/ACSAMI.6B14320/ASSET/IMAGES/LARGE/AM-2016-14320Z_0007.JPEG

Lu, Y., Zhang, L., Lin, H., 2014. The use of a microreactor for rapid screening of the reaction conditions and investigation of the photoluminescence mechanism of carbon dots. *Chem Eur J* 20, 4246–4250. https://doi.org/10.1002/chem.201304358

Lusina, A., Cegłowski, M., 2022. Molecularly imprinted polymers as state-of-the-art drug carriers in hydrogel transdermal drug delivery applications. *Polymers* 14, 640. https://doi.org/10.3390/POLYM14030640

Majedi, F.S., Hasani-Sadrabadi, M.M., Hojjati Emami, S., Shokrgozar, M.A., Vandersarl, J.J., Dashtimoghadam, E., Bertsch, A., Renaud, P., 2013. Microfluidic assisted self-assembly of chitosan based nanoparticles as drug delivery agents. *Lab Chip* 13, 204–207. https://doi.org/10.1039/c2lc41045a

Martins, C., Araújo, F., Gomes, M.J., Fernandes, C., Nunes, R., Li, W., Santos, H.A., Borges, F., Sarmento, B., 2019. Using microfluidic platforms to develop CNS-targeted polymeric nanoparticles for HIV therapy. *Eur J Pharm Biopharm* 138, 111–124. https://doi.org/10.1016/j.ejpb.2018.01.014

Min, K.I., Im, D.J., Lee, H.J., Kim, D.P., 2014. Three-dimensional flash flow microreactor for scale-up production of monodisperse PEG-PLGA nanoparticles. *Lab Chip* 14, 3987–3992. https://doi.org/10.1039/c4lc00700j

Mtibe, A., Mokhothu, T.H., John, M.J., Mokhena, T.C., Mochane, M.J., 2018. Fabrication and characterization of various engineered nanomaterials, in: *Handbook of Nanomaterials for Industrial Applications*. Elsevier, pp. 151–171. https://doi.org/10.1016/B978-0-12-813351-4.00009-2

Niculescu, A.G., Chircov, C., Bîrcă, A.C., Grumezescu, A.M., 2021. Nanomaterials synthesis through microfluidic methods: An updated overview. *Nanomaterials*. https://doi.org/10.3390/nano11040864

Orbay, S., Sanyal, A., 2023. Molecularly imprinted polymeric particles created using droplet-based microfluidics: Preparation and applications. *Micromachines* 14, 763. https://doi.org/10.3390/MI14040763

Pan, J., Chen, W., Ma, Y., Pan, G., 2018a. Molecularly imprinted polymers as receptor mimics for selective cell recognition. *Chem Soc Rev* 47, 5574–5587. https://doi.org/10.1039/C7CS00854F

Pan, L.J., Tu, J.W., Ma, H.T., Yang, Y.J., Tian, Z.Q., Pang, D.W., Zhang, Z.L., 2018b. Controllable synthesis of nanocrystals in droplet reactors. *Lab Chip* https://doi.org/10.1039/c7lc00800g

Pé Rez, N., Whitcombe, M.J., Vulfson, E.N., 2000. Molecularly imprinted nanoparticles prepared by core-shell emulsion polymerization. *J Appl Polym Sci* 77, 1851–1859. https://doi.org/10.1002/1097-4628

Poma, A., Guerreiro, A., Whitcombe, M.J., Piletska, E.V., Turner, A.P.F., Piletsky, S.A., 2013. Solid-phase synthesis of molecularly imprinted polymer nanoparticles with a reusable template – "plastic antibodies". *Adv Funct Mater* 23, 2821–2827. https://doi.org/10.1002/ADFM.201202397

Qu, P., Lei, J., Zhang, L., Ouyang, R., Ju, H., 2010. Molecularly imprinted magnetic nanoparticles as tunable stationary phase located in microfluidic channel for enantioseparation. *J Chromatogr A* 1217, 6115–6121. https://doi.org/10.1016/J.CHROMA.2010.07.063

Schejn, A., Frégnaux, M., Commenge, J.M., Balan, L., Falk, L., Schneider, R., 2014. Size-controlled synthesis of ZnO quantum dots in microreactors. *Nanotechnology* 25. https://doi.org/10.1088/0957-4484/25/14/145606

Shchukin, D.G., Sukhorukov, G.B., 2004. Nanoparticle synthesis in engineered organic nanoscale reactors. *Adv Mater*. https://doi.org/10.1002/adma.200306466

Shokoohinia, P., Hajialyani, M., Sadrjavadi, K., Akbari, M., Rahimi, M., Khaledian, S., Fattahi, A., 2019. Microfluidic-assisted preparation of PLGA nanoparticles for drug delivery purposes: Experimental study and computational fluid dynamic simulation. *Res Pharm Sci* 14, 459–470. https://doi.org/10.4103/1735-5362.268207

Shrimal, P., Jadeja, G., Patel, S., 2021. Microfluidics nanoprecipitation of telmisartan nanoparticles: effect of process and formulation parameters. *Chem Pap* 75, 205–214. https://doi.org/10.1007/s11696-020-01289-w

Shu, Y., Jiang, P., Pang, D.W., Zhang, Z.L., 2015. Droplet-based microreactor for synthesis of water-soluble Ag2S quantum dots. *Nanotechnology* 26. https://doi.org/10.1088/0957-4484/26/27/275701

Song, Y., Hormes, J., Kumar, C.S.S.R., 2008. Microfluidic synthesis of nanomaterials. *Small*. https://doi.org/10.1002/smll.200701029

Srikanth, S., Dudala, S., Jayapiriya, U.S., Mohan, J.M., Raut, S., Dubey, S.K., Ishii, I., Javed, A., Goel, S., 2021. Droplet-based lab-on-chip platform integrated with laser ablated graphene heaters to synthesize gold nanoparticles for electrochemical sensing and fuel cell applications. *Sci Rep* 11. https://doi.org/10.1038/s41598-021-88068-z

Su, Z., He, J., Zhou, P., Huang, L., Zhou, J., 2020. A high-throughput system combining microfluidic hydrogel droplets with deep learning for screening the antisolvent-crystallization conditions of active pharmaceutical ingredients. *Lab Chip* 20, 1907–1916. https://doi.org/10.1039/d0lc00153h

Susilo, S.H., Gumono, Irawan B., 2020. The effect of concentration reactant to mixing nitroglycerin using microchannel hydrodynamics focusing. *J Southwest Jiaotong Univ* 55. https://doi.org/10.35741/issn.0258-2724.55.3.51

Taddei, C., Sansone, L., Ausanio, G., Iannotti, V., Pepe, G.P., Giordano, M., Serra, C.A., 2019. Fabrication of polystyrene-encapsulated magnetic iron oxide nanoparticles via batch and microfluidic-assisted production. *Colloid Polym Sci* 297, 861–870. https://doi.org/10.1007/s00396-019-04496-4

Tammaro, O., Costagliola di Polidoro, A., Romano, E., Netti, P.A., Torino, E., 2020. A microfluidic platform to design multimodal PEG – crosslinked hyaluronic acid nanoparticles (PEG-cHANPs) for diagnostic applications. *Sci Rep* 10. https://doi.org/10.1038/s41598-020-63234-x

Tian, Z.H., Xu, J.H., Wang, Y.J., Luo, G.S., 2016. Microfluidic synthesis of monodispersed CdSe quantum dots nanocrystals by using mixed fatty amines as ligands. *Chem Eng J* 285, 20–26. https://doi.org/10.1016/j.cej.2015.09.104

Wan, L., Chen, Z., Huang, C., Shen, X., 2017. Core–shell molecularly imprinted particles. *TrAC Trends Anal Chem* 95, 110–121. https://doi.org/10.1016/J.TRAC.2017.08.010

Wang, H., Li, X., Uehara, M., Yamaguchi, Y., Nakamura, H., Miyazaki, M., Shimizu, H., Maeda, H., 2004. Continuous synthesis of CdSe-ZnS composite nanoparticles in a microfluidic reactor. *Chem Commun* 4, 48–49. https://doi.org/10.1039/b310644f

Wang, J., Zhao, H., Zhu, Y., Song, Y., 2017. Shape-controlled synthesis of CdSe nanocrystals via a programmed microfluidic process. *J Phys Chem C* 121, 3567–3572. https://doi.org/10.1021/acs.jpcc.6b10901

Wang, J.X., Zhang, Q.X., Zhou, Y., Shao, L., Chen, J.F., 2010. Microfluidic synthesis of amorphous cefuroxime axetil nanoparticles with size-dependent and enhanced dissolution rate. *Chem Eng J* 162, 844–851. https://doi.org/10.1016/j.cej.2010.06.022

Ward, K., Fan, Z.H., 2015. Mixing in microfluidic devices and enhancement methods. *J Micromech Microeng* 25. https://doi.org/10.1088/0960-1317/25/9/094001

Wehking, J.D., Gabany, M., Chew, L., Kumar, R., 2014. Effects of viscosity, interfacial tension, and flow geometry on droplet formation in a microfluidic T-junction. *Microfluid Nanofluid* 16, 441–453. https://doi.org/10.1007/s10404-013-1239-0

Yang, C.H., Wang, L.S., Chen, S.Y., Huang, M.C., Li, Y.H., Lin, Y.C., Chen, P.F., Shaw, J.F., Huang, K.S., 2016. Microfluidic assisted synthesis of silver nanoparticle–chitosan composite microparticles for antibacterial applications. *Int J Pharm* 510, 493–500. https://doi.org/10.1016/j.ijpharm.2016.01.010

Ye, L., Cormack, P.A.G., Mosbach, K., 1999. Molecularly imprinted monodisperse microspheres for competitive radioassay. *Anal Commun* 36, 35–38. https://doi.org/10.1039/A809014I

Ye, L., Mosbach, K., 2008. Molecular imprinting: Synthetic materials as substitutes for biological antibodies and receptors. *Chem Mater* 20, 859–868. https://doi.org/10.1021/CM703190W/ASSET/IMAGES/LARGE/CM-2007-03190W_0007.JPEG

Yen, B.K.H., Günther, A., Schmidt, M.A., Jensen, K.F., Bawendi, M.G., 2005. A microfabricated gas-liquid segmented flow reactor for high-temperature synthesis: The case of CdSe quantum dots. *Angew Chem* 117, 5583–5587. https://doi.org/10.1002/ange.200500792

Zhang, D., Wang, Y., Deng, J., Wang, X., Guo, G., 2020. Microfluidics revealing formation mechanism of intermetallic nanocrystals. *Nano Energy* 70. https://doi.org/10.1016/j.nanoen.2020.104565

Zhang, W., Kang, J., Li, P., Liu, L., Wang, H., Tang, B., 2016. Two-photon fluorescence imaging of sialylated glycans in vivo based on a sialic acid imprinted conjugated polymer nanoprobe. *Chem Commun* 52, 13991–13994. https://doi.org/10.1039/C6CC08211D

Zhao, C.X., He, L., Qiao, S.Z., Middelberg, A.P.J., 2011. Nanoparticle synthesis in microreactors. *Chem Eng Sci* 66, 1463–1479. https://doi.org/10.1016/j.ces.2010.08.039

Zhao, H., Wang, J.X., Wang, Q.A., Chen, J.F., Yun, J., 2007. Controlled liquid antisolvent precipitation of hydrophobic pharmaceutical nanoparticles in a microChannel reactor. *Ind Eng Chem Res* 46, 8229–8235. https://doi.org/10.1021/ie070498e

Zhu, P., Wang, L., 2017. Passive and active droplet generation with microfluidics: A review. *Lab Chip* 17, 34–75. https://doi.org/10.1039/C6LC01018K

Zimmerman, S.C., Wendland, M.S., Rakow, N.A., Zharov, I., Suslick, K.S., 2002. Synthetic hosts by monomolecular imprinting inside dendrimers. *Nature* 418(6896), 399–403. https://doi.org/10.1038/nature00877

# 6 Microfluidization of Juice Derived from Plant Products

*Divya Arora, Sanjana Kumari, and Barjinder Pal Kaur*

## 6.1 INTRODUCTION

To improve public health, it is imperative that fruit and vegetable consumption rise. Frequent consumption of fruits and vegetables reduces the risk of obesity, heart disease, and cancer, among other chronic conditions. Moreover, fruits and vegetables offer exceptional anti-inflammatory and antioxidant qualities (Wallace et al., 2020). Drinking juices made from plants (fruits, vegetables, stems, etc.) is one of the dietary recommendations for a healthy lifestyle since it facilitates the body's absorption of fibre, vitamins, minerals, and phytochemicals (Butu & Rodino, 2019); and health advantages of various plant parts may be easily accessed in the form of juices. Recently, smoothies and premium natural juices have been rapidly gaining popularity in the fast-moving world.

Naturally occurring enzymes like polyphenol oxidase (PPO), polyphenol peroxidase (POD), and pectin methyl esterase (PME) are linked to the quality and sensory attributes of the majority of juices. Juices lose their nutritional value and sensory appeal due to enzymatic browning (Singh et al., 2022b). Thus, an effective processing technique for deactivating the oxidative enzymes and preserving the juice quality is needed. The term "food processing" refers to the methods or procedures used to extend the stability and shelf life of food products. In general terms, food processing technologies can be classified into six different categories (Pandey et al., 2022), which are illustrated in Figure 6.1.

Fruit and vegetable juices have traditionally been preserved using thermal processing. Earlier, heat treatment was the most popular way to treat fresh juices since it eliminates microbes and deactivates enzymes that lead to deterioration. However, under extreme circumstances, heat processing can result in several physicochemical changes that reduce the bioavailability of several nutrients and bioactive substances, as well as disperse sensory qualities like taste, appearance, smell, and touch (Petruzzi et al., 2017). Additionally, there is a growing demand for high-quality food which is nominally processed with characteristics similar to fresh food in terms of nutritional value, flavour, texture, and colour. As a result, the need for high-quality, safety-conscious, minimally processed meals has encouraged researchers and the food industry to investigate non-thermal technologies as a means of producing food items with the fewest possible changes. Among all other techniques, pressure-based non-thermal processing technologies have gained popularity in the past decade because they extend the shelf life of the juices, preserving their organoleptic tastes, freshness, and nutritional value.

Microfluidization is one of the most promising pressure-based processing technologies in which juice preservation is done utilizing high pressures without the use of chemicals (Suhag et al., 2023). Microfluidization is usually a continuous and short-duration (<5 s) process operated at a lower temperature, thus causing minimal disruption in flavours and nutritional compounds of juices (Singh et al., 2022a). This chapter deals explicitly with the outcome of microfluidization on the juices derived from plant sources.

128

DOI: 10.1201/9781032632599-6

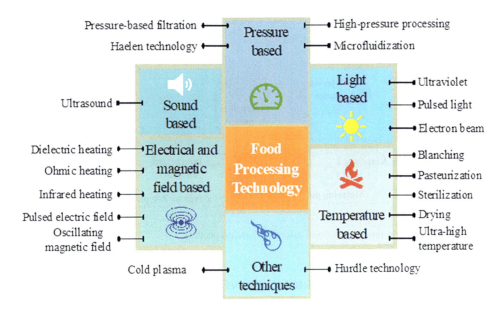

**FIGURE 6.1** General classification of food processing technologies.

## 6.2 MICROFLUIDIZATION TECHNIQUE – ADAPTATION AND SCOPE IN THE FOOD INDUSTRY

Microfluidization, a green technology, combines the advantages of impinging streams, water jets, and conventional high-pressure homogenization technologies (Li et al., 2022). Utilizing high pressure and a microfluidizer with a specially constructed interaction chamber, the microfluidization process may produce a fine emulsion with smaller particle sizes. The pharmaceutical and cosmetics industries were the first to employ ultra-fine emulsion using microfluidization (Kumar et al., 2022a). It has gradually become more well known in the food industry as a cutting-edge food processing technique. Microfluidization was initially employed in the food business by the dairy industry to homogenize milk, and it produced superior results than high-pressure homogenizers (Ozturk & Turasan, 2021). More researchers are utilizing this cutting-edge technology in the food business for a variety of additional applications because of the microfluidization technique's enormous success in the dairy sector for homogenization and size reduction. Numerous applications have been shown by recent research, including extraction (Huang et al., 2022), encapsulation (Ganesan et al., 2018), and macromolecule modification (Guo et al., 2020), other than stabilization (Wang et al., 2023) and homogenization (Oliete et al., 2019).

The food processing industry has embraced microfluidization due to its advantages, which include minimal nutritional loss, short processing times, continuous operation, and green label technology because it processes food without the use of exogenous chemicals (Suhag et al., 2023; Singh et al., 2022a). The dynamics of these precisely engineered microchannels drive this high-powered process. A microfluidizer can have an operational pressure range of up to 30,000 psi (Kavinila et al., 2023). An immediate pressure drop, impact, cavitation, high pressure, high velocity, and extreme shear rate are all experienced by the processed fluid when microchannels are precisely engineered (Ozturk & Turasan, 2022). This makes it possible to produce particles at the nanoscale, which leads to the manufacture of uniform and stable food items. Currently, the following areas are the major focus of microfluidization research (Li et al., 2022): (1) the structural alterations and physicochemical characteristics of materials treated by microfluidization; (2) the improvement of the intracellular chemicals' extraction process, such as polysaccharides, polyphenols, and flavones, by causing structural damage to the cell; and (3) inactivation of biological activity due to leakage of intracellular substance that irreversibly kills microorganisms.

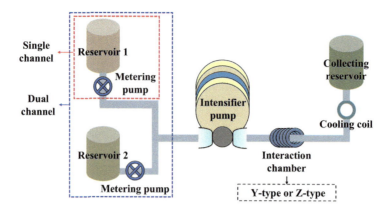

**FIGURE 6.2** Schematic representation of microfluidization process.

## 6.3 COMPONENTS AND WORKING PRINCIPLES OF MICROFLUIDIZER

An interaction chamber, a heat exchanger, a pressure intensifier pump or cooling coil, and an intake reservoir make up the microfluidizer assembly. Another essential component is an intensifier pump that runs on pneumatic power and can generate air pressure of up to 40,000 psi (Kavinila et al., 2023). The intensifier pump's primary function is to push liquid through high-pressure, high-velocity microchannels using a sequence of suction and discharge strokes. Figure 6.2 represents the schematic representation of microfluidization process. An extra part known as an auxiliary processing module (APM) can further enhance the microfluidizer's performance. The liquid feed is sent through a precisely crafted interaction chamber that is split into two or more microstreams, colliding with one another in the process of microfluidization. This process produces distinctive effects such as high pressure, high velocity, shear rate, pressure drop, vibrations, and hydrodynamic cavitation (Leyva-Daniel et al., 2020). The high pressure and velocity inside the microchannels create the shearing action between the liquid droplets and the wall of microchannels and within the liquid droplets. The liquid stream then reaches the interaction chamber, where it follows a certain path depending on the geometry of the interaction chamber. High impact force is caused by the splitting and collapsing of the liquid stream in a Y-type chamber and the zigzag form in a Z-type chamber (Kumar et al., 2022b). Further, the liquid is passed through the low-pressure zone, due to which cavitation is formed after experiencing an abrupt decline in the pressure. At the final stage, liquid interacts with the cooling coil, where it loses the excess temperature caused by the energy loss throughout the process.

## 6.4 TYPES OF MICROFLUIDIZER

Microfluidizers can be classified based on their basic components, inlet reservoir and interaction chamber. According to Bai and McClements (2016), there are two kinds of microfluidizers: single channel and double channel. These types of microfluidizers are distinguished by the number of inlet reservoirs and processing steps required. Another classification of microfluidizers, based on the type of interaction chamber, is Y- and Z-type chambers (Sahil et al., 2022). Different types of microfluidizers are discussed in detail in the following paragraphs.

### 6.4.1 Microfluidizers on the Basis of Inlet Reservoirs' Numbers

a) Single-channel microfluidizer: A single-channelled reservoir, as the name implies, consists of a single inflow reservoir. This kind of unit requires a two-step process and may not be appropriate for a process requiring the participation of two distinct phases, for example, emulsion formation. However, it is suitable for processes involving the use of a single fluid, such as homogenization or extraction.

b) Dual-channel microfluidizer: A dual-channel microfluidizer features two input reservoirs for the ability to feed two phases independently, making it an upgraded version of a single-channel microfluidizer. Emulsions may be formed in a single step thanks to this improved design. Together with the time and cost savings, this arrangement also provides the benefit of reduced waste.

### 6.4.2 Microfluidizers on the Basis of Interaction Chamber's Type

a) "Y" type: High-pressure liquid stream of a Y-type chamber splits into two parts that collide at high speeds of up to 400 m/s with one another and the microchannel walls (Sahil et al., 2022), which result in $10^7$ s$^{-1}$ shear rate (Villalobos-Castillejos et al., 2018). This design is usually meant for low-viscosity operations for liquid–liquid dispersion. This arrangement is appropriate for liposomal-based, polymer-based, and oil-in-water emulsion encapsulation. The Y-type chambers have dimensions ranging from 75 to 125 µm (Microfluidics, 2014). In the Y-type chamber, flow stabilization among high and atmospheric pressure is provided by APM, which is situated downstream of the interaction chamber (Kavinila et al., 2023).

b) "Z" type: The liquid stream in a Z-type chamber is driven into microchannels through the zigzag path, due to which it experiences collisions with the walls of flow channels. The Z-type chamber is usually meant for solid–liquid dispersion; however, it is also suitable for water-in-oil emulsions, cellular fragmentation, extraction, and dispersion. The overall dimensions of the Z-type chamber range between 87 and 1000 µm (Microfluidics, 2014), and the APM is situated at the upper side of the interaction chamber, where it acts as a pre-mixing or pre-processing unit (Kavinila et al., 2023). The visual representation of Y- and Z-type interaction chambers is shown in Figure 6.3.

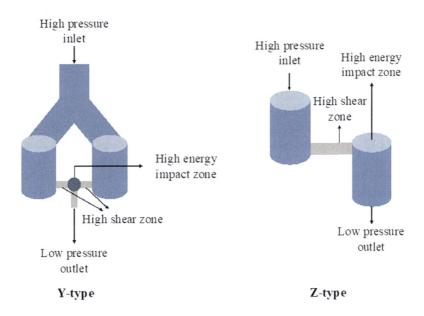

**FIGURE 6.3** Pictorial representation of "Y-" and "Z"-type interaction chamber.

## 6.5 NEED FOR MICROFLUIDIZATION OF JUICES

Plant-based juice is highly sensitive to the environment (oxidation), and its quality degrades immediately after extraction. There has always been the problem of storage of juices for later consumption (Abliz et al., 2021). Despite the homogeneous composition of the extracted juice, certain particles precipitate or turn turbid as a result of gravity or forces between the particles, which cause phase separation and lower the juice's quality (Abliz et al., 2020). The conventional thermal processing technology used to be the most economical treatment for juice processing that has proven to prolong the shelf life; however, the development of heat stress within the system results in colour change and loss of nutritional quality (Vigneshwaran et al., 2022). The food business is always looking for efficient methods that, in situations when the conventional thermal preservation method falls short, can preserve nutritional content, stability, and flavour. Microfluidization is the non-thermal and green technology emerging in the food sector for juice processing. Various studies have been conducted validating its potential as a safe and minimally processed technique. The effect of microfluidization on different properties affecting the juice quality is discussed in the following paragraphs.

## 6.6 EFFECT OF MICROFLUIDIZATION ON DIFFERENT PROPERTIES OF JUICES

So far, the need for microfluidization, its working, and types have been discussed. Now, the changes in plant-based juice after microfluidization are also of concern. In microfluidization, a homogenization technique, particle size is of prime importance, especially in plant-based juices (Li et al., 2022). Once a fruit or vegetable is processed to form juice, various factors are affected by the processing method, such as particle size, rheological properties, textural properties, and sensory properties. This represents that the physicochemical properties of juices are affected by microfluidization. It also has an impact on the polyphenol content and antioxidant activity of juices, indicating its impact on the nutritional value of the product. The effects of microfluidization on different properties of juices are discussed in the following subsections.

### 6.6.1 Effect on Physicochemical Properties

Physicochemical properties of plant-derived juices vary when they are processed into juices, and the method of this conversion plays a very important role. In a study on bilberry juice, the effects of homogenization techniques (magnetic stirring, ultrasonication, and a combination of homogenization and microfluidization) on the physicochemical properties of feed solutions (10–20%, w/w) as well as the benefits of spray-dried galactoglucomannans (GGM) and glucuronoxylans (GX) powders were examined (Halahlah et al., 2023). The results reported magnetic stirring to be the most efficient method to prepare spray-dried GGM feed solution for spray-dried microencapsulation of bilberry juice. The significant drop in the mean particle size of ginger rhizome juice from 15.85 to 12.86 μm at 137.89 MPa (1 cycle) was further supported by research on the juice that found cell disintegration. At 103.42 MPa (3 cycles), the values of the total phenolic content, antioxidant activity test 2,2-diphenyl-1-picrylhydrazyl (DPPH), and 2,2-azino-bis-3-ethylbenzothiazoline-6-sulphonic acid (ABTS) rose from 56.79 to 65.28 mg GAE/100 mL, 69.99% to 77.51% inhibition, and 74.61% to 84.74% inhibition, respectively. The release of pigments and phenolic chemicals brought on by cell rupture was identified as the reason for this rise. Table 6.1 shows that the juices derived from plant products have significant changes in physicochemical properties upon microfluidization.

### 6.6.2 Effect on Nutritional Properties

Fruits are generally perishable and due to this nature, the tendency of their juices to spoil is higher. Fruit juices on microfluidization have been witnessed to show changes in their nutritional properties. In the case of Dancy tangerine juice (Nayak et al., 2023), after microfluidization, the juice was

# Microfluidization of Juice Derived from Plant Products

## TABLE 6.1
## Changes in Physicochemical Properties of Juices from Plant Products on Microfluidization

| Plant Source | Plant Part | Changes in Physicochemical Properties | References |
|---|---|---|---|
| Bilberry juice | Fruit | According to the study, magnetic stirring is the most effective way to prepare spray-dried GGM feed solutions for spray-dried microencapsulation of bilberry juice. After being ultrasonicated and microfluidized, spray-dried GGM solutions developed a gel-like texture that rendered them inappropriate for spray drying. There are three methods for homogenizing spray-dried GX, which can be manufactured and are resistant to modifications in processing settings. Glass transition temperatures were greater for spray-dried GGM powders than for spray-dried GX. High solid ratios of wall materials to bilberry juice can be included in feed solutions because of the low viscosity of spray-dried GGM and spray-dried GX solutions. Because cellulose solutions are more viscous, spray-drying becomes difficult. After a week in storage, the spray-dried GGM, spray-dried GX, and spray-dried GGM + CMC feed solutions did not change, except for increased viscosity and anthocyanin loss. | Halahlah et al. (2023) |
| Ginger juice | Rhizome | Microscopic examination revealed cell disintegration, further supporting the considerable ($p < 0.05$) decrease in the mean particle size of ginger rhizome juice from 15.85 (control) to 12.86 $\mu$m at 137.89 MPa and one cycle. Total phenolic content, antioxidant activity assay DPPH, and ABTS values increased from 56.79 to 65.28 mg GAE/100 mL, 69.99% to 77.51% inhibition, and 74.61% to 84.74% inhibition, respectively, at 103.42 MPa and three cycles. This increase was attributed to the release of phenolic compounds and pigments caused by cell disruption. | Suhag et al. (2023) |
| Sugarcane juice | Stem | Microfluidization has a significant impact on its physicochemical properties. The absence of a discernible effect on pH indicates that microfluidization may occur concurrently with pH adjustments made to preserve sugarcane juice. The findings suggest that superior sugarcane juice with exceptional nutritional qualities might be produced at pressures more than 100 MPa but lower than 200 MPa. | Tarafdar and Kaur (2021) |
| Yam juice | Root | This study quantified DHPM's impact on yam juice. TSS, turbidity, flavonoid concentration, and non-enzymatic browning all showed significant decreases following DHPM processing; the largest declines were 35.5%, 86.2%, 20.7%, and 66.7%, respectively. Moreover, an essential positive link existed between turbidity and the average particle size, which dropped from 1944 to 358 nm. | Liu et al. (2021) |
| *Rosa roxburghii* Tratt juice | Fruit | After the juice from the *R. roxburghii* fruit is extracted, a large amount of residue is left behind, with dietary fibres making up most of this residue. Through enzymatic hydrolysis, dynamic high-pressure microfluidization, and carboxymethylation modification, respectively, RIDF was converted into enzymatic-hydrolysed RIDF, dynamic high-pressure microfluidized RIDF, and carboxymethylated RIDF in this investigation. The physicochemical and functional features of RIDF were considerably impacted by three different modification techniques. | Wang et al. (2020) |

*Note*:   ABTS = 2,2-azino-bis-3-ethylbenzothiazoline-6-sulphonic acid; CMC = carboxymethylcellulose; DHPM = dynamic high-pressure microfluidization; DPPH = 2,2-diphenyl-1-picrylhydrazyl; GAE = gallic acid equivalents; RIDF = *Rosa roxburghii* Tratt fruit; TSS = total soluble solids.

homogeneous, hesperidin and naringin concentrations increased, colour was deeper, opalescence stability was better, and PME activity was reduced. It was also reported that when granulated citrus fruits are microfluidized at high pressures, the quality of the juices improves. The total soluble solids (TSS), turbidity, flavonoid content, and non-enzymatic browning were all reduced following high-pressure microfluidization for the yam juice (Liu et al., 2021). Related studies have been reported in Table 6.2 suggesting the nutritional changes in case of different plant juices.

## TABLE 6.2
## Changes in Nutritional Properties of Juices from Plant Products on Microfluidization

| Plant Source | Plant Part | Changes in Nutritional Properties | References |
|---|---|---|---|
| Dancy tangerine juice | Fruit | The juice that had been microfluidized was assessed for several quality attributes, including pectin methyl esterase activity, cloud value, fractal dimension, lacunarity, and colour attributes. The juice in the control group had the maximum fractal dimension, suggesting an irregular cell structure, whereas the juice in the 103 MPa (3 passes) had the lowest, suggesting homogeneity. Juice that had been microfluidized had increased hesperidin and naringin concentrations, a deeper colour, and better opalescence stability. It also had lower PME activity. The greatest organoleptic quality was shown by juice that was treated once at 103 MPa. When juice is extracted from granulated citrus fruits, the high pressure used during microfluidization can significantly increase the juice's quality. | Nayak et al. (2023) |
| Sugarcane juice | Stem | Microfluidization at 100 MPa (7 cycles) or 150 MPa (3–7 cycles) can yield sugarcane juice with the least amount of polyphenol and antioxidant degradation and microbiological safety. Numerous discrepancies were observed in the antioxidant activity of sugarcane juice, as assessed by various techniques when combining our findings with those from the literature. These variations may result from a variety of causes, including the nature of technology, the research conducted on the radical scavenging assay, the testing setup used for antioxidant tests, and the intricacy of the molecular structure found in the item to be processed. | Tarafdar et al. (2021) |
| Yam juice | Root | The content of TSS, turbidity, flavonoid content, and non-enzymatic browning were all reduced following DHPM processing, with the largest decreases occurring at 35.5%, 86.2%, 20.7%, and 66.7%, respectively. Additionally, following DHPM processing, the value of $\Delta E$ increased from 4.63 (very visible) to 11.5 (excellent), indicating a considerable shift in the colour of the yam juice. | Liu et al. (2021) |
| Sea buckthorn juice | Fruit | In this investigation, sea buckthorn juice was subjected to various treatment parameters using DHPM. While there was a minor drop in TSS and total carotenoid content, treatment at higher pressures preserved more TSS and total carotenoids than treatment at lower pressures. Furthermore, DHPM did not have an effect on all-trans-α-carotene or all-trans-β-carotene. The maximum amount of carotenoids in sea buckthorn juice may be kept when the DHPM treatment parameter is 150 MPa in a single pass. The sea buckthorn juice's physical characteristics also improved under these treatment conditions. All things considered, the DHPM treatment enhanced the juice's sensory quality overall and successfully preserved its nutritional value, serving as a foundation for later studies on the bioavailability of carotenoids in sea buckthorn juice. | Abliz et al. (2021) |

*(Continued)*

# Microfluidization of Juice Derived from Plant Products

**TABLE 6.2 (CONTINUED)**

**Changes in Nutritional Properties of Juices from Plant Products on Microfluidization**

| Plant Source | Plant Part | Changes in Nutritional Properties | References |
|---|---|---|---|
| Carrot juice | Root | This study looked at how high-pressure microfluidization affected the orange carrot juice's nutritional value. Therefore, despite differences in processing conditions, no particular trends in TPC and antioxidant activity of the juice were identified. The amount of carotenoids in carrot juice was considerably increased by microfluidization nonetheless. The amounts of lutein and β-carotene have grown considerably with an increase in passes. Similarly, raising process pressure initially boosted carotenoid content considerably (up to 68.95 MPa), but subsequently increasing pressure to 103.42 MPa did not significantly increase carotenoid concentration. | Koley et al. (2020) |

*Note*: DHPM = dynamic high-pressure microfluidization; PME = pectin methyl esterase; TPC = total phenolic content; TSS = total soluble solids.

## 6.6.3 EFFECT ON ENZYMATIC AND MICROBIAL ACTIVITIES

### 6.6.3.1 Effect on Enzymes

In sapodilla jam processed by thermally assisted high hydrostatic pressure, POD was found to be more pressure resistant than PPO (Shinwari & Rao, 2021); also for loquat fruit using spectrophotometric methods PPO was more heat sensitive than POD (Zhang & Shao, 2015). According to Velázquez-Estrada et al. (2012), high-pressure homogenization of orange juice demonstrated the heat and pressure tolerance of PME. Thus, it may be concluded that various enzyme activities can benefit from high-pressure homogenization. Similarly, microfluidized Chinese pear and mushroom were observed which showed similar effects as high-pressure homogenization (Liu et al., 2009a; Liu et al., 2009b). Although the enzymes could be affected by other parameters as well, such as type of enzyme, exposure time, pH requirement, medium composition, and temperature, ultimate reaction conditions play a major role (Dos Santos Aguilar et al., 2018). Changes in the enzymatic activity of plant-produced juices have been reported in Table 6.3.

**TABLE 6.3**

**Changes in Enzymatic Activity of Juices from Plant Products on Microfluidization**

| Plant Source | Plant Part | Changes in Enzymatic Activity | References |
|---|---|---|---|
| Dancy tangerine juice | Fruit | Dancy juice's PME activity decreased in microfluidized juice independent of the amount of pressure used or passes made. Cycles and microfluidization pressure both had a major impact on PME activity. It appears that the mechanical stress exerted in the homogenization valve likely induces irreversible conformational changes in the enzyme that result in its inactivation, as seen by the reduction in PME activity acquired during microfluidization. | Nayak et al. (2023) |
| Ginger juice | Rhizome | The PPO enzyme was shown to be baroresistant because microfluidization had no noticeable impact on its relative activity. A higher microfluidization pressure of 137.89 MPa resulted in the complete reduction of aerobic bacteria, yeast, and moulds. Due to its solvent-free nature and success in microbial reduction, microfluidization may aid in the delivery of high-quality clean-labelled ginger juice. | Suhag et al. (2023) |

*(Continued)*

## TABLE 6.3 (CONTINUED)
## Changes in Enzymatic Activity of Juices from Plant Products on Microfluidization

| Plant Source | Plant Part | Changes in Enzymatic Activity | References |
|---|---|---|---|
| Orange juice | Fruit | Compared to non-homogenized orange juice, the PME activity of the homogenized orange juices decreased with increasing pressure. On increasing the number of passes, no significant effect on the PME activity was observed. Compared with the homogenization pressure treatment, the activity of PME was lower after more homogenization passes. The inactivation of PME activity prevents the loss of turbidity, increasing the commercial value of the orange juice. | Rychlik et al. (2021) |
| Peach juice | Fruit | Both conventional homogenization and DHPM processing prevented non-enzymatic browning, and the rate of inhibition rose with DHPM pressure and passing number. The early study's results seemed to suggest that DHPM processing would be more advantageous than conventional homogenization processing. | Wang et al. (2019) |

*Note*: DHPM = dynamic high-pressure microfluidization; PME = pectin methylesterase; PPO = polyphenol oxidase.

### 6.6.3.2 Effect on Microbial Activity

The microbial activity of a juice depends upon the inactivation of the microorganisms by the process using which juice is extracted, such as microfluidization. As in the case of ginger juice, aerobic bacteria, yeast, and moulds were eliminated at a higher microfluidization pressure (Suhag et al., 2023). Moreover, after microfluidization, the POD activity for sapodilla juice decreased from 100% to 40.57%, and the residual PPO activity decreased from 100% to 80.78% (Singh et al., 2022b). Similar studies are reported in Table 6.4, indicating changes in microbial properties of different plant product juices upon microfluidization.

## TABLE 6.4
## Changes in Microbial Activity of Juices from Plant Products on Microfluidization

| Plant Source | Plant Part | Changes in Microbial Activity | References |
|---|---|---|---|
| Ginger juice | Rhizome | This study looked at the effects of high-pressure microfluidization at various microfluidization cycles and pressures on the microbiological activities of ginger rhizome (*Zingiber officinale* Roscoe) juice. Aerobic bacteria, yeast, and moulds were completely eliminated at a higher microfluidization pressure of 137.89 MPa. The delivery of high-quality, clean-labelled ginger juice may be facilitated by microfluidization because of its solvent-free nature and effectiveness in microbial reduction. | Suhag et al. (2023) |
| Sapodilla juice | Fruit | The quality deterioration of sapodilla (*Manilkara achras* L.) juice is caused by a variety of oxidative enzymes, which must be inactivated using a new and ongoing green pressure processing technique. In this work, microfluidization was used to pressurize sapodilla juice at pressures between 10,000 and 30,000 pounds per square inch (psi) over the course of one to three passes or cycles. The residual PPO activity dropped from 100% to 80.78% and the POD activity dropped from 100% to 40.57% after microfluidization, according to the results. In addition, 30,000 psi (3 passes) of microfluidization yielded the lowest microbial load (2.89 log CFU/mL) as compared to the control sample (7.57 log CFU/mL) of unprocessed juice. Therefore, microfluidization may be a good option for processing juices to prevent spoiling. | Singh et al. (2022b) |

*(Continued)*

# Microfluidization of Juice Derived from Plant Products

**TABLE 6.4  (CONTINUED)**
**Changes in Microbial Activity of Juices from Plant Products on Microfluidization**

| Plant Source | Plant Part | Changes in Microbial Activity | References |
|---|---|---|---|
| Sugarcane juice | Stem | By using natural polypeptides together with three runs of microfluidization at high pressure, sugarcane juice's shelf life was increased to 56 days at 5 ± 2 °C. There were no significant microbiological loads discovered. Thus, in the case of sugarcane juice thermal processing, this hurdle technique seems to be quite satisfactory, and it may also be used in other vegetable and fruit juices. | Kohli et al. (2019) |
| Orange juice | Fruit | The purpose of this study was to evaluate the antifungal activity of plant oil extracts against pathogenic moulds (*Aspergillus niger* and *Penicillium chrysogenum*) and yeast (*Saccharomyces cerevisiae*) in combination with organic citrus fruit extract at a ratio of 1:3 (w/w). References were synthetic commercial preservatives such as potassium sorbate and sodium benzoate. For in situ analysis in orange juice, a microemulsion loaded with an antifungal formulation was created using microfluidization. During the initial 35 days of storage, the targeted fungi were completely inhibited by the antifungal formulation-loaded microemulsion, and their populations continued to remain below the detection limit. Even at 10 times the optimized concentration, antifungal formulation-loaded microemulsion had more antifungal efficacy than synthetic preservatives. By increasing the surface area of droplets caused by size decrement, microemulsion really boosted the bioactivity and bioavailability of plant oil/citrus extracts as compared to coarse emulsion. | Maherani et al. (2019) |
| Water melon juice | Fruit | The microbiological characteristics were observed in order to investigate the impact of trans-cinnamaldehyde emulsions. Trans-cinnamaldehyde emulsions were also added to actual food systems, such as watermelon juice. The best mass ratio of trans-cinnamaldehyde to Tween 20 (1:3) was used to create nano-sized trans-cinnamaldehyde emulsions, which were subsequently subjected to high-energy emulsification (10,000 rpm high-speed homogenization and 20,000 psi high-pressure homogenization). Regarding the antibacterial activity findings, the lowest concentration of 0.8 weight percent trans-cinnamaldehyde emulsions inhibited the development of *Salmonella typhimurium* and *Staphylococcus aureus*, but not that of *Escherichia coli*, in both pure water and water melon juice. | Jo et al. (2015) |

*Note*:  CFU = colony-forming units; POD = peroxidase; PPO = polyphenol oxidase.

## 6.6.4  EFFECT ON SHELF LIFE

The rate of spoiling, which can be estimated using a variety of parameters, including changes in enzymatic activity, microbial activity, oxidative rate changes, and physical parameters like colour, texture, taste, and aroma, typically determines a product's shelf life, especially juices. Concerning factors might play a role in the juices derived from microfluidization. Sugarcane juice was microfluidized at 159 MPa (1 cycle) and kept in glass bottles at 4 °C to demonstrate the safety of the fluids. The juice demonstrated stability after 10 days of treatment (Tarafdar & Kaur, 2022). Another study on quality retention using a genetic algorithm on microfluidized sugarcane juice at 124 MPa (2 passes) exhibited a shelf life of 6 days (Tarafdar et al., 2019). Another study on sugarcane juice compared thermal hurdles like pasteurization and potassium metabisulphite with non-thermal hurdles like microfluidization and natural polypeptides. The comparison revealed that microfluidization at high pressure at 5 ± 2 °C (3 passes) produced a 56-day shelf life (Kohli et al., 2019). A comparison of these studies based on changes in the shelf life of sugarcane juice is shown in Table 6.5.

## TABLE 6.5
## Changes in Shelf Life of Juices from Plant Products on Microfluidization

| Plant Source | Plant Part | Changes in Shelf Life | References |
|---|---|---|---|
| Sugarcane juice | Stem | An attempt was made to preserve sugarcane juice using microfluidization as a stand-alone method. The sugarcane juice in glass bottles was efficiently kept fresher for up to 10 days at 4 °C with microfluidization, compared to less than a day for control. The investigation's findings also suggest that glass bottles work better than HDPE upright pouches when it comes to packing microfluidized sugarcane juice. | Tarafdar and Kaur (2022) |
| Sugarcane juice | Stem | The current study examined how sugarcane juice quality parameters were affected by the microfluidization technique. The microfluidization treatment at 124 MPa (2 passes) was shown to be the most successful in maintaining the quality of sugarcane juice, according to genetic algorithm-based optimization. The juice produced under ideal circumstances showed a 6-day shelf life, suggesting that microfluidization is a viable non-thermal sugar cane juice preservation technique. | Tarafdar et al. (2019) |
| Sugarcane juice | Stem | In this work, natural polypeptides (polylysine and nisin) and microfluidization were used as non-thermal barriers to preserve sugarcane juice. The effects were contrasted with the untreated control sample and thermal barriers consisting of potassium metabisulphite and pasteurization. Each sample was kept in glass bottles at $5 \pm 2$ °C and assessed for 63 days using physicochemical, microbiological, and sensory characteristics. There was a total decrease of 23 units in PPO activity in the control sample due to both heat and non-thermal obstacles. By utilizing non-thermal hurdles, the original microbial load of $3.19 \times 106$ CFU/mL was completely reduced to 50 CFU/mL, guaranteeing a 56-day shelf life. | Kohli et al. (2019) |

*Note*:   CFU = colony forming units; HDPE = high-density polyethylene; PPO = polyphenol oxidase.

## 6.7   LIMITATIONS AND CHALLENGES

Although microfluidization has been reported to be a successful and potential technology for the processing of various juices derived from plant products in various aspects, there are still various challenges associated with the expansion and commercialization of this technology that need to be addressed. (1) The juices from different plant products differ in their consistency and viscosity; the passage of highly viscous juices may result in the blockage of the small channels. Also, the intricate configuration of the interaction chamber can result in the deposition of juice residues (Operti et al., 2021). The components of the instrument are so tiny that it hinders the cleaning process, which might cause microbial contamination of the final product. (2) Microfluidizers have many components with precise and complex designs, which add up to the manufacturing cost of the setup, which will ultimately hike the price of the final product. Thus, its high production cost is one of the concerning factors to be used in food businesses (Li et al., 2022). (3) Depending on the concentration and viscosity of the processed juice, the design of the microfluidizer and the interaction chamber's tiny size restrict the production capacity to 1–10 L/min, which makes mass production difficult (Ozturk & Turasan, 2022). (4) The juice sample, when subjected to the microfluidizer beyond certain conditions, may lead to over-processing, which sometimes shows inconvenient results. Moreover, over-processing may lead to the loss of energy and time. The beneficial or detrimental effects of treatment parameters of microfluidizers vary with different juices due to changes in the intrinsic components of the matrix (Singh et al., 2022a). Thus, the treatment conditions (number of passes, pressure value, etc.) need to be optimized for each commodity. (5) There is a high chance

Microfluidization of Juice Derived from Plant Products

that the high-velocity collision of the microstream with an intense shear rate inside the interaction chamber may result in the temperature rise of the juice sample, which may further affect their properties (Ozturk & Turasan, 2022).

## 6.8 CONCLUSIONS AND PERSPECTIVES

Plant-based juices are a healthy and easy alternative to get the benefit of essential vitamins, minerals, fibres, and phytochemicals in this fast-moving world. Juice drinking on a regular basis lowers the risk of obesity, heart disease, cancer, and other chronic diseases. However, juices are highly perishable commodities that degrade their quality when exposed to the environment and cannot be stored for longer duration without processing. Microfluidization is a time-efficient, minimally processed technique that is labelled as a green technology evident to be effective for the processing of plant-based juice. It works on hydrodynamic pressure, generating cavitation, impact, high shear, and high pressure. The microfluidization technology has drawn the attention of researchers in the food sector because of its potential to enhance the quality, safety, and stability of juices. It has shown a remarkable performance at the lab scale; however, there are a few limitations associated with the expansion of this technique that require attention. One of the most concerning challenges is the cleaning problem due to the small and complicated components of the instrument. The effect of various instrumental parameters on different juices differs in their performance. Therefore, before a product is commercialized, researchers must standardize various instrumental conditions, such as the number of passes and pressure value.

## REFERENCES

Abliz, A., Liu, J., & Gao, Y. (2020). Effect of dynamic high pressure microfluidization on physical properties of goji juice, mango juice and carrot puree. In *E3S Web of Conferences* (Vol. 189, p. 02027). EDP Sciences. https://doi.org/10.1051/e3sconf/202018902027

Abliz, A., Liu, J., Mao, L., Yuan, F., & Gao, Y. (2021). Effect of dynamic high pressure microfluidization treatment on physical stability, microstructure and carotenoids release of sea buckthorn juice. *LWT – Food Science and Technology*, *135*. https://doi.org/10.1016/j.lwt.2020.110277

Bai, L., & McClements, D. J. (2016). Development of microfluidization methods for efficient production of concentrated nanoemulsions: Comparison of single-and dual-channel microfluidizers. *Journal of Colloid and Interface Science*, *466*, 206–212. https://doi.org/10.1016/j.jcis.2015.12.039

Butu, M., & Rodino, S. (2019). Fruit and vegetable-based beverages- nutritional properties and health benefits. In *Natural Beverages* (pp. 303–338). Academic Press. https://doi.org/10.1016/B978-0-12-816689-5.00011-0

Dos Santos Aguilar, J. G., Cristianini, M., & Sato, H. H. (2018). Modification of enzymes by use of high-pressure homogenization. *Food Research International*, *109*, 120–125. https://doi.org/10.1016/J.FOODRES.2018.04.011

Ganesan, P., Karthivashan, G., Park, S. Y., Kim, J., & Choi, D. K. (2018). Microfluidization trends in the development of nanodelivery systems and applications in chronic disease treatments. *International Journal of Nanomedicine*, 6109–6121. https://doi.org/10.2147/IJN.S178077

Guo, X., Chen, M., Li, Y., Dai, T., Shuai, X., Chen, J., & Liu, C. (2020). Modification of food macromolecules using dynamic high pressure microfluidization: A review. *Trends in Food Science & Technology*, *100*, 223–234. https://doi.org/10.1016/j.tifs.2020.04.004

Halahlah, A., Piironen, V., Mikkonen, K. S., & Ho, T. M. (2023). Wood hemicelluloses as innovative wall materials for spray-dried microencapsulation of berry juice: Part 1 – Effect of homogenization techniques on their feed solution properties. *Food and Bioprocess Technology*, *16*(4), 909–929. https://doi.org/10.1007/s11947-022-02963-5

Huang, X., Li, C., & Xi, J. (2022). Dynamic high pressure microfluidization-assisted extraction of plant active ingredients: A novel approach. *Critical Reviews in Food Science and Nutrition*, 1–9. https://doi.org/10.1080/10408398.2022.2101427

Jo, Y. J., Chun, J. Y., Kwon, Y. J., Min, S. G., Hong, G. P., & Choi, M. J. (2015). Physical and antimicrobial properties of trans-cinnamaldehyde nanoemulsions in water melon juice. *LWT – Food Science and Technology*, *60*(1), 444–451. https://doi.org/10.1016/J.LWT.2014.09.041

Kavinila, S., Nimbkar, S., Moses, J. A., & Anandharamakrishnan, C. (2023). Emerging applications of microfluidization in the food industry. *Journal of Agriculture and Food Research*, 100537. https://doi.org/10.1016/j.jafr.2023.100537

Kohli, G., Jain, G., Bisht, A., Upadhyay, A., Kumar, A., & Dabir, S. (2019). Effect of non-thermal hurdles in shelf life enhancement of sugarcane juice. *LWT – Food Science and Technology*, *112*. https://doi.org/10.1016/j.lwt.2019.05.131

Koley, T. K., Nishad, J., Kaur, C., Su, Y., Sethi, S., Saha, S., Sen, S., & Bhatt, B. P. (2020). Effect of high-pressure microfluidization on nutritional quality of carrot (Daucus carota L.) juice. *Journal of Food Science and Technology*, *57*(6), 2159–2168. https://doi.org/10.1007/s13197-020-04251-6

Kumar, A., Dhiman, A., Suhag, R., Sehrawat, R., Upadhyay, A., & McClements, D. J. (2022a). Comprehensive review on potential applications of microfluidization in food processing. *Food Science and Biotechnology*, *31*, 17–36. https://doi.org/10.1007/s10068-021-01010-x

Kumar, D., Dass, S. L., Kumar, Y., & Dey, S. (2022b). Pressure-based processing technologies for food. In *Current Developments in Biotechnology and Bioengineering* (pp. 149–182). Elsevier. https://doi.org/10.1016/B978-0-323-91158-0.00015-6

Leyva-Daniel, D. E., Alamilla-Beltrán, L., Villalobos-Castillejos, F., Monroy-Villagrana, A., Jiménez-Guzmán, J., & Welti-Chanes, J. (2020). Microfluidization as a honey processing proposal to improve its functional quality. *Journal of Food Engineering*, *274*, 109831. https://doi.org/10.1016/j.jfoodeng.2019.109831

Li, Y., Deng, L., Dai, T., Li, Y., Chen, J., Liu, W., & Liu, C. (2022). Microfluidization: A promising food processing technology and its challenges in industrial application. *Food Control*, *137*, 108794. https://doi.org/10.1016/j.foodcont.2021.108794

Liu, M., Wang, R., Li, J., Zhang, L., Zhang, J., Zong, W., & Mo, W. (2021). Dynamic high pressure microfluidization (DHPM): Physicochemical properties, nutritional constituents and microorganisms of yam juice. *Czech Journal of Food Sciences*, *39*(3), 217–225. https://doi.org/10.17221/284/2020-CJFS

Liu, W., Liu, J., Liu, C., Zhong, Y., Liu, W., & Wan, J. (2009b). Activation and conformational changes of mushroom polyphenoloxidase by high pressure microfluidization treatment. *Innovative Food Science and Emerging Technologies*, *10*(2), 142–147. https://doi.org/10.1016/j.ifset.2008.11.009

Liu, W. E. I., Jianhua, L. I. U., Mingyong, X. I. E., Chengmei, L. I. U., Weilin, L. I. U., & Wan, J. I. E. (2009a). Characterization and high-pressure microfluidization-induced activation of polyphenoloxidase from Chinese pear (*Pyrus pyrifolia* Nakai). *Journal of Agricultural and Food Chemistry*, *57*(12), 5376–5380. https://doi.org/10.1021/jf9006642

Maherani, B., Khlifi, M. A., Salmieri, S., & Lacroix, M. (2019). Design of biosystems to provide healthy and safe food – Part B: Effect on microbial flora and sensory quality of orange juice. *European Food Research and Technology*, *245*(3), 581–591. https://doi.org/10.1007/s00217-018-03228-2

Microfluidics. (2014). Microfluidizer processor user guide. https://www.alfatest.it/keyportal/uploads/2017-microfluidics-chamber-user-guide.pdf. Accessed on November 10, 2023.

Nayak, S. L., Sethi, S., Saha, S., Dubey, A. K., & Bhowmik, A. (2023). Microfluidization of juice extracted from partially granulated citrus fruits: Effect on physical attributes, functional quality and enzymatic activity. *Food Chemistry Advances*, *2*. https://doi.org/10.1016/j.focha.2023.100331

Oliete, B., Potin, F., Cases, E., & Saurel, R. (2019). Microfluidization as homogenization technique in pea globulin-based emulsions. *Food and Bioprocess Technology*, *12*, 877–882. https://doi.org/10.1007/s11947-019-02265-3

Operti, M. C., Bernhardt, A., Grimm, S., Engel, A., Figdor, C. G., & Tagit, O. (2021). PLGA-based nanomedicines manufacturing: Technologies overview and challenges in industrial scale-up. *International Journal of Pharmaceutics*, *605*, 120807. https://doi.org/10.1016/j.ijpharm.2021.120807

Ozturk, O. K., & Turasan, H. (2021). Applications of microfluidization in emulsion-based systems, nanoparticle formation, and beverages. *Trends in Food Science & Technology*, *116*, 609–625. https://doi.org/10.1016/j.tifs.2021.07.033

Ozturk, O. K., & Turasan, H. (2022). Latest developments in the applications of microfluidization to modify the structure of macromolecules leading to improved physicochemical and functional properties. *Critical Reviews in Food Science and Nutrition*, *62*(16), 4481–4503. https://doi.org/10.1080/10408398.2021.1875981

Pandey, A., Tarafdar, A., Soccol, C. R., Dussap, C. G., Sirohi, R., (2022). *Current Developments in Biotechnology and Bioengineering Advances in Food Engineering*. Elsevier Science.

Petruzzi, L., Campaniello, D., Speranza, B., Corbo, M. R., Sinigaglia, M., & Bevilacqua, A. (2017). Thermal treatments for fruit and vegetable juices and beverages: A literature overview. *Comprehensive Reviews in Food Science and Food Safety*, *16*(4), 668–691. https://doi.org/10.1111/1541-4337.12270

Rychlik, M., Rawson, A., Liu, J., Zheng, J., Yu, W., Cui, J., Zhao, S., Feng, L., & Wang, Y. (2021). Effects of high-pressure homogenization on pectin structure and cloud stability of not-from-concentrate orange juice. *Frontiers in Nutrition*, *1*, 647748. www.frontiersin.org https://doi.org/10.3389/fnut.2021.647748

Sahil, Madhumita M., Prabhakar, P. K., & Kumar, N. (2022). Dynamic high pressure treatments: Current advances on mechanistic-cum-transport phenomena approaches and plant protein functionalization. *Critical Reviews in Food Science and Nutrition*, 1–26. https://doi.org/10.1080/10408398.2022.2125930

Shinwari, K. J., & Rao, P. S. (2021). Enzyme inactivation and its kinetics in a reduced-calorie sapodilla (Manilkara zapota L.) jam processed by thermal-assisted high hydrostatic pressure. *Food and Bioproducts Processing*, *126*, 305–316. https://doi.org/10.1016/J.FBP.2021.01.013

Singh, S. V., Singh, R., Singh, A., Chinchkar, A. V., Kamble, M. G., Dutta, S. J., & Singh, S. B. (2022a). A review on green pressure processing of fruit juices using microfluidization: Quality, safety and preservation. *Applied Food Research*, 100235. https://doi.org/10.1016/j.afres.2022.100235

Singh, S. V., Singh, R., Verma, K., Kamble, M. G., Tarafdar, A., Chinchkar, A. V., Pandey, A. K., Sharma, M., Kumar Gupta, V., Sridhar, K., & Kumar, S. (2022b). Effect of microfluidization on quality characteristics of sapodilla (*Manilkara achras* L.) juice. *Food Research International*, *162*. https://doi.org/10.1016/j.foodres.2022.112089

Suhag, R., Singh, S., Kumar, Y., Prabhakar, P. K., & Meghwal, M. (2023). Microfluidization of ginger rhizome (zingiber officinale roscoe) juice: Impact of pressure and cycles on physicochemical attributes, antioxidant, microbial, and enzymatic activity. *Food and Bioprocess Technology*, 1–14. https://doi.org/10.1007/s11947-023-03179-x

Tarafdar, A., & Kaur, B. P. (2021). Microfluidization-driven changes in some physicochemical characteristics, metal/mineral composition, and sensory attributes of sugarcane juice. *Journal of Food Quality*, *2021*. https://doi.org/10.1155/2021/3326302

Tarafdar, A., & Kaur, B. P. (2022). Storage stability of microfluidized sugarcane juice and associated kinetics. *Journal of Food Processing and Preservation*, *46*(6). https://doi.org/10.1111/jfpp.16561

Tarafdar, A., Kumar, Y., Kaur, B. P., & Badgujar, P. C. (2021). High-pressure microfluidization of sugarcane juice: Effect on total phenols, total flavonoids, antioxidant activity, and microbiological quality. *Journal of Food Processing and Preservation*, *45*(5). https://doi.org/10.1111/jfpp.15428

Tarafdar, A., Nair, S. G., & Pal Kaur, B. (2019). Identification of microfluidization processing conditions for quality retention of sugarcane juice using genetic algorithm. *Food and Bioprocess Technology*, *12*(11), 1874–1886. https://doi.org/10.1007/s11947-019-02345-4

Velázquez-Estrada, R. M., Hernández-Herrero, M. M., Guamis-López, B., & Roig-Sagués, A. X. (2012). Impact of ultra high pressure homogenization on pectin methylesterase activity and microbial characteristics of orange juice: A comparative study against conventional heat pasteurization. *Innovative Food Science & Emerging Technologies*, *13*(January), 100–106. https://doi.org/10.1016/J.IFSET.2011.09.001

Vigneshwaran, G., More, P. R., & Arya, S. S. (2022). Non-thermal hydrodynamic cavitation processing of tomato juice for physicochemical, bioactive, and enzyme stability: Effect of process conditions, kinetics, and shelf-life extension. *Current Research in Food Science*, *5*, 313–324. https://doi.org/10.1016/j.crfs.2022.01.025

Villalobos-Castillejos, F., Granillo-Guerrero, V. G., Leyva-Daniel, D. E., Alamilla-Beltrán, L., Gutiérrez-López, G. F., Monroy-Villagrana, A., & Jafari, S. M. (2018). Fabrication of nanoemulsions by microfluidization. In *Nanoemulsions* (pp. 207–232). Academic Press. https://doi.org/10.1016/B978-0-12-811838-2.00008-4

Wallace, T. C., Bailey, R. L., Blumberg, J. B., Burton-Freeman, B., Chen, C. O., Crowe-White, K. M., … & Wang, D. D. (2020). Fruits, vegetables, and health: A comprehensive narrative, umbrella review of the science and recommendations for enhanced public policy to improve intake. *Critical Reviews in Food Science and Nutrition*, *60*(13), 2174–2211. https://doi.org/10.1080/10408398.2019.1632258

Wang, L., Shen, C., Li, C., & Chen, J. (2020). Physicochemical, functional, and antioxidant properties of dietary fiber from Rosa roxburghii Tratt fruit modified by physical, chemical, and biological enzyme treatments. *Journal of Food Processing and Preservation*, *44*(11), e14858. https://doi.org/10.1111/jfpp.14858

Wang, N., Wang, T., Yu, Y., Xing, K., Qin, L., & Yu, D. (2023). Dynamic high-pressure microfluidization assist in stabilizing hemp seed protein-gum Arabic bilayer emulsions: Rheological properties and oxidation kinetic model. *Industrial Crops and Products*, *203*, 117201. https://doi.org/10.1016/j.indcrop.2023.117201

Wang, X., Wang, S., Wang, W., Ge, Z., Zhang, L., Li, C., Zhang, B., & Zong, W. (2019). Comparison of the effects of dynamic high-pressure microfluidization and conventional homogenization on the quality of peach juice. *Journal of the Science of Food and Agriculture*, *99*(13), 5994–6000. https://doi.org/10.1002/jsfa.9874

Zhang, X., & Shao, X. (2015). Characterisation of polyphenol oxidase and peroxidase and the role in browning of loquat fruit. *Czech Journal of Food Sciences*, *33*(2), 109–117. https://doi.org/10.17221/384/2014-CJFS

# 7 Microfluidization of Milk and Milk Products

*Anit Kumar, Kumar Sandeep, Kanchan Kumari,*
*Prem Prakash, M. A. Aftab, and Rachna Sehrawat*

## 7.1 MILK

Milk is a complete food for humans because it contains all nutritional properties such as fat, protein, carbohydrates, minerals, and vitamins in the required amounts not only for a child but also it is used as a base material for production of different milk products. Generally, milk consists of about 87–88% water and 12–13% solid contents. Dairy products such as yoghurt, dahi, paneer, cheese, buttermilk, shrikhand, and ice creams contain several health-beneficial components, that is, protein, vitamins, and minerals. Various scientists claimed that daily consumption of dairy products can decrease different types of diseases such as osteoporosis, hypertension, colon cancer, and obesity. Calcium and vitamin D are the important nourishing components of milk products (Weaver, 2003).

## 7.2 HOMOGENIZATION

Milk constitutes an emulsion of oil in water. The tendency of milk fat globules to coalesce arises from the pressure differential, where larger globules experience lower pressure compared to smaller ones, prompting the small fat globules to merge with the larger ones. Consequently, this phenomenon results in the undesired separation of milk fat rising to the surface. To counteract this effect, milk undergoes homogenization. August Gaulin in the early 20th century invented the first homogenizer. Nowadays, commercial milk containing high fat is homogenized at two different pressures, that is, 2500 psi at the first stage and 500 psi at the second stage at 60 °C that break down the larger fat globules (0.2–15 microns in diameter) into smaller fat globules (≤2 microns in diameter) due to shear stress, inertial forces, and cavitation and produce fine emulsion of oil in water. Pressure and temperature conditions of homogenization may vary according to the apparatus and type of valve. The initial stage of pressure reduces the diameter of milk fat globules, while the subsequent stage is engineered to disperse the clusters of fat globules formed during the first stage. In accordance with Stoke's law, reducing the size of milk fat globules significantly slows down the cream separation rate, primarily attributable to the density contrast between milk fat and the aqueous phase. This reduction also serves to inhibit coalescence to some degree, resulting in a more stable milk emulsion and an extended shelf life.

High-pressure homogenization (HPH) is also an advanced type of homogenization. Several HPH equipment are nowadays available in the market and some equipment producers are Microfluidics (USA), Gea Niro Soavi (Italy), Bee International (USA), and APV (UK).

## 7.3 MICROFLUIDIZATION OR HIGH-PRESSURE HOMOGENIZATION

Homogenization is a process aimed at standardizing the size of particles or fat globules, while microfluidization represents another method of homogenization. Microfluidization, a form of

142
DOI: 10.1201/9781032632599-7

Microfluidization of Milk and Milk Products

high-pressure homogenization, is renowned for generating fine emulsions. The high-shear fluid processing through a microfluidizer was patented by Cook and Lagace (1985). Initially, microfluidizer was used in cosmetic and pharmaceutical industries to produce very fine emulsions, and Paquin and Giasson (1989) first explained the applications of microfluidization in food processing.

### 7.3.1 PRINCIPLE OF OPERATION

The microfluidizer operates on the principle of forcing fluid into the interaction chamber at exceedingly high pressure and velocity. Within the chamber, the fluid is split into two streams and collides at a 180-degree angle, resulting in rapid pressure drop and impact-induced cavitation, shear, impact, and turbulence effects.

Shear force: This force arises from the interaction between the product streams and the channel walls at high velocity.

Impact force: This force is generated by collisions, occurring when the high-velocity product stream impinges upon itself. After exiting the interaction chamber, a heat exchanger restores the product stream to ambient temperature.

## 7.4 EFFECT OF HIGH-PRESSURE HOMOGENIZATION ON MILK COMPONENTS

### 7.4.1 MILK FAT

The statement discussed the capability of a microfluidizer to create stable emulsions characterized by a consistent particle size distribution, with droplet sizes reaching less than 0.1 μm. This technology plays a crucial role in ensuring that the composition of products remains consistent (Dhiman & Prabhakar, 2021). The effects of microfluidization on milk at different pressures were also discussed in various literatures. When milk is microfluidized at pressures of 35 and 103 MPa, it leads to the formation of fat droplets with a size of approximately 0.1 μm. Specifically, at moderate pressure levels, the formation of fat clusters is prevented, which is attributed to the presence of a high protein load in the milk (Mccrae, 1994). The effects of ultra-high pressure homogenization (UHPH) on milk fat globule size in relation to pressure levels were also discussed. This indicates that UHPH has been demonstrated to decrease the size of milk fat globules as the pressure increases (Hayes & Kelly, 2003). Additionally, it suggests that employing a two-stage high-pressure homogenization process is generally more effective in reducing the average fat globule size and improving rennet coagulation properties compared to a single-stage high-pressure homogenization process at equivalent total pressures, particularly notable at 200 MPa. Bucci et al. (2018) reported the effects of two-pass microfluidization on reducing particle sizes compared to single-pass microfluidization and controls. Specifically, they indicate that two-pass microfluidization resulted in significantly smaller median sizes (with statistical significance denoted by $P < 0.05$), ranging from 0.390 to 0.501 μm. These sizes were nearly 20 times smaller than the control samples. In comparison, previous research demonstrated that single-pass microfluidization at pressures of 100 or 150 MPa led to a median size value of 0.26 μm, representing a reduction of approximately 10-fold compared to the controls, indicating that multiple passes can indeed lead to additional particle size reduction. Preliminary data suggest that three or more passes do not result in further reduction. This suggests the optimal number of passes for achieving the minimum particle size. Furthermore, the results indicate that both particle sizes decrease as microfluidization pressure and inlet temperature increase. However, it's noted that lack of further reduction in particle size with increasing microfluidization pressure may be due to cluster formation (Olson et al., 2004). The passage also highlights the role of homogenization in reducing the diameters of milk fat globules and increasing their uniformity, which is beneficial in preventing cream separation during the storage of homogenized milk. This is attributed to factors such as the adsorption of milk plasma proteins on the surface of fat globules (Hayes et al., 2005). Hardham et al. (2000) reported that microfluidization of ultra-high temperature (UHT) milk had benefits in terms of reducing fat separation compared to conventional homogenization. Specifically, MF was able to maintain the stability of UHT milk for up to 9 months, whereas conventional homogenization only kept it stable for 2–3 months. Additionally, microfluidization resulted in a fat reduction of approximately 25% smaller compared to the fat reduction achieved by conventional homogenization. This highlights the effectiveness of microfluidization in enhancing the stability and quality of UHT milk products.

The milk fat globule is a complex structure crucial for maintaining milk fat as an emulsion. Inside the globule, there's liquid fat, primarily composed of triglycerides, which is surrounded by a membrane known as the milk fat globule membrane. This membrane comprises two layers of phospholipids: an internal monolayer positioned close to the core of lipids and an external bilayer. This intricate arrangement helps stabilize the fat globules within the milk emulsion, contributing to the overall structure and properties of milk. The milk fat globule membrane is composed not only of phospholipids but also of other important components such as polar lipids, cholesterol, proteins, and various minor components (Singh & Gallier, 2017). The milk fat globule membrane is associated with various types of proteins and glycoproteins, which establish different interactions within it (Garcia-Amezquita et al., 2009). The average size of milk fat globules typically ranges from 0.1 to 9 μm in diameter (Garcia-Amezquita et al., 2009).

### 7.4.2 MILK PROTEIN

HPH milk proteins, particularly casein, play a crucial role as natural emulsifying agents. As pressure is applied, milk proteins are distributed across the available surface area. This widespread coverage facilitates the stabilization of emulsions by forming a protective layer around fat droplets, preventing their coalescence and promoting a uniform dispersion in the liquid matrix. The amphiphilic nature of casein molecules allows them to interact with both fat and water phases, thus aiding in the formation and stabilization of emulsions (Serra et al., 2009; Ross et al., 2003). HPH is known to induce significant changes in the structure and composition of milk proteins. Casein micelles, which are the primary protein structures in milk, are broken down into smaller fragments under the influence of HPH. This process leads to an increase in the concentration of calcium phosphate and casein in the serum phase of the milk. Additionally, HPH treatment can cause the denaturation of various milk and whey proteins, with notable effects on proteins such as β-lactoglobulin, α-lactalbumin, and certain immunoglobulins. Denaturation refers to the alteration of a protein's native structure, which can lead to changes in its functional properties. Overall, HPH treatment of milk can result in significant modifications to the protein composition and structure, which may impact various properties of the milk and its products (Trujillo et al., 2002; Sfakianakis et al., 2014).

It is interesting to note that α-lactalbumin exhibits greater resistance to HPH compared to β-lactoglobulin. The structural changes in α-lactalbumin typically begin to occur at around 200 MPa of pressure during HPH. This indicates that α-lactalbumin requires higher pressure levels to undergo significant alterations compared to β-lactoglobulin. During HPH, the intense pressure forces induce structural rearrangements in proteins by enhancing the exposure of their hydrophobic regions (Patel et al., 2006). The relationship between pressure treatment and structural changes in milk proteins is indeed significant. At lower pressures, around 100 MPa, the denaturation of proteins tends to be reversible. This means that the proteins can unfold under pressure but may regain their native structure once the pressure is removed. However, as the pressure increases, particularly above 200 MPa, the denaturation of proteins becomes more irreversible. This suggests that the higher pressure levels induce more significant and permanent alterations to the protein structure, making it difficult for the proteins to regain their original conformation once pressure is relieved (Qi et al., 2015).

Microfluidization of Milk and Milk Products

HPH can indeed induce conformational changes in whey proteins and caseins, which may increase their susceptibility to proteolysis, leading to the release of free amino acids. These free amino acids can serve as a nutrient source for probiotic bacteria, promoting their growth and viability. By breaking down the protein structures into smaller fragments, HPH can enhance the availability of amino acids, which are essential nutrients for the growth and metabolism of probiotic bacteria. This increased availability of nutrients can create a more favorable environment for probiotics, leading to enhanced viability and activity (Patrignani et al., 2007). In their native state, whey proteins exhibit structural stability and limited interaction with caseins, fat globules, or calcium ions. However, following homogenization, whey proteins undergo denaturation, enabling them to bind with fat particles and form complexes with caseins. This homogenization process, when coupled with heat treatment, induces significant transformations in the protein structure. This includes the loss of secondary structures, particularly $\beta$-sheet and $\alpha$-helix formations, as well as a reduction in the exposed surface area of tertiary structures (Qi et al., 2015).

### 7.4.3 MILK VITAMINS AND VOLATILE COMPOUNDS

#### 7.4.3.1 Milk Vitamins

Amador-Espejo et al. (2015) conducted a study to assess the impact of UHPH on the levels of various vitamins in milk. They compared the effects of UHPH with those of traditional heat treatments. The study focused on both water- and fat-soluble vitamins present in milk. The water-soluble vitamins analyzed included nicotinamide, thiamine, pyridoxal, pyridoxamine, folic acid, cyanocobalamin, riboflavin, and total vitamin C. Additionally, the fat-soluble vitamins analyzed were retinol and $\alpha$-tocopherol. By examining the changes in the levels of these vitamins following UHPH treatment and comparing them with those resulting from heat treatments, the study aimed to provide insights into the potential effects of UHPH on the nutritional content of milk, particularly concerning its vitamin composition. Such analyses contribute to understanding the suitability of UHPH as a food processing technique and its implications for the nutritional quality of milk products. The study's findings indicate a notable decrease in the degradation of certain vitamins—nicotinamide, thiamine, riboflavin, and vitamin C—in milk treated with UHPH compared to those subjected to heat treatments. Furthermore, UHPH treatments demonstrated the ability to preserve a significant portion of the vitamin content found in raw milk. Among the various UHPH-treated samples, the one treated at 300 MPa with a temperature of 45 °C exhibited vitamin content most closely resembling that of raw fresh milk. However, it was observed that increasing the temperature during UHPH treatments led to a substantial increase in vitamin degradation, particularly affecting nicotinamide, thiamine, riboflavin, and vitamin C. The reduction in vitamin C levels observed in UHPH-treated milk compared to raw milk could be attributed not only to the pressure increment but also to the elevation in milk temperature as it passed through the valve during processing. This suggests that temperature variations during processing play a critical role in influencing vitamin degradation, alongside pressure changes. This observation is consistent with existing knowledge indicating that pressure changes have minimal effects on small molecules like vitamins. Previous studies (Trujillo et al., 2002; Barba et al., 2012) have also emphasized the limited impact of pressure on such molecules. Overall, these findings emphasize the importance of optimizing UHPH processing conditions, particularly temperature, to minimize vitamin degradation and ensure the nutritional quality of milk products. The potential benefits of employing UHPH technology in milk processing compared to traditional heat treatment methods suggested that the shorter retention time of milk at the valve after homogenization in UHPH processes may contribute to higher retention of vitamin C compared to heat-treated milk. This implies that UHPH could be a promising technology for reducing the destruction of vitamins in milk, thereby producing milk with enhanced nutritional properties.

The effects of high hydrostatic pressure (HHP) technology on low-weight molecules with high energy bonds, such as vitamins, compared to heat treatment methods were also studied by different scientists. Studies referenced suggest that the pressure increase in HHP technology has negligible effects on vitamins. For instance, research by Trujillo et al. (2002), Rastogi et al. (2007), and Barba et al. (2012) indicates minimal impact of HHP on vitamin content. Sierra et al. (2000) found no significant changes in thiamine and vitamin B6 content in milk treated with HHP at 400 MPa for 30 minutes at 23 °C. In contrast, heat treatment is noted to have significant effects on the bonds of milk vitamins, resulting in the rupture of both covalent and non-covalent bonds and altering vitamins into inactive forms. This implies that while HHP technology appears to preserve the integrity of vitamins in milk, heat treatment can lead to the degradation of these essential nutrients, diminishing the nutritional value of the milk.

Sharabi et al. (2018) observed a decrease in the degradation rate of riboflavin by up to 50% with increasing pressures. This decrease is likely attributed to an indirect effect induced by UHPH, which results in increased scattering and absorbance of wavelengths associated with riboflavin's photosensitized oxidation. This mechanism was further supported by experiments conducted in a model system. The minimal decrease in vitamin C concentration was observed immediately after UHPH treatment, and, during the shelf life, it quickly degraded. The homogenization pressure is directly related to the degradation rate of riboflavin and particle size variation with significantly (~50%) higher riboflavin degradation rate for milk treated at 40 MPa compared to the one treated at 250 MPa. In the degradation rate, no statistically significant differences were observed between samples at 200 MPa and number of cycles.

The UHPH treatment itself did not directly affect the stability of riboflavin, as indicated by our data. However, we hypothesize that an indirect effect may occur during storage, stemming from UHPH-induced alterations in particle sizes. These changes could influence the intensity of light reaching the vitamin, thereby facilitating its photodegradation. Moreover, aside from the scattering effects resulting from modifications in the size of fat globules, UHPH-induced alterations to protein structure might also play a role in this phenomenon. The structural changes induced by UHPH treatment, which can lead to protein association and denaturation, may potentially offer protection to riboflavin against photodegradation. While we posit that the reduction in particle size (along with possible UHPH-induced changes in protein structure) primarily contributes to the enhanced stability of riboflavin, the influence of changes in pH cannot be entirely discounted. Previous research has demonstrated that during processing, UHPH treatment, particularly at 300 MPa with inlet temperatures exceeding 45 °C, preserves vitamin C in milk more effectively than traditional pasteurization methods such as heating to 90 °C for 15 seconds or UHT treatment at 138 °C for 4 seconds. However, information regarding changes in vitamin C concentration during milk storage subsequent to UHPH treatment remains scarce. Therefore, we investigated the effects of UHPH processing on the stability of vitamin C in homogenized milk compared to thermal pasteurization (heating to 72 °C for 12 seconds). Our observations revealed a decrease in vitamin C content during storage across all samples. Nonetheless, samples homogenized at pressures of 100 MPa and higher exhibited higher vitamin C content after 30 hours compared to samples treated at lower pressures. The variations in pH observed during shelf life among different pressure treatments might partially elucidate the differences in vitamin C degradation. Although milk typically isn't a significant source of vitamin C in the human diet, our findings suggest a potential advantage for vitamin C preservation in milk during shelf life through UHPH treatment.

### 7.4.3.2  Volatile Compounds of Milk

Milk flavor is complex in nature and influenced by a wide rang of volatile compounds, and according to Nursten (1997), over 400 different volatile compounds have been identified in milk. Among these compounds, 47 are classified as strong contributors to the flavor of processed milk, while 46 are considered moderate contributors. Interestingly, ketones were found to be the most abundant group of volatile compounds in analyzed milk samples. Ketones are a type of organic compound characterized by a carbonyl group bonded to two other carbon atoms. They often have distinct

# Microfluidization of Milk and Milk Products

odors and flavors, which can contribute to the overall sensory profile of milk and dairy products. Understanding the composition of volatile compounds in milk is essential for dairy industry professionals to optimize processing techniques and produce dairy products with desirable flavor characteristics.

The study by Van Hekken et al. (2019) examined the concentrations of 11 selected volatile compounds in various types of milk samples—raw, thermized, pasteurized, and UHT—both before and after undergoing two different processing methods: microfluidization at 170 MPa and common two-stage homogenization at 15 MPa. Their findings indicated that microfluidization did not bring about significant alterations in the levels of volatile compounds compared to the initial milk samples. However, specific changes were observed. Heptanal showed an increase in concentration in thermized and UHT milk samples. Nonanoic acid and acetone exhibited decreased levels in raw, thermized, and pasteurized milks. Octanoic acid levels decreased in thermized and UHT milks. Moreover, the study noted that the highest concentrations of nearly all volatile compounds were found in the milk subjected to two-stage homogenization. This suggests that homogenization had a more pronounced impact on the volatile compound profiles compared to microfluidization. In summary, the study concluded that microfluidization had a minimal effect on the volatile compound profiles of milk. It's worth mentioning that despite microfluidization removing the fat droplet membrane, the process also generates temperatures adequate for inactivating lipases and alkaline phosphatases, as discussed in the study by Bucci et al. (2018).

The study conducted by Amador-Espejo et al. (2017) aimed to assess the volatile profile of whole milk treated by UHPH at 300 MPa with varying inlet temperatures (Ti) of 56 °C, 75 °C, and 85 °C. The researchers compared these results with those of UHT milk. Their findings revealed differences in the volatile compound concentrations between UHT milk and UHPH-treated milk samples. In UHT milk, ketones and dimethyl trisulfide concentrations were higher compared to any other sample. Conversely, the concentration of aldehydes in UHPH-treated milk samples was higher than that detected in heat-treated samples. This suggests that UHPH represents a promising alternative to reduce compounds associated with cooking in heat-treated milk. The study also referenced findings by Contarini and Povolo (2002), who observed an increase in 2-heptanone concentration with increasing treatment temperature in heat-treated milk (pasteurized, UHT, and in-bottle-sterilized milk). Furthermore, hexanal was identified as the aldehyde with the highest concentration among the analyzed samples, with its concentration increasing in relation to temperature and pressure increments.

Pereda et al. (2008) proposed two mechanisms to explain the observed increase in pentanal and hexanal values in UHPH-treated milk with higher pressure and temperature. Damage to milk fat globules as higher pressures in UHPH may lead to more significant damage to milk fat globules, exposing higher amounts of fatty acids to oxidation mechanisms such as light and dissolved oxygen. This increased exposure to oxidation-promoting factors could result in elevated levels of pentanal and hexanal. Temperature increase and then the rise in pressure and temperature during UHPH can cause a subsequent increase in milk temperature after leaving the homogenization valve, albeit for a short duration. This temperature elevation may lead to higher hydroperoxide formation, contributing to the increase in oxidation compounds such as aldehydes. Vazquez-Landaverde et al. (2006) provided further insight into the increase in aldehyde concentration in UHPH-treated milk due to pressure increase. Their study applied HHP to milk at varying pressures and temperatures, evaluating the volatile profile. They found that under HHP treatments, oxygen becomes more soluble, leading to an increase in the amount of hydroperoxides and subsequently higher levels of aldehydes in milk. Although HHP and UHPH operate on different principles, both technologies involve significant pressure increases (even if briefly in the case of UHPH), which can promote higher concentrations of aldehydes in milk. Additionally, in the case of UHPH-treated milk, the disruption of fat droplets by cavitation and subsequent rearrangement during homogenization may further promote the distribution of oxygen, potentially accelerating lipid oxidation and contributing to increased aldehyde levels.

## 7.5 EFFECT OF HIGH-PRESSURE HOMOGENIZATION ON PHYSICOCHEMICAL AND STRUCTURAL PROPERTIES OF MILK

### 7.5.1 RHEOLOGICAL PROPERTIES

The rheological properties of foods hold significant importance for both consumers and manufacturers. They provide crucial insights into designing processing equipment and estimating energy consumption for manufacturers. Moreover, they offer valuable information about sensory attributes, final product quality, and stability for consumers. While the steady flow curve is commonly used to assess fluid rheology, it's essential to note that many phenomena require consideration of elastic behavior as well (Augusto et al., 2012). Understanding these properties is particularly crucial for liquid foods, aiding in comprehending their behavior throughout storage, processing, and consumption phases.

The HPH treatments had a substantial impact on the microstructural and rheological characteristics of the products, with the exception of the temperature sweep. After homogenization, there was a notable decrease in particle size, transitioning from bimodal and poly-disperse distributions to mono-disperse ones. The consistency of the products decreased significantly from 91.82 to 0.51 Pa.s$^n$, while the flow behavior index increased from 0.15 to 0.36 (Gul et al., 2017).

### 7.5.2 EMULSION PROPERTIES

Homogenization, a common technological process in the dairy industry, ensures emulsion stability by reducing the size and uniformity of milk fat globules. This process not only increases fat dispersion but also alters the state of milk proteins, particularly in whole milk, where it affects fat dispersion and the absorption of milk plasma proteins, mainly casein, at the fat–plasma interface (Cano-Riuz & Richter, 1997; Dalgleish & Robson, 1985).

HPH stands out as a leading technology that enhances the aqueous solubility of functional ingredients and emulsion stability without relying on synthetic emulsifiers (Bader et al., 2011; Bouaouina et al., 2006; Grácia-Juliá et al., 2008; Paquin, 1999). Reducing drop size and achieving a uniform distribution in emulsions is crucial for improving solubility, stability, mouthfeel, and reaction intensity. HPH has proven to be highly effective in creating emulsions with submicron-sized particles and boosting protein solubility in dairy products (Bader et al., 2011).

### 7.5.3 STRUCTURAL PROPERTIES

Reconstituted skim milk powder (RSMP) underwent homogenization at varying pressures (ranging from 41 to 186 MPa) and up to six passes. As the pressure increased and more passes were applied, the average diameter of the casein micelles decreased. Various combinations of heat and pressure treatments were used, including control (C), homogenization at 186 MPa (H), heat treatment (T) at 85 °C for 10 minutes, homogenization followed by heat treatment (HT), and heat treatment followed by homogenization (TH). The impact of homogenization on reducing micelle size was observed in the order of H > TH > HT > T = C. The serum phase of homogenized milk (H) exhibited different protein size exclusion profiles compared to other treatments. UHPH led to an increase in non-sedimentable caseins (κ, αs1, and αs2) in the serum. Electron microscopy revealed the formation of smaller particles after treatments H, TH, and HT compared to C. The study concluded that UHPH could alter the structural properties of casein micelles, with the treatment sequence playing a crucial role in determining these modifications (Sandra & Dalgleish, 2005).

High-pressure homogenization at 350–400 MPa induced structural modifications in milk proteins and enhanced the solubility of whey proteins. Gamma irradiation at doses of 3 and 10 kGy cross-linked bands from α-lactalbumin and β-lactoglobulins, respectively. At doses ranging from 32 to 64 kGy, cross-linking among caseinates occurred, and the molecular size of whey proteins increased. Ultraviolet (UV) irradiation at 254 nm contributed to the stability and structural improvement of whey proteins (Syed et al., 2021).

UHP homogenization involves applying specific pressure ranges and is primarily used in the dairy industry to reduce globular size and enhance fine structures. Its temperature composition contributes to its antiseptic properties (Amador-Espejo et al., 2014).

HHP significantly impacts the technological and physicochemical properties of milk. Exposure to HHP leads to the breakdown of casein micelles into smaller particles, increasing the levels of calcium phosphate and casein in milk serum while reducing serum nitrogen and non-casein nitrogen fractions (Law et al., 1998).

The pressure from HHP leads to whey protein denaturation, affecting micelle structure and causing micelle disruption, re-aggregation, and casein micelle removal (Huppertz et al., 2004. HHP can also modify the structure of casein micelles and whey proteins by disaggregating and re-aggregating casein micelles and denaturing whey proteins.

In the food industry, homogenization is a crucial technique for reducing globular size, particularly in milk, to prevent creaming. It enhances stability in food emulsions by altering fat globule structure and increasing surface area for protein absorption, which can be achieved through homogenization processes at pressures around 20 MPa (Kielczewska et al., 2003).

## 7.6 APPLICATION OF MICROFLUIDIZATION IN MILK PROCESSING

Microfluidization has been employed as an alternative method for homogenizing milk and processing various dairy products. Microfluidization emerges as a promising processing technique where liquid foods undergo extreme pressure (150–250 MPa) as they pass through a fixed geometry interaction chamber. This process subjects products to intense shearing forces, resulting in instantaneous heating, microbial inactivation, and conformational changes in macromolecules. Previous studies have indicated that alterations to milk components due to microfluidization and similar high-pressure processes may exhibit fat-mimicking properties (Paquin, 1999), enhance color richness, reduce microbial populations (Adapa et al., 1997), and lower the cholesterol level in cow and buffalo milk (Kumar et al., 2019). Homogenization using the microfluidizer has been successfully applied in processing infant formulas (Pouliot et al., 1990), cheddar cheese (Lemay et al., 1994), mozzarella cheese (Tunick et al., 2000), fat globules, milk protein (Paquin, 1999), and ice cream (Olson et al., 2003). Pressures can reach 10–15 times greater than those of regular homogenizers. Microfluidization technology has been utilized for high-pressure homogenization of milk, demonstrating significantly smaller and more uniform particle sizes than conventional valve homogenization (Ciron et al., 2010). Ciron et al. (2010) demonstrated that microfluidization of heat-treated milk modified the microstructure of non-fat and low-fat yogurts, coinciding with greater particle size reduction in fat globules and proteins compared to the standard process. Reduced globule size facilitates better fat incorporation into the protein network (Aguilera & Kessler, 1988), while increased surface area during homogenization enhances the fat's ability to interact with casein and denatured whey proteins during acidification and subsequent gel formation (Cho et al., 1999; Lucey & Singh, 1997; Sodini et al., 2004).

One significant limitation of high-pressure treatment is the substantial cost associated with the required equipment. However, HPH is proposed as a potentially innovative milk processing technique, offering the combined benefits of homogenization and pasteurization within a single process (Hayes et al., 2005). Numerous researchers have investigated the impact of microfluidization treatment on various dairy products, that is, milk, ghee, and milk powder.

### 7.6.1 GHEE (CLEAR BUTTER FAT) AND BUTTER

Traditionally, ghee is revered as a healthful food with medicinal properties and enjoys widespread consumption across India, Pakistan, and Bangladesh. In India, it is traditionally prepared from buffalo milk and cow milk, commonly referred to as "desi ghee." Ghee is precisely defined as per Food Safety and Standards Regulation, 2011, as "pure clarified fat derived solely from milk or curd or from desi (cooking) butter or from cream to which no coloring matter or preservative has been added." Cow ghee encompasses glycerides, phospholipids, sterols, sterol esters, free fatty acids, fat-soluble vitamins, and carotenoids (Mahakalkar et al., 2014). It notably contains conjugated linoleic acid and vitamin A, purportedly serving as an immunity booster and possessing anti-carcinogenic, anti-atherogenic, anti-adipogenic, and anti-diabetic properties (Ahmad & Saleem, 2020).

Dhiman et al. (2022) investigated the impact of high-pressure microfluidization treatment (50–200 MPa, single pass) on various parameters of cow ghee, including pH, refractive index, free fatty acid values, color measurements, rheology, particle size, structural properties, and thermal properties. Microfluidization notably increased the pH values of cow ghee. The rheogram between apparent viscosity and shear rate demonstrated the shear-thinning nature of microfluidized cow ghee. Furthermore, microfluidization at 150 MPa reduced the cholesterol level in cow ghee by 39.37%, as revealed by GC–MS analysis.

Sert and Mercan (2020) investigated the microbiological, physicochemical, textural characteristics, and oxidative stability of butter produced from raw cream treated with HPH at varying pressures (0, 10, 20, 30, 50, or 70 MPa). HPH treatment increased the lightness and decreased the yellowness of butter due to smaller fat globules. The treatment also led to reduced greenness (attributed to riboflavin content) and decreased spreadability, with higher pressures resulting in smaller fat particle sizes and improved microbial quality, indicating potential for enhancing butter quality through HPH treatment.

### 7.6.2 Milk Powder

Homogenization stands as a widely adopted method in the manufacturing of powdered milk that may influence the product solubility. Sandra and Dalgleish (2005) conducted homogenization of RSMP at varied pressures (41–186 MPa) and with up to six passes, observing a reduction in the average diameter of casein micelles with higher pressures and increased passes. They employed different combinations of heat and pressure treatments: control, homogenization at 186 MPa, heat at 85 °C for 10 minutes, homogenization then heat, and heat then homogenization. Homogenization notably decreased micelle size, altering protein size exclusion profiles in the serum phase of homogenized milk. UHPH increased non-sedimentable caseins in the serum, with electron microscopy revealing the formation of smaller particles following treatments of homogenized milk, heat then homogenization, and homogenization then heat compared to the control sample. UHPH was deemed capable of modifying casein micelle structural properties, with the sequence of treatments playing a significant role in determining the nature of these modifications. High pressure primarily disrupts hydrophobic and ionic interactions believed to stabilize casein micelle structure, potentially leading to micelle disintegration through disruption of these interactions along with cavitation, turbulence, and shear effects in UHPH. Iordache and Jelen (2003) demonstrated that microfluidization of heat-treated aqueous whey protein concentrate solutions substantially increased powder solubility upon reconstitution with heat-induced viscoelastic gels produced from freeze-dried microfluidized samples. Zacaron et al. (2023) increased homogenization pressures and analyzed their effects on the particle size distribution of rehydrated whole milk powder, observing that higher pressures yielded smaller particle sizes upon rehydration without affecting sorption isotherm patterns. Mercan et al. (2018) developed new packaging materials incorporating skim milk powder produced through high-pressure homogenization, finding that increasing skim milk powder addition enhanced film density and brightness, increased hydrophobicity, and influenced mechanical properties, with higher skim milk powder content improving film elasticity and mechanical strength.

## 7.7 CONCLUSIONS AND PERSPECTIVES

This chapter summarizes the application of microfluidization or high-pressure homogenization in milk processing. Microfluidization or high-pressure homogenization has emerged as an innovative technique with diverse applications in the milk industry and successfully improved the molecular size, functional, physicochemical, thermal, nutritional, and viscosity properties of milk. Its unique operational principle and combination of multiple mechanical forces enable a range of benefits, including enzyme and spoilage microorganism inactivation, enhancement of nutritional content, improved bioavailability, and structural modification of biological macromolecules for enhanced techno-functional properties. With its capability to generate nano-sized droplets, it holds great potential in addressing challenges related to emulsion stability. However, despite the exploration of various applications and benefits at a research scale, scaling up the technology remains challenging due to constraints such as smaller dimensions of the interaction chamber and lower operating capacity. Significant efforts are required for the design and development of industry-scale microfluidizers that match the performance of research equipment. Moreover, there is a need to focus on reducing equipment and operational costs for successful implementation of the technology on an industrial scale.

## REFERENCES

Adapa, S., Schmidt, K.A. and Toledo, R. (1997) Functional properties of skim milk processed with continuous high pressure throttling. *Journal of Dairy Science*, 80: 1941–1948.

Aguilera, J.M. and Kessler, H.G. (1988) Physicochemical and rheological properties of milk-fat globules with modified membranes. *Milchwissenschaft*, 43: 411–415.

Ahmad, N. and Saleem, M. (2020) Characterisation of cow and buffalo ghee using fluorescence spectroscopy. *International Journal of Dairy Technology*, 73(1): 191–201. https://doi.org/10.1111/1471-0307.12632

Amador-Espejo, G.G., Gallardo-Chacón, J.J., Juan, B. and Trujillo, A.J. (2017) Effect of ultra-high-pressure homogenization at moderate inlet temperatures on volatile profile of milk. *Journal of Food Process Engineering*, 40(5): e12548. https://doi.org/10.1111/jfpe.12548

Amador-Espejo, G.G., Gallardo-Chacon, J.J., Nykänen, H., Juan, B. and Trujillo, A.J. (2015) Effect of ultra high-pressure homogenization on hydro- and liposoluble milk vitamins. *Food Research International* 77(1): 49–54. https://doi.org/10.1016/j.foodres.2015.04.025

Amador-Espejo, G.G., Suàrez-Berencia, A., Juan, B., Bárcenas, M.E. and Trujillo, A.J. (2014) Effect of moderate inlet temperatures in ultra-high-pressure homogenization treatments on physicochemical and sensory characteristics of milk. *Journal of Dairy Science*, 97: 659–671. https://doi.org/10.3168/jds.2013-7245

Augusto et al. (2012) Effect of high pressure homogenization (HPH) on the rheological properties of tomato juice: Time-dependent and steady-state shear. *Journal of Food Engineering*, 111(4): 570–579.

Bader, S., Bez, J., Eisner, P. (2011) Can protein functionalities be enhanced by high-pressure homogenization? – A study on functional properties of lupin proteins. *Procedia Food Science*, 1: 1359–1366. https://doi.org/10.1016/j.profoo.2011.09.201

Barba, F.J., Esteve, M.J. and Frigola, A. (2012) High pressure treatment effect on physicochemical and nutritional properties of fluid foods during storage: A review. *Comprehensive Reviews in Food Science and Food Safety*, 11:307–322.

Bouaouina, H., Desrumaux, A., Loisel, C. and Legrand, J. (2006) Functional properties of whey proteins as affected by dynamic high pressure treatment. *Internatuional Dairy Journal*, 16: 275–284. https://doi.org/10.1016/j.idairyj.2005.05.004

Bucci, A.J., Hekken, D.L. Van, Tunick, M.H., Renye, J.A. and Tomasula, P.M. (2018) The effects of microfluidization on the physical, microbial, chemical, and coagulation properties of milk. *Journal of Dairy Science*, 101: 6990–7001. https://doi.org/10.3168/jds.2017-13907

Cano-Riuz, M.E. and Richter, R.L. (1997) Effects of homogenization pressure on the milk fat globule membrane protein. *Journal of Dairy Science*, 80: 2732–2739.

Cho, Y.H., Lucey, J.A. and Singh, H. (1999) Rheological properties of acid milk gels as affected by the nature of the fat globule surface material and heat treatment of milk. *International Dairy Journal*, 9: 537–545.

Ciron, C.I.E., Gee, V.L., Kelly, A.L. and Auty, M.A.E. (2010) Comparison of the effects of high-pressure microfluidization and conventional homogenization of milk on particle size, water retention and texture of non-fat and low-fat yoghurts. *International Dairy Journal*, 20: 314–320. https://doi.org/10.1016/j.idairyj.2009.11.018

Contarini, G. and Povolo, M. (2002) Volatile fraction of milk: Comparison between purge and trap and solid phase microextraction techniques. *Journal of Agricultural and Food Chemistry*, 50(25): 7350–7355.

Cook, E.J. and Lagace, A.P. (1985) Apparatus for forming emulsions. United States Patent 4533254.

Dalgleish, D. and Robson, E. (1985) Centrifugal fraction of homogenized milks. *Journal of Dairy Research*, 52: 539–546.

Dhiman, A. and Prabhakar, P.K. (2021) Micronization in food processing: A comprehensive review of mechanistic approach, physicochemical, functional properties and self-stability of micronized food materials. *Journal of Food Engineering*, 292: 110248. https://doi.org/10.1016/j.jfoodeng.2020.110248

Dhiman, A., Suhag, R., Verma, K., Thakur, D., Kumar, A., Upadhyay, A. and Singh, A. (2022) Influence of microfluidization on physico-chemical, rheological, thermal properties and cholesterol level of cow ghee. *LWT – FoodScience and Technology*, 160: 113281. https://doi.org/10.1016/j.lwt.2022.113281

Food Safety and Standards (2011) (Food product standards and food Additives) Regulations (pp. 1–162).

Garcia-Amezquita, L.E., Primo-Mora, A.R., Barbosa-Cánovas, G.V. and Sepulveda, D.R. (2009) Effect of nonthermal technologies on the native size distribution of fat globules in bovine cheese-making milk. *Innovative Food Science and Emerging Technologies*, 10: 491–494.

Grácia-Juliá, A., René, M., Cortés-Muñoz, M., Picart, M., López-Pedemonte, T., Chevalier, D., Dumay, E. (2008) Effect of dynamic high pressure on whey protein aggregation: A comparison with the effect of continuous short-time thermal treatments. *Food Hydrocolloids*, 22: 1014–1032. https://doi.org/10.1016/j.foodhyd.2007.05.017

Gul, O., Saricaoglu, F.T., Mortas, M., Atalar, I. and Yazici, F. (2017) Effect of high pressure homogenization (HPH) on microstructure and rheological properties of hazelnut milk. *Innovative Food Science and Emerging Technologies*, 41: 411–420.

Hardham, J.F., Imison, B.W. and French, H.M. (2000) Effect of homogenisation and microfluidization on the extent of fat separation during storage of UHT milk. *Australian Journal of Dairy Technology*, 55(1): 16–22.

Hayes, M.G., Fox, P.F. and Kelly, A.L. (2005) Potential applications of high pressure homogenisation in processing of liquid milk. *Journal of Dairy Research*, 72: 25–33.

Hayes, M.G. and Kelly, A.L. (2003) High pressure homogenization of raw whole milk (a) effects on fat globule size and other properties. *Journal of Dairy Research*, 70: 297–305.

Huppertz, T., Fox, P.F. and Kelly, A.L. (2004) High pressure treatment of bovine milk: Effects on casein micelles and whey proteins. *Journal of Dairy Research*, 71: 97–106.

Iordache, M. and Jelen, P. (2003) High pressure microfluidization treatment of heat denatured whey proteins for improved functionality. *Innovative Food Science and Emerging Technologies*, 4 (4): 367–376. https://doi.org/10.1016/S1466-8564(03)00061-4

Kielczewska, K., Kruk, A., Czerniewicz, M., Warmińska, M. and Haponiuk, E. (2003) The effect of high-pressure homogenization on changes in milk colloidal and emulsifying systems. *Polish Journal of Food and Nutrition Sciences*, 12: 43–46.

Kumar, A., Badgujar, P.C., Mishra, V., Sehrawat, R., Babar, O.A. and Upadhyay, A. (2019) Effect of microfluidization on cholesterol, thermal properties and in vitro and in vivo protein digestibility of milk. *LWT – Food Science and Technology*, 116: 108523. https://doi.org/10.1016/j.lwt.2019.108523

Law, A.J., Leaver, J., Felipe, X., Ferragut, V., Pla, R. and Guamis, B. (1998) Comparison of the effects of high pressure and thermal treatments on the casein micelles in goat's milk. *Journal of Agricultural and Food Chemistry*, 46: 2523–2530. https://doi.org/10.1021/jf970904c

Lemay, A., Paquin, P. and Lacroix, C (1994) Influence of microfluidization of milk on Cheddar cheese composition, color, texture, and yield. *Journal of Dairy Science*, 77: 2870–2879.

Lucey, J.A. and Singh, H. (1997) Formation and physical properties of acid milk gels: A review. *Food Research International*, 30: 529–542.

Mahakalkar, A., Kashyap, P., Bawankar, R. and Hatwar, B. (2014). The versatility of cow ghee- an Ayurveda Pubicon. *American Journal of Drug Delivery and Therapeutics*, 1(1): 28–34.

Mccrae, C.H. (1994) Homogenization of milk emulsions: Use of microfluidizer. *International Journal of Dairy Technology*, 47: 28–31.

Mercan, E., Sert, D. and Akın, N. (2018) Determination of powder flow properties of skim milk powder produced from high-pressure homogenization treated milk concentrates during storage. *LWT*, 97: 279–288. https://doi.org/10.1016/j.lwt.2018.07.002

Nursten, H.E. (1997) The flavour of milk and dairy products: I. Milk of different kinds, milk powder, butter and cream. *International Journal of Dairy Technology*, 50(2): 48–56.

Olson, D.W., White, C.H. and Richter, R.L. (2004) Effect of pressure and fat content on particle sizes in microfluidized milk. *Journal of Dairy Science*, 87: 3217–3223.

Olson, D.W., White, C.H. and Watson, C.E. (2003) Properties of frozen dairy desserts processed by microfluidization of their mixes. *Journal of Dairy Science*, 86: 1157–1162.

Paquin, P. (1999) Technological properties of high pressure homogenizers: The effect of fat globules, milk proteins and polysaccharides. *International Dairy Journal*, 9: 329–335. https://doi.org/10.1016/S0958-6946(99)00083-7

Paquin, P. and Giasson, J. (1989) Microfluidization as an homogenization process for cream liqueur. *Lait*, 69 (6): 491–498.

Patel, H.A., Singh, H., Anema, S. and Creamer, L.K. (2006) Effects of heat and high hydrostatic pressure treatments on disulfide bonding interchanges among the proteins in skim milk. *Journal of Agricultural and Food Chemistry*, 54: 3409–3420.

Patrignani, F., Iucci, L., Lanciotti, R., Vallicelli, M., Maina, J.M., Holzapfel, W.H. and Guerzoni, M. (2007) Effect of high–pressure homogenization, nonfat milk solids, and milkfat on the technological performance of a functional strain for the production of probiotic fermented milks. *Journal of Dairy Science*, 90: 4513–4523.

Pereda, J., Ferragut, V., Miquel Quevedo, J., Guamis, B. and Trujillo, A.J. (2008) Effects of ultra-high-pressure homogenization treatment on the lipolysis and lipid oxidation of milk during refrigerated storage. *Journal of Agricultural and Food Chemistry*, 56(16): 7125–7130.

Pouliot, Y., Britten, M. and Latreille, B. (1990) Effect of high-pressure homogenization on a sterilized infant formula: Microstructure and age gelation. *Food Structure*, 9: 1–8.

Qi, P.X., Ren, D., Xiao, Y. and Tomasula, P.M. (2015) Effect of homogenization and pasteurization on the structure and stability of whey protein in milk. *Journal of Dairy Science*, 15: 22–30.

Rastogi, N.K., Raghavarao, K.S.M.S., Balasubramaniam, V.M., Niranjan, K. and Knorr, D. (2007) Opportunities and challenges in high pressure processing of foods. *Critical Reviews in Food Science and Nutrition*, 47: 69–112.

Ross, A.I.V., Griffiths, M.W., Mittal, G.S. and Deeth, H.C. (2003) Combining nonthermal technologies to control foodborne microorganisms. *International Journal of Food Microbiology*, 89: 125–138.

Sandra, S.M. and Dalgleish, D.G. (2005) Effects of ultra-high-pressure homogenization and heating on structural properties of casein micelles in reconstituted skim milk powder. *International Dairy Journal*, 15(11):1095–1104.

Serra, M., Trujillo, A.J., Guamis, B. and Ferragut, V. (2009) Flavour profiles and survival of starter cultures of yogurt produced from high-pressure homogenized milk. *International Dairy Journal*, 19: 100–106.

Sert, D. and Mercan, E. (2020) Microbiological, physicochemical, textural characteristics and oxidative stability of butter produced from high-pressure homogenisation treated cream at different pressures. *International Dairy Journal*, 111: 104825. https://doi.org/10.1016/j.idairyj.2020.104825

Sfakianakis, P., Topakas, E. and Tzia, C. (2014) Comparative study on high-intensity ultrasound and pressure milk homogenization: Effect on the kinetics of yogurt fermentation process. *Food and Bioprocess Technology*, 8: 548–557.

Sharabi, S., Okun, Z. and Shpigelman, A. (2018) Changes in the shelf life stability of riboflavin, vitamin C and antioxidant properties of milk after (ultra) high pressure homogenization: Direct and indirect effects. *Innovative Food Science & Emerging Technologies*, 47: 161–169. https://doi.org/10.1016/j.ifset.2018.02.014

Sierra, I., Vidal, V.C. and Lopez, F.R. (2000) Effect of high pressure on the vitamin B1 and B6 content of milk. *Milchwissenschaft*, 55: 365–367.

Singh, H. and Gallier, S. (2017) Nature's complex emulsion: The fat globules of milk. *Food Hydrocolloids*, 68: 81–89.

Sodini, I., Remeuf, F., Haddad, S. and Corrieu, G. (2004) The relative effect of milk base, starter, and process on yogurt texture: a review. *Critical Reviews in Food Science and Nutrition*, 44: 113–137.

Syed, Q.A., Hassan, A., Sharif, S., Ishaq, A. and Saeed, F., Icon, O., Afzaal, M., Hussain, M. and Anjum, F.M. (2021) Structural and functional properties of milk proteins as affected by heating, high pressure, Gamma and ultraviolet irradiation: A review. *International Journal of Food Properties*, 24(1): 871–884.

Trujillo, A.J., Capellas, M,. Saldo, J., Gervilla, R. and Guamis, B. (2002) Applications of high-hydrostatic pressure on milk and dairy products: A review. *Innovative Food Science and Emerging Technologies*, 3: 295–307.

Tunick, M.H., Hekken, D.L.V., Cooke, P.H., Smith, P.W. and Malin, E.L. (2000) Effect of high pressure micro-fluidization on microstructure of mozzarella cheese. *LWT – Food Science and Technology*, 33(8): 538–544. https://doi.org/10.1006/fstl.2000.0716

Van Hekken, D.L., Iandola, S. and Tomasula, P.M. (2019) Short communication: Volatiles in microfluidized raw and heat-treated milk. *Journal of Dairy Science*, 102(10): 8819–8824.

Vazquez-Landaverde, P.A., Torres, J.A. and Qian, M.C. (2006) Effect of high-pressure–moderate-temperature processing on the volatile profile of milk. *Journal of Agricultural and Food Chemistry*, 54(24): 9184–9192.

Weaver, C.M. (2003) Dairy nutrition beyond infancy. *Australian Journal of Dairy Technology*, 58: 58–60.

Zacaron, T.M., Francisquini, J.D., Perrone, Í.T. and Stephani, R. (2023) The effect of homogenisation pressure on the microstructure of milk during evaporation and drying: particle-size distribution, electronic scanning microscopy, water activity and isotherm. *Journal of Dairy Research*, 90(3): 299–305. https://doi.org/10.1017/S0022029923000456

# 8 Microfluidics-Based Food Micro- and Nanodelivery Systems

*Monika Chand, Pratima Raypa, Deepak Joshi, Nitu Rani, and Narashans Alok Sagar*

## 8.1 INTRODUCTION

The science and technology of processing small amounts of fluids through channels that have dimensions ranging from a few to hundreds of micrometers is known as MFs. A wide range of applications in various domains, such as chemical analysis, delivery systems, diagnostics, drug scanning, and microreactors, are similarly made possible by MFs. The capacity to control fluids inside micrometer-scale channels is one of the MFs primary benefits. These microfluidic (MF) devices have demonstrated a great deal of promise to lower reagent consumption, manufacturing costs, and analysis times while also improving portability and device efficiency. Droplet-based MFs allows for the creation of a variety of food-grade emulsion-based delivery systems, including solid lipid microparticles, microgels, microcapsules, multiple emulsions, and gigantic liposomes, because of their laminar regime.

The food business is increasingly concerned about food safety and quality. To produce food items that are both safe and of the highest quality, every stage of food processing is essential (Onyeaka et al., 2022). Food emulsion generation is one of the most effective ways to improve the quality of food systems (Zhang et al., 2020). Novel technologies have long been used to manufacture emulsions in food systems in addition to conventional emulsions made through homogenization, such as nanoemulsions produced through high-pressure microfluidization (Hu & Zhang, 2022). Developing methods to boost the bioavailability of bioactive compounds remains a challenge for the food industry.

One technological approach for facilitating the transportation of various substances in food and into the gastrointestinal system is to use microstructures that can encapsulate them. Applications for MFs in food safety are being dev</oped at an increasing rate, and there are an increasing number of them in the food business. An obvious illustration would be the MF chip created by Li et al. (2019) to detect nitrite in sausage using fluorometric analysis. *Salmonella enteritidis, Escherichia coli, Staphylococcus aureus*, and *Listeria monocytogenes* are four common foodborne pathogens that may be detected by using the ready-to-use chip created by Xing et al. (2023).

## 8.2 MICROFLUIDIC PLATFORMS FOR FOOD MICRO- AND NANODELIVERY SYSTEMS

MF platforms have demonstrated substantial potential for producing food-grade delivery systems, especially when it comes to encapsulating functional substances (Bianchi, De La Torre, & Costa, 2023). These platforms empower meticulous management of such systems' structural design and fabrication, utilizing food-grade materials like carbohydrates and lipids (Feng & Lee, 2019).

DOI: 10.1201/9781032632599-8

Additionally, MF devices have been employed to manipulate fluids on the nanoliter scale, rendering them well-suited for producing lipid nanoparticles and polymer nanoparticles utilized in drug delivery (Garg et al., 2016).

### 8.2.1 EMULSION-BASED

Stable emulsions with regulated droplet sizes are produced using emulsion-based MF technologies. Encapsulating lipophilic bioactives enhances their solubility and bioavailability. Manipulating stable microparticles with specific chemical and physical properties can be formed in homogeneous emulsions in microchannels (Ma et al., 2020). These systems disperse oil droplets comprising lipophilic components in an aqueous phase with a surfactant or an emulsifier to provide protection and target delivery (Costa et al., 2017). The concentration and the type of emulsifiers and the droplet size polydispersity are some of the major factors that affect the kinetic stability of the whole system against destabilization mechanisms (i.e., coalescence, creaming, flocculation, and Ostwald ripening) (Jafari et al., 2008). Recent advancements in the use of emulsion-based MF platforms for food micro- and nanodelivery systems have been significant in food systems (Bianchi, De La Torre, & Costa, 2023), highlighting that droplet-based MFs may be used to shape the properties of delivery systems based on the molecules that are entrapped. This is further supported by Odriozola-Serrano, Oms-Oliu, and MartÃãn-Belloso (2014) and Salvia-Trujillo et al. (2018), who both highlight the advantages of using delivery systems based on nanoemulsions to enhance the stability and functionality of lipophilic component (Kentish et al., 2008). The application of ultrasonics for the preparation of nanoemulsions was mentioned, which can be superior to traditional dispersion methods or at least comparable in droplet size and energy efficiency. These studies collectively underscore the promising role of emulsion-based MF platforms in enhancing the delivery of active ingredients in food products. Emulsion-based MF platforms find applications in the production of salad dressings, beverages, and other formulations where uniform dispersion of ingredients is critical.

### 8.2.2 LIPOSOME-BASED

Liposome-based MF platforms focus on the controlled formation of liposomes—lipid vesicles that can encapsulate hydrophilic and lipophilic compounds. This approach enhances bioactive ingredients' stability and targeted release, making them valuable for applications in functional foods and pharmaceuticals. The use of liposomes in food micro- and nanodelivery systems has been explored in various studies. Srinivasan et al. (2019) and Hood (2014) highlight liposomes' potential in enhancing bioavailability and controlled release of food materials. Hood (2014) further discusses the use of MF techniques for synthesizing and preparing tumor-targeted liposomes, which could be applied to food systems. Pessoa (2018) and Pessoa et al. (2018) present a method for synthesizing monodispersed PEG-stabilized liposomes using MF devices, which could improve the consistency and stability of liposome-based delivery systems. Hiyama et al. (2010) introduces a novel approach for the targeted delivery of cargo-liposomes using biomolecular-motor-based motility and DNA hybridization, which could be adapted for food applications.

### 8.2.3 PROTEIN-BASED

Protein-based MF systems are designed for the encapsulation and controlled release of proteins in food matrices. This technology helps preserve the functionality of sensitive proteins and peptides, making it applicable in the fortification of beverages, dairy products, and other protein-enhanced foods. Protein-based MF platforms have shown promising effect in the development of food micro- and nanodelivery systems (Okagu et al., 2019; Maviah et al., 2020), both demonstrate that the potential of these systems can be used to encapsulate and transport bioactive compounds, mainly hydrophobic and poorly soluble compounds. These food-based protein nanocarriers are promising for delivering hydrophobic molecules due to their varied functional properties, such as water binding capacity, foaming, and surface activity. It also exhibits low toxicity, cost-effectiveness, as well as biodegradability. In order to ensure bioavailability, adhesion, and stability in the intestinal medium, the size of protein delivery systems is critical (Okagu et al., 2019). Reduced dimensions of delivery systems can lead to higher surface area-to-volume ratios and physicochemical interactions, which in turn affect the delivery system's overall properties in the human gastrointestinal tract (Cerqueira et al., 2014). The use of natural polysaccharides in these systems further enhances their biocompatibility and stability, indicating that the interaction of proteins increases the thermal stability of proteins (Wang et al., 2025). These studies highlight how protein-based MF devices have the potential to completely transform food delivery systems.

### 8.2.4 Microfluidic Bioprinting

MF bioprinting involves the precise deposition of food materials in three-dimensional (3D) structures. This technique enables the creation of custom-designed food products with specific compositions and textures. MF bioprinting holds promise for personalized nutrition and the development of novel food formulations. MF bioprinting has the potential to revolutionize food and agriculture industries by enabling the encapsulation of nutrients and bioactive compounds, monitoring of pathogens and toxins, and improving food quality through micro–nano-filtration (Neethirajan et al., 2011; Mu et al., 2022). This technology also offers a promising approach for the development of micro/nanostructures for food delivery systems, such as microfibers/films for packaging and droplet MFs for emulsifications and encapsulations (Mu et al., 2022). MFs, particularly MF spinning technology, offers a promising approach for developing micro/nanostructures in food science and technology, with applications in producing various types of microfibers with different functionalities for food packaging, analysis, and protection of sensitive components. The use of MF devices in biosensors can enhance the detection of chemical contaminants, toxins, and pathogens in food, thereby improving food security (Farré, Kantiani, & Barceló, 2012). The application of bioprocess MFs in food and agriculture industries is an area of ongoing research and development (Marques & Szita, 2017).

## 8.3 CHARACTERISTICS OF MICROFLUIDIC FOOD MICRO- AND NANODELIVERY SYSTEMS

MFs technology in food science has transformed the creation of delivery systems, allowing for precise management of encapsulation, delivery, and morphology of bioactive compounds. This chapter examines the features of MF food micro- and nanodelivery systems, with a specific focus on antioxidant strength, bioavailability, and morphology regulation.

### 8.3.1 Antioxidant Potential

MFs technology enables the preservation of antioxidant potency in food products through innovative solutions. Researchers are improving the stability and bioavailability of antioxidants by enclosing them in micro- and nanocarriers, which helps to maximize their beneficial effects. Comunian et al. (2017) showed the precise encapsulation of delicate compounds like echium oil, which is rich in antioxidants, using MFs. Researchers successfully improved encapsulation efficiency and oxidative stability during storage by combining MFs and ionic gelation. This study underscores the potential of MF carriers in preserving antioxidants. Furthermore, MF technology enables precise delivery of extra health-promoting compounds. Mu et al. (2024) used MFs to trap food additives such as curcumin and β-carotene inside nanoparticles, creating a shielded environment that stops oxidation. This method maintains current antioxidants and enhances food items with extra advantageous elements.

MFs is used in creating functional food products. Yao et al. (2021) effectively incorporated lutein, a crucial antioxidant, into noodles through MF technology, creating a stable product ideal for storage. This novel method improves the nutritional content of common foods, providing consumers with easy access to vital antioxidants. Moreover, MF procedures can be used to produce biopolymer microgels for enclosing water-soluble proteins, as investigated by Wang (2025). The microgels provide controlled release characteristics, enabling precise delivery in the digestive tract and enhancing thermal stability. Furthermore, microgels possess anti-inflammatory properties, offering extra health advantages in addition to antioxidant protection.

MFs technology is transforming food delivery systems by developing carriers that efficiently preserve and deliver antioxidants. MF carriers have the potential to enhance the nutritional value and functionality of food products by protecting existing antioxidants and incorporating additional health-promoting compounds.

### 8.3.2 BIOAVAILABILITY

Maximizing the bioavailability of bioactive compounds is essential for enhancing their therapeutic efficacy. MFs technology provides accurate manipulation of particle size and composition, facilitating effective encapsulation and distribution of bioactive compounds in food products. Hong et al. (2020) were the first to utilize MFs to create dual-loaded liposomes with curcumin and catechin. The uniform spherical vesicles showed high encapsulation efficiency and significantly enhanced antiproliferation effects in colon cancer cells when compared to separate compounds. The novel method emphasizes the capability of MFs to improve the absorption of bioactive compounds for medical purposes.

The size of micelles is essential in influencing the bioavailability of enclosed compounds. Maravajjala et al. (2020) created adjustable-sized polymeric micelles through MF platforms. They found that smaller micelles have higher encapsulation efficiency, controlled drug release, and increased cytotoxicity in two-dimensional and 3D models. This study emphasizes the significance of particle size in enhancing the therapeutic effectiveness of encapsulated compounds.

MFs technology provides unique chances to improve the bioavailability of bioactive compounds by precisely controlling particle size, composition, and morphology. Researchers can enhance the effectiveness of bioactive compounds and transform drug delivery and biomedical applications by enclosing them in micro- and nanocarriers.

### 8.3.3 MORPHOLOGY

The structure of micro- and nanoparticles significantly influences their effectiveness and capabilities in food delivery systems. MFs technology allows for precise manipulation of particle shape, facilitating the creation of carriers with customized characteristics to improve delivery effectiveness. Chen et al. (2017) investigated an innovative approach for shaping and managing the structure of polymer particles through the use of MF chips. Researchers could control the shape of polymer particles by regulating the flow modes of droplets in MF channels. This method enabled the creation of rod-shaped particles in a confined environment and ellipsoid particles in a rolling environment, showcasing the flexibility of MFs in producing non-spherical particle shapes. Kang et al. (2016) enhanced MFs' abilities in shaping particle structure by causing phase separation in droplet templates. This novel method allowed for the creation of non-spherical particles with various equilibrium shapes, determined by thermodynamic conditions. The authors' simple model offered valuable insights into the relationship between dominant factors and particle morphologies, despite the challenging prediction of final particle properties caused by kinetic factors.

Nie et al. (2005) presented a rapid and scalable technique for continuously synthesizing polymer capsules and particles by utilizing MF reactors. Researchers can create uniform droplets and particles with specific shapes in the size range of 20–200 μm by accurately managing the emulsification process in MF channels. Combining numerous continuous MF reactors on one chip allowed for the efficient production of particles with various shapes and structures, showing great promise for applications in food delivery systems.

Ultimately, MFs technology offers exceptional chances for shaping and managing the structure of micro- and nanoparticles in food distribution systems. Researchers can design carriers with specific shapes to improve delivery effectiveness and functionality by manipulating droplet flow patterns, initiating phase separation, and regulating emulsification processes in MF channels. The progress made allows for the creation of advanced food delivery systems that have enhanced stability, bioavailability, and targeted delivery features.

## 8.4 INTEGRATED MFs SYSTEM

### 8.4.1 MACHINE LEARNING WITH MFs

MFs, a multidisciplinary field combining physics, engineering, biology, and chemistry, has seen notable progress with the introduction of machine learning (ML) methods. The merging of these fields has resulted in improved abilities in a wide range of applications, from theranostics to formulation design (Riordon et al., 2019). ML, a subset of artificial intelligence, is beneficial in MFs by utilizing extensive datasets and enhancing intricate processes that are difficult to represent using conventional approaches (McIntyre et al., 2022) (Figure 8.1).

#### 8.4.1.1 Flexibility and Optimization

MFs provides accurate manipulation of variables, making it a suitable platform for integrating ML. ML algorithms are proficient at determining the best parameters for particular applications, which helps to improve experimental design and reduce resource usage (Fukada & Seyama, 2022). The synergy improves automation in MF systems, allowing them to adjust independently using real-time feedback (Ibarz et al., 2021). Reinforcement learning is a fundamental ML method that enables iterative optimization through rewarding favorable results and adjusting system parameters accordingly (Kalashnikov et al., 2021).

#### 8.4.1.2 Advanced Algorithms for Optimization

Model-free episodic controllers (MFECs) and deep Q networks (DQNs) are commonly used algorithms in reinforcement learning for MF applications (Dressler et al., 2018). DQNs, recognized for their exceptional performance in virtual settings, excel at processing extensive datasets and attaining outstanding outcomes (Mnih et al., 2013). Conversely, MFECs provide effective learning using minimal data, making them appropriate for situations with limited resources (Blundell et al., 2016).

#### 8.4.1.3 Applications in Organ-on-a-Chip and Nanoparticle Synthesis

The combination of ML with organ-on-a-chip (OoC) technology has transformed drug screening and disease modeling. ML algorithms examine intricate cellular interactions in small-scale

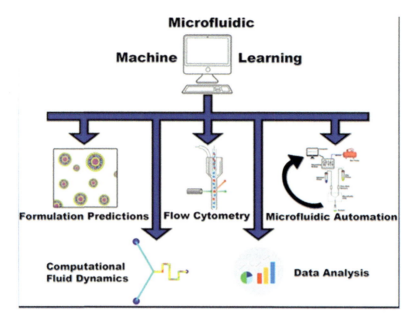

**FIGURE 8.1** Structure and functioning of MFs. (Source: Dedeloudi et al., 2023.)

environments, offering an understanding of disease advancement and reaction to treatment (Nguyen et al., 2018). ML-driven optimization has facilitated the creation of nanoparticles with accurate control over size and morphology, leading to progress in drug delivery and biomedical imaging (Tao et al., 2021).

### 8.4.1.4 Computational Fluid Dynamics (CFD) and ML

Computational fluid dynamics (CFD) is a strong tool used to know and improve fluid behavior in MF systems. CFD software utilizes mathematical models such as the Navier–Stokes equations to simulate fluid flow patterns, forecast the impacts of various parameters, and assist in the design of MF devices.

The Navier–Stokes equations (i) are fundamental in CFD simulations as they govern the conservation of mass, momentum, and energy in fluid motion:

$$\frac{\rho D \vec{v}}{Dt} = -\nabla P + \rho \vec{g} + \mu \nabla^2 \vec{v} \qquad (i)$$

The symbols used are: $\nabla$ is the gradient differential operator, $\rho$ is the fluid density, $P$ is the fluid pressure, $t$ is the time, $\nabla^2$ is the Laplacian operator, $\vec{v}$ is the kinematic viscosity vector, and $g$ is the gravitational force vector.

The equations include different parameters like fluid density, pressure, viscosity, and gravitational force. Yet, their usefulness is usually restricted to flow patterns with low Reynolds numbers, where inertial effects are insignificant.

Researchers have recently investigated combining ML methods with CFD to improve the effectiveness and precision of simulations. An approach involves utilizing ML algorithms, specifically deep learning, to analyze CFD data and enhance simulation parameters. Maulik et al. (2021) have shown that autoencoded neural networks are effective in reducing computational requirements without compromising simulation accuracy. Reduced order models (ROMs) are increasingly used in CFD applications to provide simplified representations of fluid flow phenomena. These models offer an efficient way to predict flow behavior, making them ideal for high-throughput screening and optimization studies. Brunton et al. (2020) investigated the application of clustered reduced order models (CROMs) for unsupervised learning of flow patterns in MF channels to analyze intricate fluid dynamics.

CFD is commonly used in MFs to forecast the creation and control of droplets. Lashkaripour et al. (2021) used CFD simulations in conjunction with neural networks to precisely forecast droplet size and generation rates. Researchers enhanced droplet formation processes and device performance by modifying channel shapes and flow characteristics.

Although there are advantages to combining ML with CFD, obstacles persist, especially concerning data accessibility and model generalization. Further research, like the studies conducted by Vinuesa and Brunton (2022) and Whitehouse and Boschitsch (2021), is necessary to delve deeper into the potential of ML-enhanced CFD in MFs and tackle these obstacles.

The integration of CFD and ML shows significant potential for enhancing the development and improvement of MF systems, allowing researchers to address intricate fluid dynamics challenges more effectively and precisely.

### 8.4.1.5 Data Analysis and Flow Cytometry in MFs

MFs is a prominent technology in diagnostics known for its superior accuracy and quick processing speed. Analyzing data from MFs experiments can be difficult, especially when dealing with extensive datasets. This section delves into the integration of ML with MFs to enhance data analysis and optimize assay methods. Rizkin et al. (2020) conducted a study showing how sensors placed both inside and outside MFs devices can be used to identify chemical reaction pathways. The system could adjust and improve reaction conditions in real-time by combining sensor outputs with ML algorithms. The system's ability to adapt, powered by ML, allowed it to learn and implement the best reaction conditions, resulting in improved efficiency.

MFs enables the incorporation of different sensors and detectors like soft sensors, pressure sensors, and microscopy to improve data gathering and analysis. Li et al. (2022) emphasized the utilization of pressure sensors combined with ML for MF applications. Microscopy is commonly used in flow cytometry (FC) and is essential for data analysis in MF systems.

FC is a diagnostic method that quantifies the physicochemical characteristics of cells and particles. Ahmad et al. (2022) conducted a study that compared ML-based cell characterization in FC. They used convolutional neural networks (CNNs) to sort cells based on their physicochemical properties. The study showed that ML-based methods were more effective than traditional techniques in accurately sorting cells. Yalikun et al. (2020) highlighted the significance of accurate flow rate regulation in MFs for FC applications, which allows for clearer imaging and precise cell sorting. Advanced microscopy methods like time-lapse microscopy and digital holographic microscopy have been combined with MFs–FC–ML research to improve cell analysis and categorization (Xin et al., 2021).

Research has shown the efficacy of ML in analyzing functional connectivity data and enhancing diagnostic results. Rizzuto et al. (2021) successfully classified red blood cell (RBC) shapes in cases of hemolytic anemia using MFs–FC–ML approaches. This showcases the adaptability and capacity of ML-powered facial recognition in identifying different conditions and diseases. The combination of ML with MFs and FC improves data analysis, resulting in more effective and precise diagnostic methods.

## 8.4.2 3D-Printed Microfluidic Fabrication Techniques

### 8.4.2.1 Introduction to 3D Printing in MFs

The cost-effective 3D printing technology has transformed the fabrication of MF devices by providing benefits like quick prototyping and customization. This section delves into different methods and uses of 3D printing in MFs.

### 8.4.2.2 Fabrication of Molds for PDMS-Based MFs

3D printing is a versatile method for creating molds of polydimethylsiloxane MF device (PDMS). This method combines the advantages of both 3D printing and PDMS, such as biocompatibility, oxygen permeability, and ease of production. Gross et al. (2015) and Comina et al. (2013) conducted studies showing the creation of PDMS masters through rapid prototyping methods, followed by conventional PDMS molding. These masters facilitate the creation of MF devices with complex designs at an affordable price.

### 8.4.2.3 Integration of Functional Features

Researchers have investigated incorporating different functional characteristics into 3D-printed MF devices. Comina et al. (2016) created monolithic lab-on-a-chip devices through high-resolution stereolithography (SLA) 3D printing. These devices demonstrated features like passive and active transport, lateral flow, and micro-mixing. Chan et al. (2015) created 3D-printed masters to produce interconnected MF networks in a single step, showcasing the production of operational MF chips with components such as chaotic advective mixers and peristaltic valves.

### 8.4.2.4 Applications in Biomedical Research

3D-printed MF devices are extremely versatile for biomedical purposes. Kamei et al. (2015) utilized 3D-printed soft lithography molds to study concentration gradients of growth factors and their effects on the survival and growth of embryonic stem cells. Gross et al. (2015) detailed the development and production of MF electrical cell lysis devices coated with PDMS or polystyrene to improve cell adhesion.

### 8.4.2.5 Innovations in Mold Fabrication

Novel methods for creating molds have been investigated to improve the design and performance of MF devices. Gelber and Bhargava (2015) created sacrificial molds with isomalt to make channels with circular cross-sections and simplified device structures. Further research is needed to understand how the mold material affects cells after it dissolves.

### 8.4.2.6 Microfluidic Component Fabrication

MF devices require precise fabrication of components to achieve desired functionalities, involving both passive and active elements. This chapter explores the fabrication techniques and applications of MF components, highlighting recent advancements in the field.

### 8.4.2.7 Passive Microfluidic Components

Current research has concentrated on creating passive MF parts through the use of 3D printing technology. Shallan et al. (2014) showed an economical way to create MF parts by utilizing a digital micromirror-based SLA 3D printer. Their research demonstrated the advancement of intricate MF systems, incorporating passive elements like micromixers, gradient generators, and droplet generators. Donvito et al. (2015) utilized 3D printing to create a T-junction droplet generator MF device that produced uniform droplets similar to traditional techniques. Martino et al. (2014) introduced an innovative microcapillary assembly design created through SLA 3D printing, designed for producing double emulsions. Chen et al. (2014) used SLA 3D printing to create millifluidic devices for producing multi-compartment particles with applications in food science, drug delivery, and bioassays.

# 8.4.2.8 Active Microfluidic Components

Functional MF elements, like membrane-operated valves and pumps, are essential for the automation of MF systems. Au et al. (2015) showed how to create MF pumps and valves using SLA 3D printing, allowing for pneumatic control of fluid movement. Rogers et al. (2015) reported the implementation of MF membrane valves using SLA 3D printing, exhibiting consistent performance of up to 800 actuations.

# 8.4.2.9 Modular Microfluidic Components

Utilizing a modular method for assembling MF components provides flexibility and simplifies the customization of devices. Paydar et al. (2014) suggested using 3D printing to create MF interconnects with the ability to use multiple materials, allowing for smooth connections between MF parts. Lee et al. (2014) showed how 3D-printed MF modules can be assembled into lab-on-a-chip devices by enhancing connections with metal pins and rubber O-rings. Soe et al. (2013) proposed the idea of design-by-assembly for lab-on-a-chip devices by utilizing software microfluidic modules (SoftMABs) to streamline device design and production via 3D printing.

The studies demonstrate the various uses and advantages of 3D printing in creating MF components, leading to advanced features and personalized MF platforms.

# 8.4.2.10 Complete Microfluidic Device Fabrication

The fabrication of complete MF devices using 3D printing technology has enabled the integration of complex functionalities into single-step fabricated devices, streamlining fabrication processes and enhancing performance in various applications.

## Biosensing and Detection Applications

In their study, Krejcova et al. (2014) showcased the construction of a paramagnetic particle-based optimized biosensor apparatus designed to identify and separate biological components. The device, which was created through the utilization of 3D printing technology, integrated electrochemical detection devices and enabled a streamlined two-step detection procedure to effectively isolate and detect hemagglutinin.

The fabrication of an SLA-printed MF device featuring a helical microchannel for the purpose of pathogenic bacteria detection was documented by Lee et al. (2015). Innovative design elements comprised the apparatus, such as an additional inlet for sheath flow and a trapezoidal cross-section to prevent particle accumulation and complete particle separation, respectively. Clusters of magnetic nanoparticles functionalized with antibodies were utilized to detect bacteria with exceptional sensitivity and detection limits.

A novel 3D-printed MF platform was devised by Chudobova et al. (2015) with the objective of promptly and precisely identifying methicillin-resistant *Staphylococcus aureus* (MRSA). By employing an FDM 3D printer constructed from ABS material, the platform enabled the systematic cultivation of bacteria, isolation of DNA, detection of genes, and polymerase chain reaction (PCR), thereby demonstrating its capability to detect pathogens with sensitivity and specificity.

## Integrated Sensing and Cell Culture Platforms

A MF unit was suggested by Takenaga et al. (2015) for chemical sensing. The unit was constructed using SLA 3D printing and was equipped with a light addressable potentiometric sensor (LAPS). The platform exhibited improved conditions for cellular growth and facilitated the continuous observation of cellular metabolic activity in real-time, thereby highlighting its potential for integrated sensing applications.

Anderson et al. (2013) described a MF device that was 3D-printed and intended for high-throughput drug transport research. The device's incorporation of numerous parallel channels separated by permeable polycarbonate membranes enabled the investigation of drug transport across membranes and its impact on cultured cells, thereby providing significant contributions to the understanding of drug–cell interactions.

*Electrochemical Sensing Applications*

The fabrication of reusable MF devices integrated with diverse electrodes for electrochemical sensing applications was examined by Erkal et al. (2014). The sensitivity with which these devices detected neurotransmitters, nitric oxide, RBC oxygen tension, and additional analytes demonstrated their adaptability and potential for a wide range of electrochemical applications.

A 3D-printed MF chip was devised by Vlachova et al. (2015) to electrochemically detect hydrolyzed microRNA. Constructed via fused deposition modeling (FDM) 3D printing on ABS material, the chip showcased effective nucleobase detection and featured replaceable glassy carbon electrodes; thus, it provided a potentially fruitful foundation for nucleic acid analysis.

These studies collectively demonstrate the adaptability and effectiveness of 3D printing in the fabrication of complete MF devices, thereby creating novel opportunities for the development of sophisticated applications in biosensing, detection, and cell culture.

### 8.4.3 SENSORS WITH MFS FOR QUALITY CONTROL

In recent years, the integration of MF chip technology with advanced sensing techniques has emerged as a powerful approach for quality control in various industries, particularly food safety and environmental monitoring. This section explores the application of MFs in conjunction with sensors for detecting contaminants, pathogens, heavy metals, and food additives, providing rapid, sensitive, and cost-effective solutions for quality assurance (Table 8.1).

#### 8.4.3.1 Detection of Pesticide Residue

Pesticides play a crucial role in crop protection, but their residue in food presents health risks. Various MF-based techniques have been developed for rapid pesticide residue detection:

- Enzyme Inhibition Method: Hossain et al. (2009) manufacture a paper chip using inkjet printing to detect organophosphorus pesticide residues by measuring acetylcholinesterase activity.
- Spectral Detection Method: Liu et al. (2015) devised a paper-based detector using the luminol–$H_2O_2$ chemiluminescence system to detect dichlorvos in vegetables.
- Chromatographic Detection Method: Wang et al. (2014) introduced a colorimetric microdevice based on plug MF technology for detecting organophosphorus pesticides.

These methods offer advantages such as portability, automation, and low cost, addressing the limitations of traditional detection methods.

#### 8.4.3.2 Detection of Pathogenic Bacteria

Foodborne pathogens pose significant risks to food safety. MF systems have been developed for rapid and sensitive detection of bacteria:

- Paper-Based Microfluidic Chip: Jokerst et al. (2012) created a paper-based microfluidic chip that can detect *Salmonella*, *E. coli* O157:H7, and *L. monocytogenes* in meat products.
- 3D Microfluidic Magnetic Preconcentrator: Park et al. (2017) enhanced the sensitivity of current detection systems by creating a preconcentrator for detecting *E. coli* O157: H7.
- Carbon Nanotube Multilayer Biosensor: Li et al. (2017) integrated loop-mediated isothermal amplification with a biosensor to achieve highly sensitive detection of *E. coli* O157:H7.

## TABLE 8.1
### Role of Microfluidic in Detecting Contaminants and Pathogens

| Target | Detection Method | LOD | Application | Chip Material | References |
|---|---|---|---|---|---|
| OPS | Colorimetric | — | Food, beverage | Paper | Hossain et al. (2009) |
| DDV | CL method | 3.60 ng/mL | Tomato, cucumber, and cabbage | Paper | Wei et al. (2014) |
| DDV | CL method | 0.80 ng/mL | Vegetable | Paper | Liu et al. (2015) |
| OPS | Colorimetric method | 33 nM, 90 nM | — | — | Wang et al. (2014) |
| Chlorpyrifos | EIS | 0.041 mg/L | Vegetable real samples | PDMS | Guo et al. (2014) |
| Trichlorfon | Colorimetric method | 1.65 µg/mL | — | Paper | Yang et al. (2018) |
| OPS | Semi-quantitative method | | — | Paper | Deng et al. (2019) |
| Salmonella, E. coli O157: H7, L. and monocytogenes | Colorimetric method | $10^4$, $10^6$ and $10^8$ CFU/mL | Ready-to-eat meat | Paper | Jokerst et al. (2012) |
| E. coli O157: H7 | Spectrophotometry | 10 CFU/mL | Blood | Plastic | Park et al. (2017) |
| E. coli O157: H7 | LAMP | 1 CFU/mL | — | Glass | Li et al. (2017) |
| Salmonella | LAMP | 50 cells | Pork | — | Sun et al. (2015) |
| Salmonella | Electrochemical | $10^3$ CFU/mL | Food extract and borate buffer | Silicon | Kim et al. (2015) |
| Salmonella | Optical immunoassay | 10.00 CFU/mL | Poultry packaging (fresh) | PDMS | Fronczek et al. (2013) |
| Cu (II) | — | 0.06 mg/L | Tailing water and tap water | Paper | Jayawardane et al. (2013) |
| Cd (II), Pb (II) | Electrochemical | 2.30 µg/L, 2.00 µg/L | Carbonated beverages | Paper | Shi et al. (2012) |
| Ag (I), Hg (II) | Fluorescence | 47.0 nM, 121.0 nM, | — | Paper | Zhang et al. (2015) |
| Cu (II) | Colorimetric | 0.30 ng/mL | Groundwater, drinking water, rice, tomato | Paper | Chaiyo et al. (2015) |
| Pb (II) | Electrochemical | 10 µM | — | PDMS | Fan et al. (2012) |
| Pb (II) | Electrochemical | 95 nM | — | — | Francesca et al. (2018) |
| Hg (II), Cu (II) | Fluorescence | 0.056 µg/L, 0.035 µg/L, | Water samples | Paper | Ji et al. (2017) |
| Tartrazine, sunset yellow | Surface-enhanced Raman spectroscopy | $10^{-4}$ M, $10^{-5}$ M | Orange juice, grape juice | Paper | Zhu et al. (2015) |
| Nitrate, nitrite | Colorimetric | 19.00 µM, 1.00 µM, | Mineral water, tap water, pond water | Paper | Jayawardane et al. (2013) |
| Nitrite, sulfite | Fluorescence | | — | Polymer | Fujii et al. (2004) |
| Nitrite | Colorimetric | 5.60 µM | Sausage, ham, preservative water | Paper | Cardoso et al. (2015) |
| Benzoic acid | — | — | Commercial food samples | Paper | Liu et al. (2018) |
| Glucose | Electrochemical | 0.18 mM | Carbonated drinks | Paper | Lawrence et al. (2014) |

These MF platforms offer rapid detection, high sensitivity, and potential for on-site analysis, addressing the limitations of conventional methods.

### 8.4.3.3 Detection of Heavy Metal

Heavy metals in food can cause various health issues. MF-based sensors have been developed for sensitive detection of heavy metal contaminants:

- Disposable Paper-Based Sensor: Jayawardane et al. (2013) created a paper-based sensor to detect copper (II) in water samples, with sensitivity similar to atomic absorption spectrophotometry.
- Electrochemical Paper-Based MF Chip: Shi et al. (2012) developed a MF chip capable of simultaneously detecting lead (Pb) and cadmium (Cd) in beverages with sensitivity similar to conventional techniques.

These sensors provide rapid, sensitive, and cost-effective detection of heavy metals, addressing the challenges of conventional detection techniques.

### 8.4.3.4 Detection of Food Additives

The addition of food additives requires careful monitoring to ensure safety. MF-based systems have been developed for detecting various additives:

- Surface-Enhanced Raman Spectroscopy (SERS): In their study, Zhu, Li, and Yang (2015) devised a paper-based chip that exhibited heightened sensitivity in the detection of pigments such as tartrazine and sunset yellow in beverages.
- Amperometric Glucose Biosensor: Lawrence et al. (2014) developed a screen-printed paper-based chip to detect glucose in carbonated drinks, yielding precise results similar to high-performance liquid chromatography.

These MF platforms offer rapid, cost-effective, and selective detection of food additives, ensuring food safety and quality control.

## 8.5 APPLICATIONS OF MICROFLUIDIC-BASED DELIVERY SYSTEM

MF-based delivery systems are a new emerging technology within the realm of food science, with the potential to exhibit a diverse array of morphologies that can significantly impact their stability and functional efficacy. Employing these strategies in the food sector offers numerous benefits, such as minimizing dispersion, achieving consistent product uniformity, and streamlining the synthesis process, thereby reducing costs and time required for scale-up (Ushikubo et al., 2015). Adopting this approach enables swift and even transfer of heat and mass, leading to significant enhancement in yield and size distribution while simultaneously mitigating the generation of unwanted by-products. Moreover, the potential for solvent recycling in synthesis adds to the cost efficiency and environmentally friendly technology (Capretto et al., 2013). Numerous research studies explore various applications of MF systems, encompassing areas such as encapsulation, food packaging, drug delivery, and food safety, as delineated in the accompanying table. These applications span a range of functionalities, including emulsions, liposomes, niosomes, microgels, and contributions to enhancing food safety measures, as mentioned in Table 8.2. Conventional methods like colloid mills, homogenization, and high-speed ultrasonic mixers do not offer control over particle size and might lead to elevated dispersion. In addition, heat or shear pressures produced during homogenization can cause some compounds to lose their function or undergo denaturation, as seen in proteins. Conversely, through a gentler procedure, MF devices provide more uniform monodisperse droplets and high encapsulation efficiency. Furthermore, Ushikubo (2015) states that double emulsions are a

# TABLE 8.2
## Application of MFs in the Food System

| Type of Structure | Application | Composition of Emulsions | Reference |
| --- | --- | --- | --- |
| Emulsions O/W | To formulate plant protein stabilized emulsion | Hexadecane/pea protein solution | Hinderink et al. (2020) |
| Emulsions O/W | To quantify emulsion coalescence | $\beta$-Lactoglobulin/whey protein isolate/ oxidized whey protein isolate | Muijlwijk et al. (2017) |
| Emulsions O/W | To formulate encapsulated microparticles using chrysalis oil as a source of $\alpha$-linolenic acid ethyl ester | Tween 20 and silkworm pupa oil isolate as water phase, whereas caffeic acid-modified chitosan in acetate buffer as oil phase | Bai et al. (2019) |
| Emulsions O/W | Transport of astaxanthin used as a bioactive compound | Sodium dodecylsulfate in deionized water as water phase: astaxanthin in soybean oil as oil phase | Ulianova et al. (2020) |
| Emulsion W/O/W | To improve perception of sweetness | NaCl and sucrose in water as water phase while polyglycerol polyricinoleate in sunflower oil as oil phase: sucrose and waxy rice starch gelatinized in water as water phase | Al Nuumani et al. (2020) |
| Emulsion W/O/W | To study double-emulsion stability | Tween 20, MilliQ water with SDS as water phase: sunflower oil with PGPR: and $\beta$-lactoglobulin in imidazole buffer with water as water phase | Hughes et al. (2013) |
| Solid lipid particles | To form fat particles and alter their rheology in order to incorporate them into food or use them for drug delivery | Palm oil and Tween 20 solution | Kim and Vanapalli (2013) |
| Solid lipid particles | To encapsulate and protect the fish oil for incorporation into food | Fish oil and protein aqueous solution (solutions of gelatin, casein, and soy protein) as water phase | Comunian et al. (2018) |
| Microgels | To encapsulate the probiotic *Lactiplantibacillus plantarum* CIDCA 83114 | Pluronic acid-poly(acrylic acid) copolymer as water phase and okara oil | Quintana et al. (2021) |
| Microgels | To encapsulate the bioactive molecules | Protein and dextran: protein, Na-alginate, and PEG: paraffin oil | Sun et al. (2019) |
| Microgels | To encapsulate echium oil by ionic cross-linking microfluidic gelation | Echium oil: sodium alginate and calcium–EDTA complex: corn oil and soy lecithin | Comunian et al. (2017) |
| Microgels | To produce gellan gum microcapsules with hydrophobic compounds | Tween 20: sunflower oil, polyoxyethylene gellan gum, low-acyl gellan gum: calcium acetate and polyglycerol–polyglycerol– polyricinoleate with sunflower oil | Michelon, Leopércio, and Carvalho (2020) |
| Liposomes | To form $\beta$-carotene-filled liposomes (double emulsion) | PVA and dextran: PVA: soy lecithin and $\beta$-carotene in chloroform/hexane; ethyl acetate/hexane or ethyl acetate/pentane | Michelon et al. (2019) |
| Liposomes | To improve the bioavailability of curcumin and catechin by encapsulating in liposomes | Curcumin, dipalmitoylphosphatidylcholine, and cholesterol dissolved in isopropanol: catechin and HEPES buffer | Hong et al. (2020) |
| Niosomes | To encapsulate curcumin | Tween 85 or Span 80 with cholesterol and curcumin in ethanol: distilled water | Obeid et al. (2019) |

*(Continued)*

## TABLE 8.2 (CONTINUED)
## Application of MFs in the Food System

| Type of Structure | Application | Composition of Emulsions | Reference |
|---|---|---|---|
| Hydrogel | To synthesize double emulsion-based microfluidic production of hydrogel microspheres with varied chemical functionalities toward biomolecular conjugation | This method used spontaneous dewetting of the oil phase upon polymerization and transfer into aqueous solution, resulting in poly(ethylene glycol)-based microspheres containing primary amines chitosan, or carboxylates acrylic acid for chemical functionality | Liu et al. (2018) |
| Alginate | To formulate injectable alginate-based formulation for immobilizing enzymes into microfluidic systems | Alginate GDL mixture containing β-glucosidase was injected into the microchannel prior to gelation | Akay et al. (2017) |
| Food packaging | To release quercetin and tocopherol, packing polymers with varying degrees of hydrophobicity are manipulated | Ethylene vinyl alcohol (EVOH), ethylene vinyl acetate (EVA), low-density polyethylene (LDPE), and polypropylene (PP) polymers | Chen et al. (2012) |
| Food packaging | To produce food packaging films that are active and perform significantly using a hydrophilic/hydrophobic approach | Konjac glucomannan/polylactic acid, trans-cinnamic acid | Lin, Ni, and Pang (2019) |
| Food packaging | To formulate strategy to provide green pathway and facile for the construction of promising antibacterial food packaging | Konjac glucomannan/poly(ε-caprolactone) Silver nanoparticles | Lin, Ni, and Pang (2020) |
| Food packaging | To develop biocompatible food packaging films | Ethyl cellulose/polyvinylpyrrolidone | Rao et al. (2022) |
| Food packaging | To incorporate LAE in packaging film constructed with renewable polymer materials for controlling bacterial contamination in foods | Zein matrices: ethyl-$N^\alpha$-dodecanoyl-L-arginate hydrochloride (LAE) | Kashiri et al. (2016) |
| Food packaging | To encapsulate bacteriophages in whey protein films to extend storage and release | Model bacteriophage, T4 bacteriophage, in WPI-based edible protein films | Vonasek, Le, and Nitin (2014) |
| Food safety | To rapidly detect clenbuterol in milk using microfluidic paper-based ELISA | The microfluidic paper-based analytical device (μPAD) was combined with ELISA | Muijlwijk et al. (2017) |
| Food safety | To develop a droplet-based milli fluidic device for rapid phase diagram determination of liquid–liquid phase separated system | Protein–polysaccharide mixture made of β-lactoglobulin (BLG) and Gum Arabic (GA) | Amine et al. (2017) |
| Food safety | To develop a sensor for detecting unwanted ions in food | Microfluidic aptamer-based sensor mercury and lead ions | Huang et al. (2021) |
| Food safety | To develop hydrogel microfibers with core–shell GO-AgNPs/BC (graphene oxide–silver nanoparticles/bacterial cellulose) that have long-lasting and controlled-release antibacterial properties | Graphene oxide/bacterial cellulose Silver nanoparticles | Chen et al. (2016) |

successful encapsulation method for hydrophilic and hydrophobic compounds. One such instance of a double emulsion is W/O/W, in which an oil droplet forms inside a water droplet (Hughes et al., 2013). The double emulsion has low oil content, so it is used to manufacture low-fat goods in the baking and confectionery industries. They also aid in lowering the amount of sodium in food, hiding flavors, and halting oxidation (Muschiolik & Dickinson, 2017).

Emulsion-based structure/emulsion synthesis is used to produce new food structures that provide organoleptic quality or textural properties to different foods. Emulsion-based structures or emulsion synthesis techniques are employed to create novel food structures, imparting distinct textures or enhancing the organoleptic attributes of various food products (McClements, 2012). Elevated dispersion might result from conventional procedures such as homogenization, high-speed or ultrasonic mixers, and colloid mills that do not offer enough control over particle size. In addition, heat or shear pressures produced during homogenization can cause some compounds to lose their function or undergo denaturation, as seen in proteins. Conversely, through a gentler procedure, MF devices provide more uniform monodisperse droplets and high encapsulation efficiency. One such example is microgel, a microscopic 3D network comprising hydrophilic polymers with both solid and liquid properties. Microgels hold significant appeal in the food sector due to their ability to lower energy density and fat content in processed foods, enhancing mouthfeel. They also contribute to reducing cream formation during transportation and storage, stabilizing emulsions, and facilitating the encapsulation of nutraceutical compounds for targeted release (Shewan & Stokes, 2013). Water-in-oil (W/O) emulsion is essential for the synthesis of microgels by MFs, as it would act as a template for producing the microgel (Sagnelli et al., 2017). Other examples of emulsion-based MF systems are liposomes and niosomes. Hydrophobic molecules are difficult to combine into lipophilic compounds and have a low bioavailability in the GIT. In order to overcome this obstacle, liposomes are used to release and transport bioactive substances (Toniazzo & Pinho, 2016). Liposomes, composed of phospholipids organized into bilayers to create spherical vesicles, can be produced using the flow-focusing method featuring cross-shaped topology within MF devices (Ushikubo et al., 2015). Another approach involves the formation of niosomes which are vesicles that form closed bilayers in an aqueous solution by self-assembling non-ionic surfactants (Ge et al., 2019). These structures can trap hydrophilic substances in the aqueous spaces between the bilayers and the lipophilic substance inside the surfactant bilayer (Fuciños et al., 2023). Niosomes hold certain advantages over phospholipid compounds frequently utilized in liposomes that involve higher storage stability and lower production cost (Kopermsub, Mayen, & Warin, 2011; Marianecci et al., 2014). It has been utilized for a number of purposes in the food sector, for example, encapsulation of capsaicin, curcumin, carotenoids, lactic acid, gallic acid, anthocyanins, or for yogurt fortification using iron (Gutiérrez et al., 2016).

Food packaging: The food packaging sector has shown promise in using MF devices to monitor and regulate food safety and quality along the whole food supply chain. Antimicrobial packaging has been developed using a unique MF spinning technology based on trans-cinnamic acid, polylactic acid, and konjac glucomannan (Lin et al., 2019). The packaging property demonstrated a sizable surface area, strong mechanical characteristics, and effective antibacterial action against *Staphylococcus aureus* and *Escherichia coli*. Building of reactors for the production of microparticles.

Another application of MFs is to encapsulate bioactive compounds by synthesizing solid particles. Solid lipid particles are formed by encapsulating bioactive compounds in lipid to create an emulsion. The food sector needs these solid lipid particles to make butter or ice cream. They exhibit fat-soluble vitamin delivery and preservation, as well as ω-3 oil delivery, and are essential for mouthfeel and food flavor (Kim & Vanapalli, 2013).

These applications demonstrate the versatility of micro- and nanodelivery systems in addressing various challenges and enhancing the overall quality, functionality, and appeal of food products in the industry.

## 8.6 CONCLUSIONS AND PERSPECTIVES

MF technology revolutionized the delivery methods in various fields, especially food. The creation of MF delivery systems utilizing platforms, such as protein-based, MF bioprinting, and nanoparticle-based systems, is covered in detail in this chapter. The chapter deals with the quality characteristics of food materials, including antioxidant capacity, bioavailability, structure, stability, and shelf life.

Additionally, the adaptability of MFs in enhancing food delivery systems is highlighted by its integration with technologies like 3D printing, ML, and quality control sensors. Useful applications, including flavor encapsulation, targeted nutrition administration, controlled release of bioactives, and distribution of functional components, show evident advantages of MF-based delivery systems in a variety of sectors.

Future possibilities for delivery systems utilizing MF technology seem promising. It is anticipated that with further research and development, manufacturing methods will continue to evolve, as functions and integration with other technologies will be strengthened and applications will spread across many industries. MF-based delivery systems have enormous potential to solve problems in healthcare, food delivery, and other areas, opening the door to more effective, efficient, and long-lasting solutions.

## REFERENCES

Ahmad, A., Sala, F., Paiè, P., Candeo, A., D'Annunzio, S., Zippo, A., ... & Rousseau, D. (2022). On the robustness of machine learning algorithms toward microfluidic distortions for cell classification via on-chip fluorescence microscopy. *Lab on a Chip*, 22(18), 3453–3463.

Akay, S., et al. (2017). An injectable alginate-based hydrogel for microfluidic applications. *Carbohydrate Polymers*, 161, 228–234. https://doi.org/10.1016/j.carbpol.2017.01.004

Al Nuumani, R., et al. (2020). In-vitro oral digestion of microfluidically produced monodispersed W/O/W food emulsions loaded with concentrated sucrose solution designed to enhance sweetness perception. *Journal of Food Engineering*, 267, 109701. https://doi.org/10.1016/j.jfoodeng.2019.109701

Amine, C., et al. (2017). Droplets-based millifluidic for the rapid determination of biopolymers phase diagrams. *Food Hydrocolloids*, 70, 134–142. https://doi.org/10.1016/j.foodhyd.2017.03.035

Anderson, K.B., et al. (2013). A 3D printed fluidic device that enables integrated features. *Analytical Chemistry*, 85, 5622–5626.

Au, A.K., et al. (2015). 3D-printed microfluidic automation. *Lab Chip*, 15, 1934–1941.

Bai, Z.-Y., et al. (2019). Generation of α-linolenic acid ethyl ester microparticles from silkworm pupae oil by microfluidic droplet. *Waste and Biomass Valorization*, 10(12), 3781–3791. https://doi.org/10.1007/s12649-018-00572-y

Bianchi, J.R.D.O., De La Torre, L.G., & Costa, A.L.R. (2023). Droplet-based MFs as a platform to design food-grade delivery systems based on the entrapped compound type. *Foods*, 12(18), 3385. https://doi.org/10.3390/foods12183385

Blundell, C., Uria, B., Pritzel, A., Li, Y., Ruderman, A., Leibo, J.Z., ... & Hassabis, D. (2016). Model-free episodic control. arXiv preprint arXiv:1606.04460.

Brunton, S.L., Noack, B.R., & Koumoutsakos, P. (2020). Machine learning for fluid mechanics. *Annual Review of Fluid Mechanics*, 52, 477–508.

Capretto, L., et al. (2013). Microfluidic and lab-on-a-chip preparation routes for organic nanoparticles and vesicular systems for nanomedicine applications. *Advanced Drug Delivery Reviews*, 65(11–12), 1496–1532. https://doi.org/10.1016/j.addr.2013.08.002

Cardoso, T.M.G., Garcia, P.T., & Coltro, W.K.T. (2015). Colorimetric determination of nitrite in clinical, food and environmental samples using microfluidic devices stamped in paper platforms. *Analytical Methods*, 7, 7311–7317.

Cerqueira, M.A., et al. (2014). Design of Bio-nanosystems for oral delivery of functional compounds. *Food Engineering Reviews*, 6(1–2), 1–19. https://doi.org/10.1007/s12393-013-9074-3

Chaiyo, S., Siangproh, W., Apilux, A., & Chailapakul, O. (2015). Highly selective and sensitive paper-based colorimetric sensor using thiosulfate catalytic etching of silver nanoplates for trace determination of copper ions. *Analytica Chimica Acta*, 866, 75–83.

Chan, H.N., et al. (2015). Direct, one-step molding of 3D-printed structures for convenient fabrication of truly 3D PDMS microfluidic chips. *MFs Nanofluidics*, 19, 9–18.

Chen, C., et al. (2016). Rapid fabrication of composite hydrogel microfibers for weavable and sustainable antibacterial applications. *ACS Sustainable Chemistry & Engineering*, 4(12), 6534–6542. https://doi.org/10.1021/acssuschemeng.6b01351

Chen, Q.L., Liu, Z., & Shum, H.C. (2014). Three-dimensional printing-based electro-millifluidic devices for fabricating multi-compartment particles. *BioMFs*, 8, 064112.

Chen, R., Chen, X., Jin, X., & Zhu, X. (2017). Morphology design and control of polymer particles by regulating the droplet flowing mode in microfluidic chips. *Polymer Chemistry* 8(19), 2953–2958.

Chen, X., et al. (2012). Release kinetics of tocopherol and quercetin from binary antioxidant controlled-release packaging films. *Journal of Agricultural and Food Chemistry*, 60(13), 3492–3497. https://doi.org/10.1021/jf2045813

Chudobova, D., et al. (2015). 3D-printed chip for detection of methicillin-resistant Staphylococcus aureus labeled with gold nanoparticles. *Electrophoresis*, 36, 457–466.

Comina, G., Suska, A., & Filippini, D. (2013). PDMS lab-on-a-chip fabrication using 3D printed templates. *Lab Chip*, 14, 424–430.

Comina, G., Suska, A., & Filippini, D. (2016). Towards autonomous lab-on-a-chip devices for cell phone biosensing. *Biosensors and Bioelectronics*, 77, 1153–1167.

Comunian, T.A., Ravanfar, R., De Castro, I.A., Dando, R., Favaro-Trindade, C.S., & Abbaspourrad, A. (2017). Improving oxidative stability of echium oil emulsions fabricated by MFs: Effect of ionic gelation and phenolic compounds. *Food Chemistry*, 233, 125–134. https://doi.org/10.1016/j.foodchem.2017.04.085

Comunian, T.A., et al. (2018). Water-in-oil-in-water emulsion obtained by glass microfluidic device for protection and heat-triggered release of natural pigments. *Food Research International*, 106, 945–951. https://doi.org/10.1016/j.foodres.2018.02.008

Costa, A.L.R., et al. (2017). Emulsifier functionality and process engineering: Progress and challenges. *Food Hydrocolloids*, 68, 69–80. https://doi.org/10.1016/j.foodhyd.2016.10.012

Dedeloudi, A., Weaver, E., & Lamprou, D.A. (2023). Machine learning in additive manufacturing & MFs for smarter and safer drug delivery systems. *International Journal of Pharmaceutics*, 636, 122818. https://doi.org/10.1016/j.ijpharm.2023.122818

Deng, S., Yang, T., Zhang, W., Ren, C., Zhang, J., Zhang, Y., & Cui, T., Yue, W. (2019). Rapid detection of trichlorfon residues by a microfluidic paper-based phosphorus-detection chip (μPPC). *New Journal of Chemistry*, 43, 7194–7197.

Donvito, L., et al. (2015). Experimental validation of a simple, low-cost, T-junction droplet generator fabricated through 3D printing. *Journal of Micromechanics and Microengineering*, 25, 035013.

Dressler, O.J., Howes, P.D., Choo, J., & deMello, A.J. (2018). Reinforcement learning for dynamic microfluidic control. *ACS Omega*, 3(8), 10084–10091.

Erkal, J.L., et al. (2014). 3D printed microfluidic devices with integrated. *Lab on a Chip*, 14(12), 2023–2032.

Fan, C., He, S., Liu, G., Wang, L., & Song, S. (2012). A portable and power-free microfluidic device for rapid and sensitive lead ($Pb^{2+}$) detection. *Sensors*, 12, 9467–9475.

Farré, M., Kantiani, L., & Barceló, D. (2012). Microfluidic devices, in *Chemical Analysis of Food: Techniques and Applications*. Elsevier, pp. 177–217. https://doi.org/10.1016/B978-0-12-384862-8.00007-8

Feng, Y., & Lee, Y. (2019). Microfluidic assembly of food-grade delivery systems: Toward functional delivery structure design. *Trends in Food Science & Technology*, 86, 465–478. https://doi.org/10.1016/j.tifs.2019.02.054

Francesca, S.M., Ivan, S., Donatella, P., Daniel, F., & Rositsa, Y. (2018). Epitaxial graphene sensors combined with 3D printed microfluidic chip for heavy metals detection. *Proceedings*, 2, 982.

Fronczek, C.F., You, D.J., & Yoon, J.-Y. (2013). Single-pipetting microfluidic assay device for rapid detection of Salmonella from poultry package. *Biosensors and Bioelectronics*, 40, 342–349.

Fuciños, C., et al. (2023). MFs potential for developing food-grade microstructures through emulsification processes and their application. *Food Research International*, 172, 113086. https://doi.org/10.1016/j.foodres.2023.113086

Fujii, S.I., Tokuyama, T., Abo, M., & Okubo, A. (2004). Fluorometric determination of sulfite and nitrite in aqueous samples using a novel detection unit of a microfluidic device. *Analytical Sciences*, 20, 209–212.

Fukada, K., & Seyama, M. (2022). Microfluidic devices controlled by machine learning with failure experiments. *Analytical Chemistry*, 94(19), 7060–7065.

Garg, S., et al. (2016). MFs: A transformational tool for nanomedicine development and production. *Journal of Drug Targeting*, 24(9), 821–835. https://doi.org/10.1080/1061186X.2016.1198354

Ge, X., Han, Q. L., Zhong, M., & Zhang, X. M. (2019). Distributed Krein space-based attack detection over sensor networks under deception attacks. *Automatica*, 109, 108557.

Gelber, M.K., & Bhargava, R. (2015). Monolithic multilayer MFs via sacrificial molding of 3D-printed isomalt. *Lab on a Chip*, 15, 1736–1741.

Gross, B.C., et al. (2015). Polymer coatings in 3D printed fluidic device channels for improved cellular adherence prior to electrical lysis. *Analytical Chemistry*, 87, 6335–6341.

Guo, Y.M., et al. (2014). A PDMS microfluidic impedance immunosensor for sensitive detection of pesticide residues in vegetable real samples. *International Journal of Electrochemical Science*, 10, 4155–4164.

Gutiérrez, G., et al. (2016). Iron-entrapped niosomes and their potential application for yogurt fortification. *LWT*, 74, 550–556. https://doi.org/10.1016/j.lwt.2016.08.025

Hiyama, S., Moritani, Y., Gojo, R., Takeuchi, S., & Sutoh, K. (2010). Biomolecular-motor-based autonomous delivery of lipid vesicles as nano-or microscale reactors on a chip. *Lab on a Chip*, 10(20), 2741–2748.

Hinderink, E.B.A., et al. (2020). Microfluidic investigation of the coalescence susceptibility of pea protein-stabilised emulsions: Effect of protein oxidation level. *Food Hydrocolloids*, 102, 105610. https://doi.org/10.1016/j.foodhyd.2019.105610

Hong, S.C., et al. (2020). Microfluidic assembly of liposomes dual-loaded with catechin and curcumin for enhancing bioavailability. *Colloids and Surfaces A: Physicochemical and Engineering Aspects*, 594, 124670. https://doi.org/10.1016/j.colsurfa.2020.124670

Hood, R. R. (2014). *Pharmacy-on-a-chip: Microfluidic synthesis and preparation of tumor-targeted liposomes* (Doctoral dissertation, University of Maryland, College Park).

Hossain, S.M.Z., et al. (2009). Reagentless bidirectional lateral flow bioactive paper sensors for detection of pesticides in beverage and food samples. *Analytical Chemistry*, 81, 9055–9064.

Hu, C., & Zhang, W. (2022). Micro/nano emulsion delivery systems: Effects of potato protein/chitosan complex on the stability, oxidizability, digestibility and β-carotene release characteristics of the emulsion. *Innovative Food Science and Emerging Technologies*, 77, 102980.

Hughes, E., et al. (2013). Microfluidic preparation and self-diffusion PFG-NMR analysis of monodisperse water-in-oil-in-water double emulsions. *Journal of Colloid and Interface Science*, 389(1), 147–156. https://doi.org/10.1016/j.jcis.2012.07.073

Ibarz, J., Tan, J., Finn, C., Kalakrishnan, M., Pastor, P., & Levine, S. (2021). How to train your robot with deep reinforcement learning: Lessons we have learned. *The International Journal of Robotics Research*, 40(4–5), 698–721.

Jafari, S.M., et al. (2008). Re-coalescence of emulsion droplets during high-energy emulsification. *Food Hydrocolloids*, 22(7), 1191–1202. https://doi.org/10.1016/j.foodhyd.2007.09.006

Jayawardane, B.M., Coo, L.D., Cattrall, R.W., & Kolev, S.D. (2013). The use of a polymer inclusion membrane in a paper-based sensor for the selective determination of Cu (II). *Analytica Chimica Acta*, 803, 106–112.

Ji, Q., Li, B., Wang, X., Zhong, Z., & Chen, L. (2017). Three-dimensional paper-based microfluidic chip device for multiplexed fluorescence detection of $Cu^{2+}$ and $Hg^{2+}$ ions based on ion imprinting technology. *Sensors and Actuators B: Chemical*, 251, 224–233.

Jokerst, J.C., Adkins, J.A., Bisha, B., Mentele, M.M., & Henry, C.S. (2012). Development of a paper-based analytical device for colorimetric detection of select foodborne pathogens. *Analytical Chemistry*, 84, 2900–2907.

Kalashnikov, D., Varley, J., Chebotar, Y., Swanson, B., Jonschkowski, R., Finn, C., … & Hausman, K. (2021). Mt-opt: Continuous multi-task robotic reinforcement learning at scale. arXiv preprint arXiv:2104.08212.

Kamei, K.I., et al. (2015). 3D printing of soft lithography mold for rapid production of polydimethylsiloxane-based microfluidic devices for cell stimulation with concentration gradients. *Biomedical Microdevices*, 17(1), 1–8.

Kang, Z., Kong, T., Lei, L., Zhu, P., Tian, X., & Wang, L. (2016). Engineering particle morphology with microfluidic droplets. *Journal of Micromechanics and Microengineering*, 26(7), 075011.

Kashiri, M., et al. (2016). Novel antimicrobial zein film for controlled release of lauroyl arginate (LAE). *Food Hydrocolloids*, 61, 547–554. https://doi.org/10.1016/j.foodhyd.2016.06.012

Kentish, S., et al. (2008). The use of ultrasonics for nanoemulsion preparation. *Innovative Food Science & Emerging Technologies*, 9(2), 170–175. https://doi.org/10.1016/j.ifset.2007.07.005

Kim, G., Moon, J.-H., Moh, C.-Y., & Lim, J.-G. (2015). A microfluidic nano-biosensor for the detection of pathogenic Salmonella. *Biosensors and Bioelectronics*, 67, 243–247.

Kim, J., & Vanapalli, S.A. (2013). Microfluidic production of spherical and nonspherical fat particles by thermal quenching of crystallizable oils. *Langmuir*, 29(39), 12307–12316. https://doi.org/10.1021/la401338m

Kopermsub, P., Mayen, V., & Warin, C. (2011). Potential use of niosomes for encapsulation of nisin and EDTA and their antibacterial activity enhancement. *Food Research International*, 44(2), 605–612. https://doi.org/10.1016/j.foodres.2010.12.011

Krejcova, L., et al. (2014). 3D printed chip for electrochemical detection of influenza virus labeled with CdS quantum dots. *Biosensors and Bioelectronics*, 54, 421–427.

Lashkaripour, A., Rodriguez, C., Mehdipour, N., Mardian, R., McIntyre, D., Ortiz, L., … & Densmore, D. (2021). Machine learning enables design automation of microfluidic flow-focusing droplet generation. *Nature Communications*, 12(1), 25.

Lawrence, C.S.K., Tan, S.N., & Floresca, C.Z. (2014). A "green" cellulose paper based glucose amperometric biosensor. *Sensors and Actuators B: Chemical*, 193, 536–541.

Lee, K.G., et al. (2014). 3D printed modules for integrated microfluidic devices. *RSC Advances*, 4, 32876–32880.

Lee, W., et al. (2015). 3D-printed microfluidic device for the detection of pathogenic bacteria using size-based separation in helical channel with trapezoid cross-section. *Scientific Reports*, 5, 7717.

Li, T., Zhu, F., Guo, W., Gu, H., Zhao, J., Yan, M., & Liu, S. (2017). Selective capture and rapid identification of E. coli O157: H7 by carbon nanotube multilayer biosensors and microfluidic chip-based LAMP. *RSC Advances*, 7(48), 30446–30452.

Li, Y., Wu, G., Song, G., Lu, S. H., Wang, Z., Sun, H., … & Wang, X. (2022). Soft, pressure-tolerant, flexible electronic sensors for sensing under harsh environments. *ACS Sensors*, 7(8), 2400–2409.

Li, W., Shi, Y., Hu, X., Li, Z., Huang, X., Holmes, M., Gong, Y, Shi, J and Zou, X. (2019). Visual detection of nitrite in sausage based on a ratiometric fluorescent system. *Food Control*, 106, 106704.

Lin, W., Ni, Y., & Pang, J. (2019). Microfluidic spinning of poly (methyl methacrylate)/konjac glucomannan active food packaging films based on hydrophilic/hydrophobic strategy. *Carbohydrate Polymers*, 222, 114986. https://doi.org/10.1016/j.carbpol.2019.114986

Lin, W., Ni, Y., & Pang, J. (2020). Size effect-inspired fabrication of konjac glucomannan/polycaprolactone fiber films for antibacterial food packaging. *International Journal of Biological Macromolecules*, 149, 853–860. https://doi.org/10.1016/j.ijbiomac.2020.01.242

Liu, W., Kou, J., Xing, H., & Li, B. (2015). Based chromatographic chemiluminescence chip for the detection of dichlorvos in vegetables. *Biosensors and Bioelectronics*, 52, 76–81.

Liu, E.Y., et al. (2018). High-throughput double emulsion-based microfluidic production of hydrogel microspheres with tunable chemical functionalities toward biomolecular conjugation. *Lab on a Chip*, 18(2), 323–334. https://doi.org/10.1039/C7LC01088E

Ma, Z., et al. (2020). Comparative study of oil-in-water emulsions encapsulating fucoxanthin formulated by microchannel emulsification and high-pressure homogenization. *Food Hydrocolloids*, 108, 105977. https://doi.org/10.1016/j.foodhyd.2020.105977

Maravajjala, K. S., Swetha, K. L., Sharma, S., Padhye, T., & Roy, A. (2020). Development of a size-tunable paclitaxel micelle using a microfluidic-based system and evaluation of its in-vitro efficacy and intracellular delivery. *Journal of Drug Delivery Science and Technology*, 60, 102041.

Marianecci, C., et al. (2014). Niosomes from 80s to present: The state of the art. *Advances in Colloid and Interface Science*, 205, 187–206. https://doi.org/10.1016/j.cis.2013.11.018

Martino, C., Berger, S., Wootton, R. C., & deMello, A. J. (2014). A 3D-printed microcapillary assembly for facile double emulsion generation. *Lab on a Chip*, 14(21), 4178–4182.

Marques, M.P., & Szita, N. (2017). Bioprocess MFs: Applying microfluidic devices for bioprocessing. *Current Opinion in Chemical Engineering*, 18, 61–68. https://doi.org/10.1016/j.coche.2017.09.004

Maulik, R., Lusch, B., Balaprakash, P. (2021) Reduced-order modeling of advection-dominated systems with recurrent neural networks and convolutional autoencoders. *Physics of fluids* 33(3):037106.

Maviah, M.B.J., et al. (2020). Food protein-based nanodelivery systems for hydrophobic and poorly soluble compounds. *AAPS PharmSciTech*, 21(3), 101. https://doi.org/10.1208/s12249-020-01641-z

McClements, D.J. (2012). Crystals and crystallization in oil-in-water emulsions: Implications for emulsion-based delivery systems. *Advances in Colloid and Interface Science*, 174, 1–30. https://doi.org/10.1016/j.cis.2012.03.002

McIntyre, D., Lashkaripour, A., Fordyce, P., & Densmore, D. (2022). Machine learning for microfluidic design and control. *Lab on a Chip*, 22(16), 2925–2937.

Michelon, M., Leopércio, B.C., & Carvalho, M.S. (2020). Microfluidic production of aqueous suspensions of gellan-based microcapsules containing hydrophobic compounds. *Chemical Engineering Science*, 211, 115314. https://doi.org/10.1016/j.ces.2019.115314

Michelon, M., et al. (2019). Single-step microfluidic production of W/O/W double emulsions as templates for β-carotene-loaded giant liposomes formation. *Chemical Engineering Journal*, 366, 27–32. https://doi.org/10.1016/j.cej.2019.02.021

Mnih, V., et al. (2013). Playing atari with deep reinforcement learning. arXiv preprint arXiv:1312.5602.

Mu, R., et al. (2022). Recent trends of MFs in food science and technology: Fabrications and applications. *Foods*, 11(22), 3727. https://doi.org/10.3390/foods11223727

Mu, X., et al. (2024). Microfluidic formulation of food additives-loaded nanoparticles for antioxidation. *Colloids and Surfaces B: Biointerfaces*, 234, 113739.

Muijlwijk, K., et al. (2017). Coalescence of protein-stabilised emulsions studied with MFs. *Food Hydrocolloids*, 70, 96–104. https://doi.org/10.1016/j.foodhyd.2017.03.031

Muschiolik, G., & Dickinson, E. (2017). Double emulsions relevant to food systems: Preparation, stability, and applications. *Comprehensive Reviews in Food Science and Food Safety*, 16(3), 532–555. https://doi.org/10.1111/1541-4337.12261

Neethirajan, S., et al. (2011). MFs for food, agriculture and biosystems industries. *Lab on a Chip*, 11(9), 1574. https://doi.org/10.1039/c0lc00230e

Nguyen, M., De Ninno, A., Mencattini, A., Mermet-Meillon, F., Fornabaio, G., Evans, S.S., & Parrini, M.C. (2018). Dissecting effects of anti-cancer drugs and cancer-associated fibroblasts by on-chip reconstitution of immunocompetent tumor microenvironments. *Cell Reports*, 25(13), 3884–3893.

Nie, Z., Xu, S., Seo, M., Lewis, P.C., & Kumacheva, E. (2005). Polymer particles with various shapes and morphologies produced in continuous microfluidic reactors. *Journal of the American Chemical Society*, 127(22), 8058–8063. https://doi.org/10.1021/ja042494w

Obeid, M.A., et al. (2019). Microfluidic manufacturing of different niosomes nanoparticles for curcumin encapsulation: Physical characteristics, encapsulation efficacy, and drug release. *Beilstein Journal of Nanotechnology*, 10, 1826–1832. https://doi.org/10.3762/bjnano.10.177

Odriozola-Serrano, I., Oms-Oliu, G., & MartÃán-Belloso, O. (2014). Nanoemulsion-based delivery systems to improve functionality of lipophilic components. *Frontiers in Nutrition*, 1. https://doi.org/10.3389/fnut.2014.00024

Okagu, O.D., et al. (2019) Protein-based nanodelivery systems for food applications, in *Encyclopedia of Food Chemistry*. Elsevier, pp. 719–726. https://doi.org/10.1016/B978-0-08-100596-5.21864-7

Onyeaka, H., Passaretti, P., Miri, T., & Al-Sharify, Z.T. (2022). The safety of nanomaterials in food production and packaging. *Current Research in Food Science*, 5, 763–774.

Park, C., Lee, J., Kim, Y., Kim, J., Lee, J., & Park, S. (2017). 3D-printed microfluidic magnetic preconcentrator for the detection of bacterial pathogen using an ATP luminometer and antibody-conjugated magnetic nanoparticles. *Journal of Microbiological Methods*, 132, 128–133.

Paydar, O., et al. (2014). Characterization of 3D-printed microfluidic chip interconnects with integrated O-rings. *Sensors and Actuators A*, 205, 199–203.

Pessoa, A.C.S.N., et al. (2018). Tailoring the synthesis of monodisperse PEG-stabilized liposomes via microfluidic devices, in *Blucher Chemical Engineering Proceedings. XXII Congresso Brasileiro de Engenharia Química*, São Paulo, SP: Editora Blucher, pp. 4795–4798. https://doi.org/10.5151/cobeq2018-CO.181

Quintana, G., et al. (2021). Microencapsulation of Lactobacillus plantarum in W/O emulsions of okara oil and block-copolymers of poly(acrylic acid) and pluronic using microfluidic devices. *Food Research International*, 140, 110053. https://doi.org/10.1016/j.foodres.2020.110053

Rao, J., et al. (2022). Facile microfluidic fabrication and characterization of ethyl cellulose/PVP films with neatly arranged fibers. *Carbohydrate Polymers*, 292, 119702. https://doi.org/10.1016/j.carbpol.2022.119702

Riordon, J., Sovilj, D., Sanner, S., Sinton, D., & Young, E.W. (2019). Deep learning with MFs for biotechnology. *Trends in Biotechnology*, 37(3), 310–324.

Rizkin, B.A., Shkolnik, A.S., Ferraro, N.J., & Hartman, R.L. (2020). Combining automated microfluidic experimentation with machine learning for efficient polymerization design. *Nature Machine Intelligence*, 2(4), 200–209.

Rizzuto, V., Mencattini, A., Álvarez-González, B., Di Giuseppe, D., Martinelli, E., Beneitez-Pastor, D., ... & Samitier, J. (2021). Combining MFs with machine learning algorithms for RBC classification in rare hereditary hemolytic anemia. *Scientific Reports*, 11(1), 13553.

Rogers, C.I., et al. (2015). 3D printed microfluidic devices with integrated valves. *BioMFs*, 9, 016501.

Sagnelli, D., et al. (2017). Cross-linked amylose bio-plastic: A transgenic-based compostable plastic alternative. *International Journal of Molecular Sciences*, 18(10), 2075. https://doi.org/10.3390/ijms18102075

Salvia-Trujillo, L., et al. (2018). Emulsion-based nanostructures for the delivery of active ingredients in foods. *Frontiers in Sustainable Food Systems*, 2, 79. https://doi.org/10.3389/fsufs.2018.00079

Shallan, A.I., et al. (2014). Cost-effective three-dimensional printing of visibly transparent microchips within minutes. *Analytical Chemistry*, 86, 3124–3130.

Shewan, H.M., & Stokes, J.R. (2013). Review of techniques to manufacture micro-hydrogel particles for the food industry and their applications. *Journal of Food Engineering*, 119(4), 781–792. https://doi.org/10.1016/j.jfoodeng.2013.06.046

Shi, J., Bi, L., Zheng, H., Tang, F., Wang, W., & Xing, H. (2012). Electrochemical detection of Pb and Cd in paper-based microfluidic devices. *Journal of the Brazilian Chemical Society*, 23, 1124–1130.

Soe, A.K., Fielding, M., & Nahavandi, S. (2013). Lab-on-a-chip turns soft: Computer-aided, software-enabled MFs design. In *2013 IEEE/ACM International Conference on Advances in Social Networks Analysis and Mining (ASONAM)*. IEEE.

Srinivasan, V., et al. (2019). Liposomes for nanodelivery systems in food products, in R.N. Pudake, N. Chauhan, and C. Kole (eds) *Nanoscience for Sustainable Agriculture*. Cham: Springer International Publishing, pp. 627–638. https://doi.org/10.1007/978-3-319-97852-9_24

Sun, H., et al. (2019). Monodisperse alginate microcapsules with spatially confined bioactive molecules via microfluid-generated W/W/O emulsions. *ACS Applied Materials & Interfaces*, 11(40), 37313–37321. https://doi.org/10.1021/acsami.9b12479

Sun, Y., Quyen, T.L., Hung, T.Q., Chin, W.H., Wolff, A., & Bang, D.D. (2015). A lab-on-a-chip system with integrated sample preparation and loop-mediated isothermal amplification for rapid and quantitative detection of Salmonella spp. in food samples. *Lab on a Chip*, 15(9), 1898–1904.

Takenaga, S., Kaji, H., Nishizawa, M., Takahashi, K., Tokeshi, M., Baba, Y., & Aoyagi, Y. (2015). Fabrication of biocompatible lab-on-chip devices for biomedical applications by means of a 3D-printing process. *Physica Status Solidi A*, 212(6), 1347–1352.

Tao, H., Wu, T., Kheiri, S., Aldeghi, M., Aspuru-Guzik, A., & Kumacheva, E. (2021). Self-driving platform for metal nanoparticle synthesis: Combining MFs and machine learning. *Advanced Functional Materials*, 31(51), 2106725.

Toniazzo, T., & Pinho, S.C. (2016). Lyophilized liposomes for food applications: Fundamentals, processes, and potential applications, in J.M. Lakkis (ed.) *Encapsulation and Controlled Release Technologies in Food Systems*. 1st edn. Wiley, pp. 78–96. https://doi.org/10.1002/9781118946893.ch4

Ulianova, Yu.V., et al. (2020). Production of O/W emulsions containing astaxanthin by microfluidic devices. *Nanotechnologies in Russia*, 15(1), 63–68. https://doi.org/10.1134/S1995078020010103

Ushikubo, F.Y., et al. (2015). Designing food structure using MFs. *Food Engineering Reviews*, 7(4), 393–416. https://doi.org/10.1007/s12393-014-9100-0

Vinuesa, R., & Brunton, S.L. (2022). Enhancing computational fluid dynamics with machine learning. *Nature Computational Science*, 2(6), 358–366.

Vlachova, J., Horky, M., Kalcher, K., & Vytřas, K. (2015). A 3D microfluidic chip for electrochemical detection of hydrolysed nucleic bases by a modified glassy carbon electrode. *Sensors*, 15, 2438–2452.

Vonasek, E., Le, P., & Nitin, N. (2014). Encapsulation of bacteriophages in whey protein films for extended storage and release. *Food Hydrocolloids*, 37, 7–13. https://doi.org/10.1016/j.foodhyd.2013.09.017

Wang, J., Suzuki, H., & Satake, T. (2014). Coulometric microdevice for organophosphate pesticide detection. *Sensors and Actuators B: Chemical*, 204, 297–301.

Wang, M., et al. (2025). Effects of microgels fabricated by microfluidic on the stability, antioxidant, and immunoenhancing activities of aquatic protein. *Journal of Future Foods*, 5(1), 57–67. https://doi.org/10.1016/j.jfutfo.2024.01.005

Wei, L., Juan, K., Huizhong, X., & Baoxin, L. (2014). Paper-based chromatographic chemiluminescence chip for the detection of dichlorvos in vegetables. *Biosensors and Bioelectronics*, 52, 76–81.

Whitehouse, G.R., & Boschitsch, A.H. (2021). Investigation of grid-based vorticity-velocity large eddy simulation off-body solvers for application to overset CFD. *Computers & Fluids*, 225, 104978.

Xin, L., Xiao, W., Che, L., Liu, J., Miccio, L., Bianco, V., … & Pan, F. (2021). Label-free assessment of the drug resistance of epithelial ovarian cancer cells in a microfluidic holographic flow cytometer boosted through machine learning. *ACS Omega*, 6(46), 31046–31057.

Xing, G., Li, N., Lin, H., Shang, Y., Pu, Q., & Lin, J.M. (2023). Microfluidic biosensor for one-step detection of multiplex foodborne bacteria ssDNA simultaneously by smartphone. *Talanta*, 253, 123980.

Yalikun, Y., Ota, N., Guo, B., Tang, T., Zhou, Y., Lei, C., … & Tanaka, Y. (2020). Effects of flow-induced microfluidic chip wall deformation on imaging flow cytometry. *Cytometry Part A*, 97(9), 909–920.

Yang, N., Wang, P., Xue, C.Y., Sun, J., Mao, H.P., & Oppong, P.K. (2018). A portable detection method for organophosphorus and carbamates pesticide residues based on multilayer paper chip. *Journal of Food Process Engineering*, 41, e12867.

Yao, Y., Lin, J.J., Chee, X.Y., Liu, M.H., Khan, S.A., & Kim, J.E. (2021). Encapsulation of lutein via microfluidic technology: Evaluation of stability and in vitro bioaccessibility. *Foods*, 10(11), 2646.

Zhang, R., Belwal, T. Li, L., Lin, X., Xu, Y., & Luo, Z. (2020). Recent advances in polysaccharides stabilized emulsions for encapsulation and delivery of bioactive food ingredients: A review. *Carbohydrate Polymers*, 242, 116388.

Zhang, Y., Zuo, P., & Ye, B.C. (2015). A low-cost and simple paper-based microfluidic device for simultaneous multiplex determination of different types of chemical contaminants in food. *Biosensors and Bioelectronics*, 68, 14–19.

Zhu, Y., Li, Z., & Yang, L. (2015). Designing of the functional paper-based surface-enhanced Raman spectroscopy substrates for colorants detection. *Materials Research Bulletin*, 63, 199–204.

# 9 Microfluidization of Cereals-Based Products

*Jithender Bhukya, R. Nisha, Sophia Chanu Warepam, Harsh Dadhaneeya, Raj Singh, and C. Nickhil*

## 9.1 INTRODUCTION

Cereal-based products have been staple foods in human diets for centuries, and their processing methods have evolved over time to meet the demands of growing populations and changing consumer preferences (Láng et al., 2013). Traditional processing methods for cereals often involve a series of steps such as cleaning, milling, stone grinding, fermentation, and baking. The primary objective is to transform raw cereals into edible and palatable products while preserving their nutritional value (Thielecke et al., 2021). Stone grinding involved the use of millstones to crush grains, preserving their nutritional integrity while imparting a distinct texture (Boukid, 2021). Milling, on the other hand, harnessed the power of mechanical processes to refine cereals into finer particles, unlocking possibilities for diverse culinary applications (Yishak, 2014). Fermentation, on the other hand, is employed to enhance flavor, texture, and nutritional content by introducing microorganisms that interact with the cereal components (Verni et al., 2019). Although these traditional methods have been effective, they may have limitations in terms of efficiency, precision, and the ability to retain certain nutritional elements.

In recent years, the food processing industry has witnessed the emergence of innovative techniques to address the shortcomings of traditional methods (Hassoun et al., 2023). One such advancement is MF, a modern processing technique that offers a more controlled and efficient approach to the production of cereal-based products. MF involves the use of microfluidics, a field at the intersection of physics, chemistry, and engineering that deals with the behavior and manipulation of fluids at the microscale (Chiozzi et al., 2022; Mu et al., 2022). In cereal processing, MF leverages microfluidic principles to precisely control the flow, mixing, and interaction of ingredients at a microscopic level. This technique allows for improved homogeneity in product composition, enhanced texture, and better retention of nutritional elements (Mert, 2020). The fundamental principles of MF involve the manipulation of fluid dynamics, surface tension, and interfacial phenomena at the microscale to achieve desired processing outcomes (Bhukya et al., 2021; Bhukya et al., 2023; Maurya et al., 2023). By providing a more tailored and efficient processing platform, MF represents a promising avenue for the advancement of cereal-based product manufacturing in the modern era.

Microfluidization (MF)represents a significant leap forward in food processing technology, offering numerous advantages over conventional methods. One key advantage is the precise control it provides over particle size and distribution (Mert, 2020). Conventional methods often struggle to achieve uniformity in particle size, leading to variations in product texture (Mishra et al., 2023). Also compared to conventional methods like grinding and blending, microfluidization minimizes the risk of agglomeration and ensures consistent quality, ultimately enhancing the sensory attributes

DOI: 10.1201/9781032632599-9

of food products (Soukoulis & Bohn, 2018; Nisha et al., 2023). Furthermore, MF enhances homogenization, ensuring a more uniform distribution of ingredients. This is particularly crucial for products requiring consistent quality, where the even distribution of components significantly impacts the final product's stability and appearance (Kumar et al., 2022). Additionally, MF operates under controlled high-pressure conditions, which not only contributes to the reduction of particle size but also enables the incorporation of bioactive compounds. This capability is crucial in the context of the growing consumer demand for functional foods, as MF allows for the precise incorporation of vitamins, minerals, and antioxidants, enhancing the nutritional profile of cereals-based products in a targeted and efficient manner (Ozturk & Turasan, 2021; Boukid et al., 2021; Leszczyńska et al., 2023).

In the realm of cereals-based products, MF offers a spectrum of potential benefits that positively influence product quality. Improved texture is a notable advantage, as MF allows for precise control over particle size, resulting in cereals with a smoother and more appealing mouthfeel (McClements et al., 2019). This is particularly crucial for breakfast cereals, where texture plays a significant role in consumer acceptance. The enhanced homogenization achieved through MF contributes to a more consistent nutritional profile in cereals-based products (Boukid et al., 2021). Nutrients are uniformly distributed, minimizing the risk of nutrient degradation during processing. This is especially important for fortified cereals, where the even distribution of added vitamins and minerals is critical for meeting nutritional standards. Furthermore, MF operation at lower temperatures helps in preserving the heat-sensitive vitamins and enzymes present in cereals, contributing to an improved overall nutritional profile (Mishra et al., 2023; Li et al., 2022). The controlled processing conditions of MF also play a role in extending the shelf life of cereals-based products, reducing the risk of spoilage and maintaining product freshness (Skendi et al., 2020).

## 9.2 MICROFLUIDIZATION TECHNIQUES FOR CEREAL FOODS

MF techniques play a crucial role in processing cereal foods, enabling precise control over particle size and distribution. Several microfluidization techniques are employed in the cereal industry.

### 9.2.1 HIGH-PRESSURE HOMOGENIZATION

High-pressure homogenization (HPH) is a widely utilized MF technique that involves subjecting a fluid to elevated pressures to achieve desired particle size reduction and homogenization (Malik et al., 2023). In the cereal foods, HPH is applied to disrupt cell walls, release intracellular components, and create emulsions or dispersions with improved stability. This technique is particularly effective in reducing particle size, thereby enhancing the sensory attributes and mouthfeel of cereal-based products like sauces, dressings, and beverages (Goldstein & Reifen, 2022; Chandran et al., 2023).

### 9.2.2 SHEAR-INDUCED MICROFLUIDIZATION

Shear-induced MF is another technique that capitalizes on mechanical forces to manipulate the structure and properties of cereals. It involves subjecting the material to intense shear forces, inducing deformation and alignment of particles. This method is especially beneficial for producing gels and improving the texture of cereal products. Shear-induced MF has been employed in the development of cereal bars, snacks, and bakery products to achieve a desirable balance between crispiness and chewiness (Shahbazi et al., 2018; Dekkers et al., 2016).

### 9.2.3 Microfluidic Encapsulation

Microfluidic encapsulation, on the other hand, focuses on the controlled encapsulation of bioactive compounds within micro-sized droplets or particles. In cereal foods, this technique is employed for the encapsulation of vitamins, flavors, and functional ingredients. By encapsulating these components, their stability during processing and storage is enhanced, and their controlled release into the final product can be achieved. Microfluidic encapsulation is particularly valuable in fortifying cereals with essential nutrients without compromising the product's sensory attributes (Siddiqui et al., 2023; Li et al., 2023; Katouzian & Jafari, 2016).

### 9.2.4 Ultrasonication

Ultrasonication is a MF technique employed in cereal food processing, utilizing high-frequency sound waves to induce cavitation and microstreaming. This gentle yet effective method contributes to particle size reduction and improved homogenization, enhancing the overall texture and stability of cereal-based products (Welti-Chanes et al., 2017; Leong et al., 2017).

### 9.2.5 Jet Cutting

Jet cutting, another MF approach, involves the use of high-velocity jets to disintegrate and homogenize cereal components. This technique excels in achieving precise and uniform particle size reduction, contributing to improved product quality and consistency. Jet cutting is particularly valuable in cereal processing where maintaining uniformity is crucial for the desired textural properties in products such as cereals, snacks, and baked goods (Bala et al., 2023; Cano-Sarmiento et al., 2015; Lukinac & Jukić, 2022).

### 9.2.6 Nano Emulsification

Nano emulsification is a microfluidization technique that focuses on creating stable nano-sized emulsions. In cereal foods, this method is employed to encapsulate and deliver bioactive compounds, flavors, or nutrients with enhanced stability and bioavailability. Nano emulsification contributes to the development of functional and fortified cereal products, meeting consumer demands for both convenience and health benefits (Modarres-Gheisari et al., 2019; Rousta et al., 2021).

## 9.3 CEREAL-BASED PRODUCT CHARACTERISTICS

### 9.3.1 Overview of Cereal Grains and Their Components

Cereals pertaining to the Poaceae or Gramineae family, and the primary varieties include barley, millet, maize, oats, rice, rye, sorghum, and wheat (Sopade, 2017). Cereals are cultivated significantly due to their economic value as a commodity and their role in supplying food and energy globally, surpassing all other types of crops. Therefore, cereal grains may be referred to as staple crops (Luithui et al., 2019). Products made from cereal have essential components of the human diet for a year, supplying numerous flavors, textures, and nutritional profiles to complement the human diet. Understanding the unique characteristics that delineate these goods is essential for the advancement of products, the delight of consumers, and general fulfillment in the market (Zamaratskaia et al., 2021).

Wheat is a trendy grain available in different varieties like hard red wheat, soft red wheat, and durum wheat. Its high gluten content is crucial in bread, pasta, and pastries because it imparts elasticity and structure to these baked items. Wheat is a vibrant source of complex carbohydrates, B vitamins, and minerals such as iron and magnesium (Olakanmi et al., 2022). Corn, or maize, is a responsive cereal grain utilized in various forms, mainly cornmeal, popcorn, and corn syrup (Mandal et al., 2023). Corn is delightful and easily digested by humans, monogastric, and ruminant animals. It is considered one of the most efficient sources of metabolizable energy compared to other grains (Loy & Lundy, 2019). Corn is a crucial component of various diets, and its by-products are extensively employed in the food sector to manufacture multiple goods (Aktaş & Akın, 2020). Rye is the sole crop after wheat that has a substantial quantity of gluten proteins. As a result, it is extensively utilized by food manufacturers worldwide to produce items such as bread, biscuits, rusk, flakes, and beer. Rye bread is famous among those who want a darker and more compact substitute for conventional wheat bread (Kaur et al., 2021). Oats are known for their various advantages, especially in enhancing cardiovascular health. Oats are used for making oatmeal biscuits, granola bars, and morning breakfasts (Leszczyńska et al., 2023). Barley is a resilient cereal grain with a firm texture and a somewhat nutty taste. It's frequently utilized for making soups, stews, and salads. Barley is high in beta-glucans, a soluble fiber that helps lower cholesterol levels (Collar, 2014).

Cereal grains comprise three significant components: bran, endosperm, and germ. Each component contributes a distinct nutrient set (Bader Ul Ain et al., 2020). The germ or embryo, which is also the innermost layer, encompasses the necessary genetic material for the growth and development of the plant. The germ is formed through the union of male and female gametes (Evers & Millar, 2002). Vitamins, good fats, minerals, and phytochemicals are abundant in the embryo. Another component, starchy endosperm, is a dense solid mass positioned at the grain's core. It is the most significant morphological component in all crops and the most valuable element (Ratnavathi, 2019). This layer acts as a reservoir for energy in the growth and expansion process. The outermost layer, known as the bran, contains significant dietary fiber, antioxidants, and B vitamins, and it additionally protects the grain. Advancements in science and genetic resources have greatly accelerated studies on the development and growth of cell walls in cereal grains. When the cellular structure of cell walls is investigated during grain development, it is clear that there are considerable differences in the walls of various grain regions and even between adjoining cells (Burton & Fincher, 2014). The implementation of novel MF techniques can be utilized in the cereal food processing sector to address the existing research gaps and achieve a consistently balanced nutritional product (Figure 9.1).

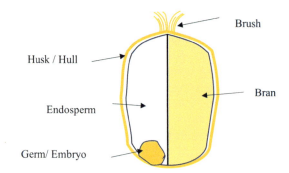

**FIGURE 9.1** Wheat grain structure.

## 9.3.2 Impact of Composition during Processing

The nutritional content of cereal grains has a major effect on the shaping procedure and the quality of the finished product. Every step, from selecting raw materials to producing the final product, impacts its functional, nutritional, and sensory qualities. The two essential characteristics of food that enhance its commercial value for customers are its nutritional content and functional value (Aryee and Boye, 2014). Various conventional techniques for processing and preparing food cannot retain the micronutrients and other plant-based compounds in the final product at certain levels. This processing technique is designed to enhance the physicochemical accessibility of micronutrients (Luithui et al., 2019). The effect of composition in grain processing is both challenging and broad. The varied components of cereal grains impact our favorite food products' sensory, nutritional, and functional elements from the first milling stages to the finished product on the market.

## 9.3.3 Conventional Processing Techniques

Traditional grain processing has served as the foundation and backbone of the food processing sector for many years. This section acts as a bridge between the extensive ancient practices of conventional cereal processing technologies and the promising opportunities introduced by MF. This section provides readers with a thorough comprehension of this crucial sector's past, present, and future.

Traditional techniques for reducing particle size entail the use of milling and grinding. These techniques have plenty of drawbacks, which might involve the production of heterogeneous particles, the product adhering on a surface, a high energy requirement and poor performance, and the product deterioration due to mechanical and thermal factors (Fahim et al., 2014). Grinding has the capability to reduce the size of food by breaking it down. This can be achieved using compression, attrition, impact, and shearing forces created within the equipment. Because of the higher shear and impact on the interior cohesiveness of product, the food matrix structure is ruptured. Particle size distribution is highly significant in food processing sector as it permits efficient mixing, pumping, transportation, and the functionalization of food materials. The particle size distribution of flour granules is extremely important for determining the quality and functional qualities of the finished product (Dhiman & Prabhakar, 2021). Conventional processing methods cause the destruction or loss of bioactive components such as antioxidants and phytochemicals, which could lead to a reduction in the potential health benefits related to these substances (Ktenioudaki et al., 2015). The impact of grain processing on phytochemical components is illustrated in Table 9.1. Conventional processing techniques frequently provide inadequate control of the rheological characteristics of cereals, specifically in relation to viscosity and texture. This limitation may make it more difficult to create products with accurate textures or satisfy a wide range of customer preferences. The effect of cereal processing on phytochemical compounds is given in Table 9.1.

Modern advances in cereal processing are essential for tackling numerous problems and fulfilling changing consumer, food industry, and global population demands. The market for more nutritional and functional foods is expanding (Aryee and Boye, 2014). Processing advancements may help cereals preserve and even improve their nutritional value. The increasing concern lies in the ecological consequences of the food sector. Advancement in grain processing might boost sustainability by decreasing water and energy use and lowering waste. Modern customers have diverse tastes, ranging from gluten-free and whole-grain alternatives to items with distinct textures and flavors. Advances in cereal processing may respond to this variety by allowing for personalization of ingredients, textures, and nutritional value, assuring a greater market appeal (Zamaratskaia et al., 2021). Technological progress in automation, precision engineering, and process optimization could end up in reduced manufacturing costs, hence, enhancing the affordability and availability of cereals to a wider demographic.

**TABLE 9.1**

**Effect of Cereal Processing on Phytochemical Compounds**

| Process | Effected Compounds | Used Cereal | Final Product | Effect on Compound | References |
|---------|-------------------|-------------|---------------|-------------------|------------|
| Milling | Phenolic acids | Wheat bread and durum | Flours with different rates of extracting | Reduce | Liyana-Pathirana and Shahidi (2007) |
| | Tocols | Red and white rice | Flours with different rates of extracting | Reduce | Finocchiaro et al. (2007) |
| Mixing/ kneading | Phenolic acids | Mixture (oat, rye, buckwheat, and wheat) | Multigrain bread | Reduce | Angioloni and Collar (2011) |
| | Carotenoids | Wheat flour | Pasta, bread | Reduce | Hidalgo et al. (2010) |
| Fermentation | Tocols | Rye | Bread | Reduce | Katina et al. (2007) |
| | Lignans | Wheat | Bread | Minor impact | Liukkonen et al. (2003) |
| Baking | Carotenoids | Wheat flour | Pasta and bread | Reduce | Hidalgo et al. (2010) |
| | Phenolic content | Mixture (amaranth, oat fiber, and soy flour) | Biscuits | Reduce | Vitali et al. (2009) |

## 9.4 APPLICATION OF MICROFLUIDIZATION IN CEREALS-BASED PRODUCTS

Microfluidization stands out with distinct advantages compared to conventional grinding and milling techniques in the domain of food processing. It involves exposing food to extremely high pressure, rapid shear forces, strong impact, sudden pressure changes, and cavitation, resulting in a uniform distribution of particles and emulsions. This technology is now widely employed in various food items, including gums, cereals, fruit juices, proteins, and their derivatives. A succinct summary of the research on cereal microfluidization is shown in Table 9.2. As evidenced by various studies, it yields diverse enhancements in food properties. Stewart and Slavin (2009) studied how the size of wheat bran particles affects its impact. Wang et al. (2012) studied the MF process's effect on wheat bran's physicochemical characteristics, while Yan et al. (2013) observed improved emulsifying properties of wheat gluten through MF. Kasemwong et al. (2011) utilized microfluidization to change the size of starch particles, examining its effects on the thermal properties and structure of cassava starch. In their research, Zhao et al. (2018) investigated the influence of ultrafine treatment on the bioaccessibility, phenolic profiles, and functional attributes of insoluble dietary fiber (IDF) extracted from rice bran. Additionally, Wang et al. (2018) investigated the MF process's influence on rice bran's IDF, concentrating on the vitro digestibility and the sorption characteristics of Pb(II).

### 9.4.1 Utilizing Microfluidization in Bakery Product Manufacturing

Microfluidization is utilized to enhance the quality and characteristics of bakery product goods such as cake, biscuits, and pastries through processes. The bakery products of MF have underscored the functional advantages associated with generating fibrous structures from cereal brans which contribute to unique and crucial functional as well as nutritional attributes to products, encompassing aspects such as cooking performance, color, dietary fiber content, texture, and shelf life. Microfluidized wheat bran was used in biscuit samples as an attempt to reduce fat content. While fat is commonly used in biscuits for various beneficial features, such as contributing to sensory appeal, rheological properties, uniform grain, providing tenderness, longer shelf life, moisture retention,

**TABLE 9.2**

**Overview of Studies on Microfluidization of Cereals**

| Sample | Research Purpose | Material and Methodology | Key Finding | References |
|---|---|---|---|---|
| Rice starch | To evaluate the potential use of modified insoluble dietary fiber as a fiber-rich ingredient in starchy foods, and additionally investigate the impacts of dynamic high-pressure microfluidization on the physicochemical characteristics of insoluble dietary fiber | • Interaction Chambers: 75 µm<br>• Pressure: 80, 120, and 170 MPa<br>• Passes: 3 passes<br>• Solid to water ratio: 1:35 | • The incorporation of modified insoluble dietary fiber improved the viscosity of rice starch paste<br>• Microfluidization significantly reduced the size of insoluble dietary fiber particles and improved its water-retaining capacity | Li et al. (2018) |
| Corn bran | To find a relationship between the antioxidant and physiochemical characteristics of maize bran and the procedure pressure and the total quantity of passes throughout the microfluidization procedures. Additionally, identify the processing conditions that yield the highest values for these qualities under the experimental variable range | • Interaction chambers: 200 µm<br>• Pressure: 124.1–158.7 MPa<br>• Passes: 1–5 passes<br>• Solid to water ratio: 1:9<br>• Experimental design: Central composite | • Swelling capacity, phenolic content, water-holding capacity, DPPH radical scavenging activity, surface reactive, and oil-holding capacity of microfluidized corn bran displayed linear positive relationships to the pressure and negative quadratic relationships to the total number of passes<br>• The antioxidant activity exhibited a negative quadratic correlation with both procedure variables | He et al. (2016) |
| Wheat bran | To evaluate the effectiveness of the microfluidization procedure to increase the bound phenolic compounds' antioxidant activity in wheat bran | • Interaction chambers: Z-shaped having diameters 200 and 87 µm and pressures 159 and 172 MPa, respectively.<br>• Passes: 1–3<br>• Solid to water ratio: 1:50 | • The microfluidization technique enhances the physicochemical qualities of wheat bran by minimizing particle size and altering its microstructure<br>• The technique considerably improved hydrolyzable phenolics, alkaline, and surface reactivity by 20%, 60%, and 280%, respectively | Wang et al. (2013) |
| Whole-grain oat pulp | To incorporate the manufacturing process of oat pulp in order to accomplish effective crushing, dietary fiber, homogenizing, emulsification of starch, and other components | • Pressures: 60, 90, 120, and 150 MPa.<br>• Solid to water ratio: 1:10 | • Dynamic high-pressure microfluidizer processing substantially lowered the mean particle size and repose time of the water particles in oat pulp while increasing the viscosity | Jiang et al. (2023) |

*(Continued)*

**TABLE 9.2 (CONTINUED)**

**Overview of Studies on Microfluidization of Cereals**

| Sample | Research Purpose | Material and Methodology | Key Finding | References |
|---|---|---|---|---|
| Whole corn slurry beverages | To determine how effectively industry-scale microfluidizer system in converting sweetcorn grains into a suspension of ultrafine particles. Additionally, ISMS was evaluated for its impact on the physical, chemical, and nutritional aspects of corn slurry | • Pressures: 60, 90, and 120 MPa<br>• Passes: 1<br>• Solid to water ratio: 1:2 | Increasing the pressure enhanced the physiological stability of corn slurry, which was associated with a decrease in particle size, a rise in viscosity, better particulate matter dispersion, and the formation of a network with reduced dimensions. These findings imply that microfluidizer system might be a novel processing method for producing whole maize drinks with higher nutritional content, longer shelf life, and superior functional performance | Guo et al. (2021) |
| Wheat bran fibers substitute in biscuits | To highlight the impact of smaller wheat bran fiber on the rheology, texture, and overall quality of biscuits when employed as a fat substitute | • Interaction chambers: Y-shaped having diameters 200 and 100 μm.<br>• Pressures: 15,000 psi | These fibers were adequate for the dough's workability when used to make low-fat biscuits, their usage increased the fiber content and/or decreased the size of the fibers, which produced harder biscuits with a reduced spread ratio | Erinc et al. (2018) |
| Gluten-free corn breads | To assess the possible use of microfluidization as a milling procedure for corn gluten, as well as to investigate potential enhancements using hydrocolloids and pH changes | • Interaction chambers: Y-shaped having diameters 200 and 100 μm with pressures 500 and 1250 bar, respectively.<br>• Passes: 1–3<br>• Solid to water ratio: 1:3 | With these treatments, there was an enlargement in surface area and the breakdown of large blocks of corn meal. Corn meal's ability to retain water was further enhanced, and, as a result, it became suitable for use as the primary component in gluten-free bread recipes | Ozturk and Mert (2018) |

and improved texture, health concerns related to saturated and trans-fat consumption have made fat reduction a public health issue. Despite efforts to diversify nutrition with rich ingredients, challenges persist in maintaining product quality when reducing fats (I. D. Mert, 2020). Erinc et al. (2018) have studied the substitution of wheat bran fibers of varying sizes for fat in biscuit formulations. For short dough products like biscuits and cookies, which traditionally include a significant amount of fat to enhance tenderness, the study observed a preference for avoiding strong gluten network formation. While an elevated water content typically leads to softer dough, the inclusion of wheat bran fiber allowed for increased water content. Nevertheless, the research found that the dough did not become softer with the addition of 20% and 30% dry fiber, which was due to the fibers holding onto water. The study revealed that smaller-sized fibers, microfluidized exhibited higher water-holding capacity, thereby reducing the plasticizing effect of water and limiting dough softening. Even though larger fibers improved texture, having more fiber or making the fibers smaller made the biscuits harder and reduced their spreading, though they remained suitable for making low-fat biscuits. Similarly, Mert et al. (2014) examined the differences in microstructure by comparing standard wheat bran with microfluidized wheat bran processed through three passes using SEM images (Figure 9.2). They also assessed how microfluidized wheat bran influenced the thickness and flow properties of cake batter as well as the quality of the resulting cake products. MF involves subjecting the wheat bran to high-pressure homogenization, resulting in the breakdown of the bran's fibrous structure of wheat bran through high-pressure homogenization releasing more phenolic compounds and increasing free phenolic content, enabling it to hold more water and stay suspended in water. Microfluidized wheat bran settles more slowly compared to ball-milled wheat bran and regular.

The greater surface area of fibers in microfluidized wheat bran creates a strong structure, improving the batter's thickness, resistance to flow, and elasticity. Adding microfluidized wheat bran to batter results in cakes that are firmer, require more cutting force, and are harder, but they have less stickiness compared to cakes with regular bran. The presence of fibers restricts water's plasticizing effect, elevating yield stress and consistency coefficient. According to research, micro-fluidized treated wheat bran improves the physical properties of fibrous products by delaying the increase in firmness during staling and reducing moisture loss. Microfluidized wheat bran can be used as an effective substitute for other constituents in bakery products.

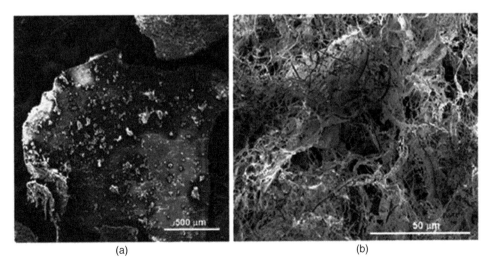

**FIGURE 9.2** SEM images comparing (a) regular wheat bran and (b) microfluidized (three passes) wheat bran samples. (Adapted from the work of Mert et al. (2014) with permission.)

## 9.4.2 MICROFLUIDIZATION TECHNIQUES FOR CEREAL-BASED DIETARY FIBER PRODUCTION AND PRECISION SIZE REDUCTION

Dietary fiber has a profound impact on health, influencing diabetes, gastrointestinal health, obesity, insulin sensitivity, and heart disease. Unlike traditional grinding methods, microfluidization prevents fiber collapse, leading to improved fiber properties such as hydration, oil retention, digestibility, and antioxidant activity. However, high fiber intake may lead to reduced product quality and increased gas production in the gut. Modifying fiber properties is crucial to addressing these potential issues while maximizing the health benefits of dietary fiber. MF offers an effective approach for enhancing food production with elevated dietary fiber content, surface area, and phenolic compound exposure in fibers, leading to improved functional and nutritional properties in dietary fiber and cereal bran, potentially increasing overall yield. Wheat gluten is widely employed in the food industry for its significant contribution to the wheat starch industry and its abundance in plant proteins (Mert, 2020). To broaden its food applications, it is crucial to enhance various functional properties like solubility, emulsifying, and foaming abilities. De Bondt et al. (2020) have found that microfluidization significantly impacts wheat bran physicochemical properties, with pressure and passes exerting the most influence reducing the median particle size to 14.8 μm through effective processing and improving the extraction of starch and arabinoxylan for better results by enhancing their extractability. The study emphasized that initial particle size had minimal significance. However, lower bran concentration, higher pressure, and increased passes during MF resulted in smaller particles, enhanced water retention capacity, improved extractability, and increased stability in a 2% wheat bran suspension and viscosity. Additionally, Ortiz de Erive et al. (2020) sought to investigate microfluidized corn bran into bread formulations on specific loaf volume, microstructure, and texture. The findings showed that bread formulated with a mix of flours containing elevated levels of microfluidized corn bran had a decreased specific loaf volume compared to regular white bread. Examination under electron microscopy revealed disturbances in the incomplete starch gelatinization and gluten network in the composite bread, impacting its quality adversely. When the moisture levels in the dough, featuring 18%, 20%, and 22% bran incorporation, surpassed established standards (38.3%, 38.6%, and 38.8%), the resulting bread approached the control's specific loaf volume, microstructure, and texture. Further research is necessary to comprehend the impact of increased moisture on the shelf life and microbial growth of bran-enriched bread.

Microfluidization has gained popularity as a technique for size reduction in recent times. In this process, a fluid is split into two microstreams that collide at high speeds. This collision exposes the microstreams to strong shear and impact forces, effectively breaking down and evenly spreading larger particles into fibrous formations (I. D. Mert, 2020). The influence of dietary fibers, whether originating from wheat bran or alternative sources, varies based on fiber size. High-pressure microfluidization reduces particle size and creates a more relaxed, porous structure in dietary fibers (Guo et al., 2020). Other researchers have employed high-pressure microfluidization-treated dietary fibers in food systems. Achieving a reduction of wheat bran below 40 μm using conventional milling proves challenging due to its relatively low density and soft structure. Liu et al. (2010) and Wang et al. (2012) confirmed effective size reduction. Likewise, according to the findings of Sun et al. (2016), examination via scanning electron microscopy (SEM) demonstrated the initial irregular shape and compact layering of IDF but after microfluidization, the fiber structure changed resulting in enhancing surface porosity and decreased particle size. Nevertheless, differences in porous surfaces may arise among dietary fibers sourced from various origins, each characterized by unique chemical compositions. Similarly, Rosa-sibakov et al. (2015) explored the impact of breaking down wheat bran in different ways and how it is affected due to its resilience under high moisture conditions. Three varieties of wheat bran, including peeled bran, aleurone-rich fraction, and standard bran, have been processed using wet grinding, dry grinding, MF, and enzymatic degradation. The results showed that wet grinding improved particle stability by disrupting the physical structure size

Microfluidization of Cereals-Based Products 187

down to 10–16 μm, reducing gravitational sedimentation. Enzymatic treatment significantly improved the solubility of bran preparations, elevating it from an initial range of 18–24% to a higher level of 40–50%. However, the improved solubility did not correlate with enhanced particle stability. The microfluidization of aleurone yielded and peeled bran heightened viscosity and enhanced stability, possibly attributed to improved particle homogenization and modified microstructure. In brief, the processed wheat bran demonstrated promise for beverage applications. The combination of MF and wet grinding exhibited potential for producing beverages with wheat bran and warrants further investigation. Similarly, Wang et al. (2012) investigated how MF affects the chemical and physical properties of wheat bran. The study showed that MF made wheat bran particles have a larger specific surface area and higher bulk density, with more passes leading to smaller particle size and greater surface area, although the effects occurred more gradually. The study also utilized confocal laser scanning to reveal the isolated structural components of wheat bran after MF. Moreover, MF resulted in enhanced water retention, and the swelling capacity of wheat bran is linked to a decrease in particle size. The hydration of wheat bran fibers flies on both swelling and water binding.

### 9.4.3 Microfluidization in Protein Modification

In the food industry, certain protein-rich by-products and proteins, traditionally limited to animal feed due to their lack of functionality, are gaining attention for enhancement and broader application in human food products. Improving the functional properties of these materials has become a focus, with MF emerging as a preferred environmentally friendly method. Various factors including processing pressure, the number of passes, the initial concentration of the protein, the category of chamber used, and different pre-treatment methods can impact the efficiency and results of MF protein modification. This process alters the conformational structure of proteins, creating micro- or nanosized particles that increase surface area, improve solubility and diffusion, and foster new structures. These changes enhance protein functionality through improved interactions between different protein groups (Guo et al., 2020). High-pressure microfluidization modifications in protein properties are linked to alterations in aggregated state, structure, and conformation. The initial impact involves the separation of large particles, resulting in modifications to the protein's quaternary structure. This process may induce unfolding and denaturation by increased α-helix intensity, maximum emission fluorescence intensity, exposed sulfhydryl, altered UV absorbance, and total sulfhydryl contents in studies on trypsin. Moreover, alterations in secondary structure, including a reduction and elevation in random coil and α-helix, have been observed across different processing pressures (Sun et al., 2016). While detailed reports on high-pressure microfluidization-induced alterations in the primary structure are lacking, the aggregation has been observed under high pressures and sulfhydryl/disulfide interchange reactions, suggesting a complex interplay of hydrophobic interactions, electrostatic and hydrogen bond interactions in the modification process (Gong et al., 2019). Figure 9.3 displays a potential schematic representation depicting the structural changes in the protein. These findings contribute significantly to our understanding of the physicochemical and functional attributes of proteins modified by MF.

Ozturk and Mert (2019) utilized microfluidization to enhance both the water-holding capacity and emulsion-forming properties of corn gluten meal (CGM). They conducted microfluidization at pressures ranging from 500 to 1250 bar, maintaining a temperature of 25 °C, and implementing pH adjustments to 6, 8, and 10. The application of both microfluidization (MF) and microfluidization with pH adjustment (MF-pH) demonstrated an enhancement in emulsion structure. This improvement was characterized by an increase in molecular surface area, a decrease in particle size, and the formation of tissues and microspores, resulting in overall advancements in emulsifying properties. While pH adjustments (MF-pH6 and MF-pH8) enhanced the effects of MF, intensive pH adjustment (MF-pH10) led to structural impairment and protein denaturation. The Herschel–Bulkley model explained flow behaviors, demonstrating elastic gel-like behavior and shear thinning in all formulations. The creation of smaller particles through microfluidization and microfluidization with pH

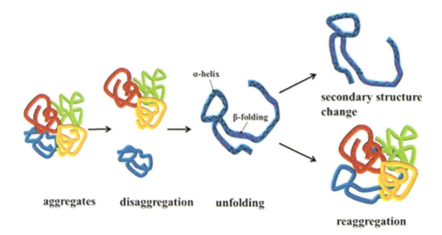

**FIGURE 9.3** Schematic of the potential structural changes in the protein following high-pressure microfluidization. (Adapted from the work of Guo et al. (2020) with permission.)

adjustment modifications in CGM structure contributed to more stable emulsions. In conclusion, MF is an efficient process for modifying CGM, offering the potential to transform an underused substance into a valuable asset within the food industry.

### 9.4.4 Advancing Bioavailability and Digestibility through Microfluidization

The restricted bioavailability of phenolics in cereal kernels, which are present as bound compounds limiting enzyme accessibility, has prompted recent research into alternative methods beyond conventional techniques to boost their bioavailability and enhance their health benefits. As determined by Mert et al. (2014), the impacts of conventional microfluidized wheat bran and untreated wheat bran on a bakery product's quality and storage characteristics were examined. The MF process caused the creation of a distinctly separated fibrous structure with greater content of methanol-soluble free phenolic compound resulting in higher surface area and increased water holding capacity. Similarly, this investigation demonstrates that the antioxidant functional compounds within phenolic elements present in wheat bran exhibit antioxidant capacity when entrapped or bound to the fiber matrix and disclosed to the liquid stage. Microfluidization substantially increases the accessibility and antioxidant capacity of phenolic compounds, increasing with the extent of treatment by 20%, 60%, and 280%, respectively. While not breaking covalent linkages, MF increases alkaline, antioxidant capacity, and acid hydrolyzable phenolic contents, revealing an underestimation in antioxidant capacity and total phenolic content using conventional methods based on solvent extraction and hydrolysis. Some researchers also used dietary fibers treated with high-pressure microfluidization to comprehend the fiber solubility impacts on the physiological utilization of these fibers in humans.

In a study by Chen et al. (2013), MF of IDFs from oats and peach resulted in reduced particle size, increased soluble fiber composition, and improved physicochemical properties, leading to enhanced restraint pancreatic lipase and lowered postprandial serum availability. It is important to note that the transformation from insoluble to soluble fiber could influence the digestion process, as soluble fibers are primarily absorption in the small intestine while insoluble fibers proceed to the colon. The difference in digestion profiles between soluble and insoluble fibers can impact physiological benefits. Soluble fibers interact with L-cells, activating the gut-brain axis, while insoluble fibers are utilized by the colon's gut microbiome. It highlights the importance of exploring how this solubility transition affects fiber utilization in human physiology. Similarly, Zheng et al. (2021) investigate the structural properties, rheological, and in vitro digestibility of rice starch modified by dynamic high-pressure microfluidization (DHPM) and high methoxyl pectin (HMP). The incorporation of hydroxypropyl methylcellulose (HMP) increased viscosity and elasticity, particularly with a 10% HMP addition. Conversely, applying debranched high-amylose maize starch (DHPM) led to reduced viscosity and elasticity under increasing pressure. In the 100 MPa DHPM treatment, enhanced ordered and crystalline structures were observed compared to starch-HMP blends, resulting in a compact surface and decreased digestibility. However, intense shear force at 200 MPa DHPM disrupted the crystalline and semi-crystalline structure of starch.

## 9.5 PROCESS OPTIMIZATION AND CONTROL PARAMETERS

### 9.5.1 MICROFLUIDIZATION PROCESS PARAMETERS

First, pressure is a critical parameter in microfluidization, influencing the degree of shear forces applied to the cereal mixture. Higher pressures generally result in finer particle sizes, but an optimum pressure range exists for each specific application. Excessive pressure may lead to increased energy consumption and equipment wear (Bucci et al., 2018; Li et al., 2022). Second, the temperature control is vital in microfluidization to prevent thermal degradation of sensitive components in cereals, such as enzymes or vitamins. It also impacts the viscosity of the mixture, influencing the flow behavior during processing (Bucci et al., 2018; Li et al., 2022). Finally, the flow rate determines the residence time of the product within the microfluidization chamber. Balancing the flow rate is essential to ensure efficient particle size reduction while avoiding over-processing that could negatively impact the product's sensory attributes (Tang et al., 2013; Li et al., 2022).

### 9.5.2 OPTIMIZATION STRATEGIES FOR DESIRED OUTCOMES

#### 9.5.2.1 Response Surface Methodology

Response Surface Methodology (RSM) is a statistical approach widely utilized in microfluidization optimization. It involves designing experiments based on a mathematical model to explore the effects of multiple variables simultaneously. By systematically varying parameters within predefined ranges, researchers can identify optimal conditions that result in the desired outcomes, whether its particle size reduction, homogeneity, or texture modification (Shen et al., 2014; Umokaso et al., 2022).

#### 9.5.2.2 Factorial Design

Factorial design is another valuable optimization strategy, enabling the systematic examination of individual parameters and their interactions. By varying factors at different levels, researchers can discern the impact of each variable on the final product. This method provides insights into how combinations of factors influence outcomes, aiding in the identification of optimal conditions for microfluidization processes (Dos Santos et al., 2022; Ouakhssase & Ait Addi, 2022).

#### 9.5.2.3 Multivariate Analysis

Techniques such as principal component analysis (PCA) and partial least squares (PLS) regression facilitate a deeper understanding of complex interactions between multiple variables. Multivariate analysis helps identify patterns and correlations within data sets, guiding the optimization process by pinpointing key factors that significantly influence desired outcomes. This approach is particularly useful when dealing with intricate relationships between various parameters and their impact on product quality (Abdi & Williams, 2013).

#### 9.5.2.4 Sequential Optimization

In some cases, a stepwise or sequential optimization approach may be employed. This involves optimizing one parameter at a time, starting with the most influential, and progressively fine-tuning others. Sequential optimization can be beneficial when there is limited prior knowledge about the process or when resources are constrained, allowing for a more focused and efficient optimization process (Saraiva et al., 2021; Osaba et al., 2021).

### 9.5.2.5 Adaptive Control Strategies

Adaptive control strategies involve real-time adjustments during the microfluidization process based on continuous monitoring of critical parameters. Sensors and feedback systems provide data on particle size, viscosity, or other relevant characteristics, enabling dynamic adjustments to maintain optimal conditions. This real-time adaptability is particularly crucial when dealing with variations in raw materials or unexpected changes in processing conditions (Patel et al., 2021; Gómez-López et al., 2022).

### 9.5.3 QUALITY CONTROL MEASURES

Quality control is imperative in microfluidization to ensure that the desired outcomes are consistently achieved. Robust quality control measures encompass various analytical techniques and testing methodologies to monitor key product attributes.

### 9.5.3.1 Particle Size Analysis

Particle size analysis is a fundamental quality control measure in microfluidization. Techniques such as laser diffraction or dynamic light scattering are commonly employed to assess the particle size distribution. Monitoring and controlling particle size are critical for achieving the desired texture, mouthfeel, and stability of cereal-based products (Alldrick, 2017; Mustač et al., 2023).

### 9.5.3.2 Rheological Analysis

Rheological analysis helps evaluate the flow and deformation characteristics of microfluidized cereal products. Measuring parameters such as viscosity and elasticity provides insights into the texture and overall sensory properties of the final product. Rheological analysis is essential for ensuring that the processed cereals meet the desired consistency and mouthfeel (Varzakas, 2016; Mert, 2020).

### 9.5.3.3 Microstructural Examination

Microstructural examination using techniques like SEM or transmission electron microscopy (TEM) allows visual inspection of the internal structure of microfluidized cereals-based product. This provides valuable information on the morphology of particles and the degree of homogeneity achieved. Microstructural analysis aids in validating the effectiveness of the microfluidization process in modifying the cereal matrix (Holopainen-Mantila & Raulio, 2016).

### 9.5.3.4 Chemical Composition Analysis

Monitoring changes in the chemical composition of cereals-based product during microfluidization is crucial for assessing the impact on nutritional components, flavor compounds, and other sensitive constituents. Techniques such as chromatography and spectroscopy are employed to analyze the chemical composition and ensure that the processing does not compromise the nutritional quality or safety of the final product (Mustač et al., 2023; Pandhi et al., 2023).

### 9.5.4 MONITORING AND ADJUSTING PARAMETERS FOR CONSISTENCY

Consistency in microfluidized cereal products is achieved through continuous monitoring of key parameters and immediate adjustments as needed. Real-time monitoring technologies and feedback control systems play a pivotal role in maintaining optimal conditions and preventing deviations that could impact product quality.

### 9.5.4.1 In-Line Sensors

In-line sensors are deployed directly within the microfluidization process to provide real-time data on critical parameters. Sensors may measure parameters such as pressure, temperature, and particle size, allowing for continuous monitoring without interrupting the process. Immediate feedback from in-line sensors enables prompt adjustments to maintain consistency (Helal et al., 2019; Luo et al., 2017).

### 9.5.4.2 Automation and Control Systems

Automation systems, coupled with advanced control algorithms, facilitate the automatic adjustment of process parameters based on sensor feedback. These systems ensure that the microfluidization process remains within predefined ranges, optimizing conditions for consistent product quality. Automation is particularly advantageous in large-scale production where manual adjustments may be impractical (Chauhan et al., 2023; Reklaitis et al., 2017).

### 9.5.4.3 Data Logging and Analysis

Continuous data logging and analysis provide a historical record of the microfluidization process. This historical data can be analyzed to identify trends, correlations, or deviations, helping refine the process and enhance long-term consistency. Data analysis tools contribute to a deeper understanding of the relationships between parameters and product outcomes (Xu et al., 2023; Ozturk & Turasan, 2022).

### 9.5.5 Ensuring Reproducibility in Large-Scale Production

Scaling up microfluidization processes for large-scale production introduces challenges related to reproducibility and consistency. Strategies to ensure the successful transition from laboratory-scale to industrial-scale production include.

### 9.5.5.1 Scale-Up Studies

Comprehensive scale-up studies are conducted to understand how changes in equipment size, processing rates, and residence times impact the microfluidization process. These studies involve testing multiple batches at various scales to validate the scalability of the process (Bondu & Yen, 2022).

### 9.5.5.2 Process Validation Protocols

Rigorous process validation protocols are implemented to verify that the microfluidization process consistently meets predefined quality standards. This involves conducting thorough testing, including particle size analysis, rheological testing, and other relevant assessments, to confirm reproducibility (Tušek et al., 2021; Atıl, 2021).

### 9.5.5.3 Standard Operating Procedures

Developing and adhering to standardized operating procedures ensure consistency in the execution of the microfluidization process. Standard Operating Procedures (SOPs) outline the specific steps, parameters, and quality control measures to be followed during production, minimizing variability between batches (Dhotre & Sathe, 2023).

### 9.5.5.4 Quality Management Systems

Implementing robust quality management systems, such as Good Manufacturing Practice (GMP) or International Organization for Standardization (ISO) standards, helps ensure that the microfluidization process adheres to industry best practices. These systems provide a framework for quality assurance and control in large-scale production environments (Bondu & Yen, 2022; Kumar et al., 2022).

## 9.6 CHALLENGES AND FUTURE DIRECTIONS

### 9.6.1 CHALLENGES AND OPPORTUNITIES

The current landscape of microfluidization for cereal processing reveals both challenges and significant opportunities. Challenges stem from the inherent diversity of cereals, presenting difficulties in achieving consistent outcomes across various grain types. Adapting microfluidization to account for differences in particle size, composition, and rheological properties requires fine-tuning of processing parameters and customized equipment designs. Preserving the nutritional integrity of cereals during microfluidization remains a critical challenge, demanding comprehensive investigations to minimize nutrient degradation. Scaling up microfluidization processes for industrial applications poses efficiency concerns, necessitating research into maintaining the technology's effectiveness and cost-effectiveness in larger-scale production. However, these challenges also present opportunities for further development. Researchers are actively exploring innovative equipment designs and processing parameters tailored to different cereal types, aiming for more standardized and efficient processing. There is a growing emphasis on leveraging microfluidization to create value-added cereal products with improved sensory attributes and enhanced nutritional profiles, aligning with consumer preferences for healthier options. Collaborative efforts between academia, research institutions, and industry stakeholders are pivotal, fostering knowledge exchange and technology transfer to bridge the gap between fundamental research and practical applications. The integration of sustainable practices, such as minimizing energy consumption and optimizing water usage, represents an additional opportunity for advancing microfluidization in cereal processing.

### 9.6.2 CONSIDER THE ECONOMIC FEASIBILITY AND SCALABILITY OF MICROFLUIDIZATION PROCESSES

The microfluidization process demonstrates strong economic feasibility and scalability, making it an attractive technology across various industries. Economically, microfluidization offers significant advantages by enhancing production efficiency and reducing costs. The high-pressure homogenization employed in microfluidization results in smaller particle sizes, improving the uniformity and stability of products in pharmaceuticals, nutraceuticals, and food industries. This reduction in particle size not only enhances the bioavailability of active ingredients but also allows for efficient utilization of raw materials, minimizing waste, and optimizing resource usage. Additionally, the process contributes to the overall improvement of product quality, a crucial factor for industries aiming to meet stringent regulatory standards and consumer expectations. From a scalability perspective, microfluidization stands out as a versatile and adaptable technology suitable for diverse production scales. Its modular equipment design facilitates seamless integration into existing production lines, enabling companies to transition from lab-scale research to large-scale industrial manufacturing without significant modifications. This scalability is particularly beneficial in industries where production volumes may vary or experience rapid growth, such as pharmaceuticals and biotechnology. The ability to scale up without compromising efficiency ensures that companies can meet market demands while maintaining consistent product quality. Overall, the economic feasibility and scalability of microfluidization make it a compelling choice for industries looking to improve production processes, enhance product quality, and remain competitive in dynamic markets.

## 9.7 CONCLUSIONS AND PERSPECTIVES

Microfluidization of cereal-based goods opens up new opportunities for improving quality, nutritional content, and overall consumer pleasure. This cutting-edge technology uses high-pressure homogenization to reduce particle size, increasing texture and flavor while extending shelf life. The ability of microfluidization to encapsulate bioactive molecules improves nutrition absorption while also addressing health and taste preferences. Despite its potential, addressing scalability and cost concerns is critical for wider implementation in large-scale food production. Essentially, microfluidization is a smart and unique approach that has the potential to reinvent cereal-based products for a more nutritious and enjoyable culinary experience.

Current research and development trends in microfluidization for cereal processing reflect a dynamic landscape with a strong emphasis on optimizing the technology for enhanced efficiency, nutritional preservation, and sustainability. Researchers are actively exploring innovative microfluidization techniques tailored to the unique characteristics of various cereals, aiming to achieve consistent and controlled particle size reduction while preserving the nutritional integrity of the grains. This involves not only fine-tuning processing parameters but also customizing equipment designs to accommodate the diverse properties of different cereals. Furthermore, there is a growing interest in leveraging microfluidization to create value-added cereal products with improved sensory attributes and enhanced nutritional profiles. Computational modeling and simulation are emerging as essential tools, aiding in the prediction and optimization of fluid dynamics within microfluidic systems, and contributing to the design of more efficient processes. Sustainability is a prominent theme, with researchers focusing on minimizing energy consumption, exploring alternative environmentally friendly materials for microfluidic devices, and optimizing water usage in cereal processing. Collaboration between academia, research institutions, and industry is thriving, fostering knowledge exchange and the translation of research findings into practical applications. Overall, the current trends in microfluidization for cereal processing underscore a holistic approach, integrating technological advancements, sustainability considerations, and nutritional enhancement to propel the evolution of microfluidization as a transformative technology in the cereal industry.

The future of microfluidization in cereal processing promises transformative applications and innovations that could reshape the entire landscape of this industry. One notable avenue is the development of precisely engineered cereal products with enhanced nutritional profiles, improved textures, and extended shelf life. Microfluidization offers the potential to tailor particle sizes and distributions, influencing the structural and textural attributes of cereals. This opens the door to the creation of novel cereal-based snacks, breakfast items, and functional foods that meet evolving consumer preferences for health and convenience. Additionally, the technology may play a pivotal role in addressing global food security challenges by optimizing the processing of cereals for improved nutrient retention and bioavailability. Furthermore, microfluidics could contribute to sustainable practices in cereal processing by minimizing waste, reducing energy consumption, and optimizing water usage. As research advances, the integration of microfluidization with precision agriculture and smart manufacturing technologies could enable real-time monitoring and control of cereal processing, enhancing overall efficiency and ensuring product consistency. Collaborations between food scientists, engineers, and nutritionists will likely drive these innovations, paving the way for a future where MF revolutionizes the cereal industry, offering not only improved product quality but also solutions to broader challenges in food production and sustainability. The convergence of MF with emerging technologies holds the potential to create a new era of precision in cereal processing, meeting the demands of a rapidly evolving and health-conscious consumer market.

## REFERENCES

Abdi, H., & Williams, L. J. (2013). Partial least squares methods: Partial least squares correlation and partial least square regression. *Computational Toxicology: Volume II* (pp. 549–579).

Aktaş, K., & Akın, N. (2020). Influence of rice bran and corn bran addition on the selected properties of tarhana, a fermented cereal based food product. *LWT, 129*, 109574.

Alldrick, A. J. (2017). Food safety aspects of grain and cereal product quality. In *Cereal Grains* (pp. 393–424). Woodhead Publishing.

Angioloni, A., & Collar, C. (2011). Polyphenol composition and "in vitro" antiradical activity of single and multigrain breads. *Journal of Cereal Science, 53*(1), 90–96.

Aryee, A. N., & Boye, J. I. (2014). Current and emerging trends in the formulation and manufacture of nutraceuticals and functional food products. In *Nutraceutical and Functional Food Processing Technology* (pp 1–53). John Wiley & Sons, Ltd. 1.

Atıl, G. U. (2021). Developing enteral feeding formulations with different protein sources, rheological characterization and microstructural analysis of these formulations (Master's thesis, Middle East Technical University).

Bader Ul Ain, H., Saeed, F., Kashif, M., Mushtaq, Z., Imran, A., Ahmad, A., & Tufail, T. (2020). Effect of cereal endospermic cell wall on farinographic and mixographic characteristics of wheat flour. *Journal of Food Processing and Preservation, 44*(11), e14899.

Bala, M., Tushir, S., Garg, M., Meenu, M., Kaur, S., Sharma, S., & Mann, S. (2023). Wheat milling and recent processing technologies: Effect on nutritional properties, challenges, and strategies. In *Wheat Science* (pp. 219–256). CRC Press.

Bhukya, J., Mohapatra, D., & Naik, R. (2023). Hydrodynamic cavitation processing of ascorbic acid treated precooled sugarcane juice for physiochemical, bioactive, enzyme stability, and microbial safety. *Journal of Food Process Engineering, 46*(6), e14209.

Bhukya, J., Naik, R., Mohapatra, D., Sinha, L. K., & Rao, K. V. R. (2021). Orifice based hydrodynamic cavitation of sugarcane juice: Changes in Physico-chemical parameters and Microbiological load. *LWT, 150*, 111909.

Bondu, C., & Yen, F. T. (2022). Nanoliposomes, from food industry to nutraceuticals: Interests and uses. *Innovative Food Science & Emerging Technologies, 81*, 103140.

Boukid, F. (Ed.). (2021). *Cereal-Based Foodstuffs: The Backbone of Mediterranean Cuisine*. Springer Nature.

Boukid, F., Rosell, C. M., & Castellari, M. (2021). Pea protein ingredients: A mainstream ingredient to (re) formulate innovative foods and beverages. *Trends in Food Science & Technology, 110*, 729–742.

Bucci, A. J., Van Hekken, D. L., Tunick, M. H., Renye, J. A., & Tomasula, P. M. (2018). The effects of microfluidization on the physical, microbial, chemical, and coagulation properties of milk. *Journal of Dairy Science, 101*(8), 6990–7001.

Burton, R. A., & Fincher, G. B. (2014). Evolution and development of cell walls in cereal grains. *Frontiers in Plant Science, 5*, 456.

Cano-Sarmiento, C., Alamilla-Beltrán, L., Azuara-Nieto, E., Hernández-Sánchez, H., Téllez-Medina, D. I., Jiménez-Martínez, C., & Gutiérrez-López, G. F. (2015). High shear methods to produce nano-sized food related to dispersed systems. In *Food Nanoscience and Nanotechnology* (pp.145–161). Springer International Publishing: Cham, Switzerland.

Chandran, A. S., Suri, S., & Choudhary, P. (2023). Sustainable plant protein: A recent overview of sources, extraction techniques and utilization ways. *Sustainable Food Technology, 1*(4), 466–483.

Chauhan, R., Minocha, N., Coliaie, P., Singh, P. G., Korde, A., Kelkar, M. S., & Singh, M. R. (2023). Emerging microfluidic platforms for crystallization process development. *Chemical Engineering Research and Design, 197*, 908–930.

Chen, J., Gao, D., Yang, L., & Gao, Y. (2013). Effect of microfluidization process on the functional properties of insoluble dietary fiber. *Food Research International, 54*(2), 1821–1827.

Chiozzi, V., Agriopoulou, S., & Varzakas, T. (2022). Advances, applications, and comparison of thermal (pasteurization, sterilization, and aseptic packaging) against non-thermal (ultrasounds, UV radiation, ozonation, high hydrostatic pressure) technologies in food processing. *Applied Sciences, 12*(4), 2202.

Collar, C. (2014). Barley, maize, sorghum, millet, and other cereal grains. In *Bakery Products Science and Technology*, 107–126. New York: Wiley.

De Bondt, Y., Rosa-Sibakov, N., Liberloo, I., Roye, C., Van de Walle, D., Dewettinck, K., & Courtin, C. M. (2020). Study into the effect of microfluidisation processing parameters on the physicochemical properties of wheat (Triticum aestivum L.) bran. *Food Chemistry, 305*, 125436.

de Erive, M. O., Wang, T., He, F., & Chen, G. (2020). Development of high-fiber wheat bread using microfluidized corn bran. *Food Chemistry, 310*, 125921.

Dekkers, B. L., Nikiforidis, C. V., & van der Goot, A. J. (2016). Shear-induced fibrous structure formation from a pectin/SPI blend. *Innovative Food Science & Emerging Technologies, 36*, 193–200.

Dhiman, A., & Prabhakar, P. K. (2021). Micronization in food processing: A comprehensive review of mechanistic approach, physicochemical, functional properties and self-stability of micronized food materials. *Journal of Food Engineering, 292*, 110248.

Dhotre, I., & Sathe, V. (2023). Optimizing lemongrass essential oil extraction using microwave technology: Exploring the influence of sparger, tri-spiral condenser, and ultrasonication. In *Biomass Conversion and Biorefinery* (pp. 1–14). Springer.

Dos Santos, T. R., dos Santos Melo, J., Dos Santos, A. V., Severino, P., Lima, Á. S., Souto, E. B., ... & Cardoso, J. C. (2022). Development of a protein-rich by-product by 23 factorial design: characterization of its nutritional value and sensory analysis. *Molecules, 27*(24), 8918.

Erinc, H., Mert, B., & Tekin, A. (2018). Different sized wheat bran fibers as fat mimetic in biscuits: Its effects on dough rheology and biscuit quality. *Journal of Food Science and Technology, 55*, 3960–3970.

Evers, T., & Millar, S. (2002). Cereal grain structure and development: Some implications for quality. *Journal of Cereal Science, 36*(3), 261–284.

Fahim, T. K., Zaidul, I. S. M., Bakar, M. A., Salim, U. M., Awang, M. B., Sahena, F., ... & Sohrab, M. H. (2014). Particle formation and micronization using non-conventional techniques-review. *Chemical Engineering and Processing: Process Intensification, 86*, 47–52.

Finocchiaro, F., Ferrari, B., Gianinetti, A., Dall'Asta, C., Galaverna, G., Scazzina, F., & Pellegrini, N. (2007). Characterization of antioxidant compounds of red and white rice and changes in total antioxidant capacity during processing. *Molecular Nutrition & Food Research, 51*(8), 1006–1019.

Goldstein, N., & Reifen, R. (2022). The potential of legume-derived proteins in the food industry. *Grain & Oil Science and Technology, 5*(4), 167–178.

Gómez-López, V. M., Pataro, G., Tiwari, B., Gozzi, M., Meireles, M. Á. A., Wang, S., & Morata, A. (2022). Guidelines on reporting treatment conditions for emerging technologies in food processing. *Critical Reviews in Food Science and Nutrition, 62*(21), 5925–5949.

Gong, K., Chen, L., Xia, H., Dai, H., Li, X., Sun, L., ... & Liu, K. (2019). Driving forces of disaggregation and reaggregation of peanut protein isolates in aqueous dispersion induced by high-pressure microfluidization. *International Journal of Biological Macromolecules, 130*, 915–921.

Guo, X., Chen, M., Li, Y., Dai, T., Shuai, X., Chen, J., & Liu, C. (2020). Modification of food macromolecules using dynamic high pressure microfluidization: A review. *Trends in Food Science & Technology, 100*, 223–234.

Guo, X., McClements, D. J., Chen, J., He, X., Liu, W., Dai, T., & Liu, C. (2021). The nutritional and physicochemical properties of whole corn slurry prepared by a novel industry-scale microfluidizer system. *LWT, 144*, 111096.

Hassoun, A., Jagtap, S., Trollman, H., Garcia-Garcia, G., Abdullah, N. A., Goksen, G., & Lorenzo, J. M. (2023). Food processing 4.0: Current and future developments spurred by the fourth industrial revolution. *Food Control, 145*, 109507.

He, F., Wang, T., Zhu, S., & Chen, G. (2016). Modeling the effects of microfluidization conditions on properties of corn bran. *Journal of Cereal Science, 71*, 86–92.

Helal, N. A., Elnoweam, O., Eassa, H. A., Amer, A. M., Eltokhy, M. A., Helal, M. A., & Nounou, M. I. (2019). Integrated continuous manufacturing in pharmaceutical industry: Current evolutionary steps toward revolutionary future. *Pharmaceutical Patent Analyst, 8*(4), 139–161.

Hidalgo, A., Brandolini, A., & Pompei, C. (2010). Carotenoids evolution during pasta, bread and water biscuit preparation from wheat flours. *Food Chemistry, 121*(3), 746–751.

Holopainen-Mantila, U., & Raulio, M. (2016). Cereal grain structure by microscopic analysis. In *Imaging Technologies and Data Processing for Food Engineers* (pp. 1–39) Springer International Publishing: Cham, Switzerland.

Jiang, P., Kang, Z., Zhao, S., Meng, N., Liu, M., & Tan, B. (2023). Effect of dynamic high-pressure microfluidizer on physicochemical and microstructural properties of whole-grain oat pulp. *Foods, 12*(14), 2747.

Kasemwong, K., Ruktanonchai, U. R., Srinuanchai, W., Itthisoponkul, T., & Sriroth, K. (2011). Effect of high-pressure microfluidization on the structure of cassava starch granule. *Starch-Stärke, 63*(3), 160–170.

Katina, K., Liukkonen, K. H., Kaukovirta-Norja, A., Adlercreutz, H., Heinonen, S. M., Lampi, A. M., & Poutanen, K. (2007). Fermentation-induced changes in the nutritional value of native or germinated rye. *Journal of Cereal Science, 46*(3), 348–355.

Katouzian, I., & Jafari, S. M. (2016). Nano-encapsulation as a promising approach for targeted delivery and controlled release of vitamins. *Trends in Food Science & Technology, 53*, 34–48.

Kaur, P., Sandhu, K. S., Purewal, S. S., Kaur, M., & Singh, S. K. (2021). Rye: A wonder crop with industrially important macromolecules and health benefits. *Food Research International, 150*, 110769.

Ktenioudaki, A., Alvarez-Jubete, L., & Gallagher, E. (2015). A review of the process-induced changes in the phytochemical content of cereal grains: The breadmaking process. *Critical Reviews in Food Science and Nutrition, 55*(5), 611–619.

Kumar, A., Dhiman, A., Suhag, R., Sehrawat, R., Upadhyay, A., & McClements, D. J. (2022). Comprehensive review on potential applications of microfluidization in food processing. *Food Science and Biotechnology, 31*, 17–36.

Láng, L., Rakszegi, M., & Bedő, Z. (2013). Cereal production and its characteristics. In *Engineering Aspects of Cereal and Cereal-Based Products* (pp. 1–20). Taylor & Francis Group, LLC.

Leong, T., Juliano, P., & Knoerzer, K. (2017). Advances in ultrasonic and megasonic processing of foods. *Food Engineering Reviews*, 9(3), 237–256.

Leszczyńska, D., Wirkijowska, A., Gasiński, A., Srednicka-Tober, D., Trafiałek, J., & Kazimierczak, R. (2023). Oat and oat processed products technology, composition, nutritional value, and health. *Applied Sciences*, 13(20), 11267.

Li, Y., Deng, L., Dai, T., Li, Y., Chen, J., Liu, W., & Liu, C. (2022). Microfluidization: A promising food processing technology and its challenges in industrial application. *Food Control*, 137, 108794.

Li, Y. O., González, V. P. D., & Diosady, L. L. (2023). Microencapsulation of vitamins, minerals, and nutraceuticals for food applications. In *Microencapsulation in the Food Industry* (pp. 507–528). Academic Press.

Li, Y. T., Wang, R. S., Liang, R. H., Chen, J., He, X. H., Chen, R. Y., … & Liu, C. M. (2018). Dynamic high-pressure microfluidization assisting octenyl succinic anhydride modification of rice starch. *Carbohydrate Polymers*, 193, 336–342.

Liu, W., Zhang, Z. Q., Liu, C. M., Xie, M. Y., Tu, Z. C., Liu, J. H., & Liang, R. H. (2010). The effect of dynamic high-pressure microfluidization on the activity, stability and conformation of trypsin. *Food Chemistry*, 123(3), 616–621.

Liukkonen, K. H., Katina, K., Wilhelmsson, A., Myllymaki, O., Lampi, A. M., Kariluoto, S., & Poutanen, K. (2003). Process-induced changes on bioactive compounds in whole grain rye. *Proceedings of the Nutrition Society*, 62(1), 117–122.

Liyana-Pathirana, C. M., & Shahidi, F. (2007). The antioxidant potential of milling fractions from breadwheat and durum. *Journal of Cereal Science*, 45(3), 238–247.

Loy, D. D., & Lundy, E. L. (2019). Nutritional properties and feeding value of corn and its coproducts. In *Corn* (pp. 633–659). AACC International Press.

Luithui, Y., Baghya Nisha, R., & Meera, M. S. (2019). Cereal by-products as an important functional ingredient: Effect of processing. *Journal of Food Science and Technology*, 56, 1–11.

Lukinac, J., & Jukić, M. (2022). Barley in the production of cereal-based products. *Plants*, 11(24), 3519.

Luo, S., Wang, Y., Wang, G., Wang, K., Wang, Z., Zhang, C., & Liu, T. (2017). CNT enabled co-braided smart fabrics: A new route for non-invasive, highly sensitive & large-area monitoring of composites. *Scientific Reports*, 7(1), 44056.

Malik, T., Sharma, R., Ameer, K., Bashir, O., Amin, T., Manzoor, S., & Mohamed Ahmed, I. A. (2023). Potential of high-pressure homogenization (HPH) in the development of functional foods. *International Journal of Food Properties*, 26(1), 2509–2531.

Mandal, S., Singh, V. K., Chaudhary, D., Kaur, A., Kumar, R., Panwar, A., & Kaushik, P. (2023). From grain to gain: Revolutionizing maize nutrition. Preprint.org. 1, 1–21.

Maurya, V. K., Shakya, A., Bashir, K., Jan, K., & McClements, D. J. (2023). Fortification by design: A rational approach to designing vitamin D delivery systems for foods and beverages. *Comprehensive Reviews in Food Science and Food Safety*, 22(1), 135–186.

McClements, D. J., Newman, E., & McClements, I. F. (2019). Plant-based milks: A review of the science underpinning their design, fabrication, and performance. *Comprehensive Reviews in Food Science and Food Safety*, 18(6), 2047–2067.

Mert, B., Tekin, A., Demirkesen, I., & Kocak, G. (2014). Production of microfluidized wheat bran fibers and evaluation as an ingredient in reduced flour bakery product. *Food and Bioprocess Technology*, 7, 2889–2901.

Mert, I. D. (2020). The applications of microfluidization in cereals and cereal-based products: An overview. *Critical Reviews in Food Science and Nutrition*, 60(6), 1007–1024.

Mishra, S., Singh, R., Upadhyay, A., Mishra, S., & Shukla, S. (2023). Emerging trends in processing for cereal and legume-based beverages: A review. *Future Foods*, 100257. https://doi.org/10.1016/j.fufo.2023.100257

Modarres-Gheisari, S. M. M., Gavagsaz-Ghoachani, R., Malaki, M., Safarpour, P., & Zandi, M. (2019). Ultrasonic nano-emulsification – A review. *Ultrasonics Sonochemistry*, 52, 88–105.

Mu, R., Bu, N., Pang, J., Wang, L., & Zhang, Y. (2022). Recent trends of microfluidics in food science and technology: Fabrications and applications. *Foods*, 11(22), 3727.

Mustač, N. Č., Pastor, K., Kojić, J., Voučko, B., Ćurić, D., Rocha, J. M., & Novotni, D. (2023). Quality assessment of 3D-printed cereal-based products. *LWT*, 184, 115065.

Nisha, R., Nickhil, C., Pandiarajan, T., Pandiselvam, R., Jithender, B., & Kothakota, A. (2023). Chemical, functional, rheological and structural properties of broken rice–barnyard millet–green gram grits blend for the production of extrudates. *Journal of Food Process Engineering*, 46(5), e14324.

Olakanmi, S. J., Jayas, D. S., & Paliwal, J. (2022). Implications of blending pulse and wheat flours on rheology and quality characteristics of baked goods: A review. *Foods, 11*(20), 3287.

Osaba, E., Villar-Rodriguez, E., Del Ser, J., Nebro, A. J., Molina, D., LaTorre, A., & Herrera, F. (2021). A tutorial on the design, experimentation and application of metaheuristic algorithms to real-world optimization problems. *Swarm and Evolutionary Computation, 64*, 100888.

Ouakhssase, A., & Ait Addi, E. (2022). Mycotoxins in food: A review on liquid chromatographic methods coupled to mass spectrometry and their experimental designs. *Critical Reviews in Food Science and Nutrition, 62*(10), 2606–2626.

Ozturk, O. K., & Mert, B. (2018). The effects of microfluidization on rheological and textural properties of gluten-free corn breads. *Food Research International, 105*, 782–792.

Ozturk, O. K., & Mert, B. (2019). Characterization and evaluation of emulsifying properties of high pressure microfluidized and pH shifted corn gluten meal. *Innovative Food Science & Emerging Technologies, 52*, 179–188.

Ozturk, O. K., & Turasan, H. (2021). Applications of microfluidization in emulsion-based systems, nanoparticle formation, and beverages. *Trends in Food Science & Technology, 116*, 609–625.

Ozturk, O. K., & Turasan, H. (2022). Latest developments in the applications of microfluidization to modify the structure of macromolecules leading to improved physicochemical and functional properties. *Critical Reviews in Food Science and Nutrition, 62*(16), 4481–4503.

Pandhi, S., Mahato, D. K., & Kumar, A. (2023). Overview of green nanofabrication technologies for food quality and safety applications. *Food Reviews International, 39*(1), 240–260.

Patel, D., Zhang, Y., Dong, Y., Qu, H., Kozak, D., Ashraf, M., & Xu, X. (2021). Adaptive perfusion: An in vitro release test (IVRT) for complex drug products. *Journal of Controlled Release, 333*, 65–75.

Ratnavathi, C. V. (2019). Grain structure, quality, and nutrition. In *Breeding Sorghum for Diverse End Uses* (pp. 193–207). Woodhead Publishing.

Reklaitis, G. V., Seymour, C., & García-Munoz, S. (Eds.). (2017). *Comprehensive Quality by Design for Pharmaceutical Product Development and Manufacture.* John Wiley & Sons.

Rosa-Sibakov, N., Sibakov, J., Lahtinen, P., & Poutanen, K. (2015). Wet grinding and microfluidization of wheat bran preparations: Improvement of dispersion stability by structural disintegration. *Journal of Cereal Science, 64*, 1–10.

Rousta, L. K., Bodbodak, S., Nejatian, M., Yazdi, A. P. G., Rafiee, Z., Xiao, J., & Jafari, S. M. (2021). Use of encapsulation technology to enrich and fortify bakery, pasta, and cereal-based products. *Trends in Food Science & Technology, 118*, 688–710.

Saraiva, M., Jitaru, P., & Sloth, J. J. (2021). Speciation analysis of Cr (III) and Cr (VI) in bread and breakfast cereals using species-specific isotope dilution and HPLC-ICP-MS. *Journal of Food Composition and Analysis, 102*, 103991.

Shahbazi, M., Majzoobi, M., & Farahnaky, A. (2018). Impact of shear force on functional properties of native starch and resulting gel and film. *Journal of Food Engineering, 223*, 10–21.

Shen, N., Wang, Q., Qin, Y., Zhu, J., Zhu, Q., Mi, H., & Huang, R. (2014). Optimization of succinic acid production from cane molasses by Actinobacillus succinogenes GXAS137 using response surface methodology (RSM). *Food Science and Biotechnology, 23*, 1911–1919.

Siddiqui, S. A., Farooqi, M. Q. U., Bhowmik, S., Zahra, Z., Mahmud, M. C., Assadpour, E., & Jafari, S. M. (2023). Application of micro/nano-fluidics for encapsulation of food bioactive compounds-principles, applications, and challenges. *Trends in Food Science & Technology, 136*, 64–75.

Skendi, A., Zinoviadou, K. G., Papageorgiou, M., & Rocha, J. M. (2020). Advances on the valorisation and functionalization of by-products and wastes from cereal-based processing industry. *Foods, 9*(9), 1243.

Sopade, P. A. (2017). Cereal processing and glycaemic response. *International Journal of Food Science & Technology, 52*(1), 22–37.

Soukoulis, C., & Bohn, T. (2018). A comprehensive overview on the micro-and nano-technological encapsulation advances for enhancing the chemical stability and bioavailability of carotenoids. *Critical Reviews in Food Science and Nutrition, 58*(1), 1–36.

Stewart, M. L., & Slavin, J. L. (2009). Particle size and fraction of wheat bran influence short-chain fatty acid production in vitro. *British Journal of Nutrition, 102*(10), 1404–1407.

Sun, C., Dai, L., Liu, F., & Gao, Y. (2016). Dynamic high pressure microfluidization treatment of zein in aqueous ethanol solution. *Food Chemistry, 210*, 388–395.

Tang, S. Y., Shridharan, P., & Sivakumar, M. (2013). Impact of process parameters in the generation of novel aspirin nanoemulsions–comparative studies between ultrasound cavitation and microfluidizer. *Ultrasonics Sonochemistry, 20*(1), 485–497.

Thielecke, F., Lecerf, J. M., & Nugent, A. P. (2021). Processing in the food chain: Do cereals have to be processed to add value to the human diet? *Nutrition Research Reviews, 34*(2), 159–173.

Tušek, A. J., Šalić, A., Valinger, D., Jurina, T., Benković, M., Kljusurić, J. G., & Zelić, B. (2021). The power of microsystem technology in the food industry – Going small makes it better. *Innovative Food Science & Emerging Technologies, 68*, 102613.

Umokaso, M. M., Efiuvwevwere, B. J., & Ire, F. S. (2022). Optimization of fermentation parameters for cereal-porridge production using response surface methodology (RSM). *GSC Biological and Pharmaceutical Sciences, 18*(2), 203–214.

Varzakas, T. (2016). Quality and safety aspects of cereals (wheat) and their products. *Critical Reviews in Food Science and Nutrition, 56*(15), 2495–2510.

Verni, M., Rizzello, C. G., & Coda, R. (2019). Fermentation biotechnology applied to cereal industry by-products: Nutritional and functional insights. *Frontiers in Nutrition, 6*, 42.

Vitali, D., Dragojević, I. V., & Šebečić, B. (2009). Effects of incorporation of integral raw materials and dietary fibre on the selected nutritional and functional properties of biscuits. *Food Chemistry, 114*(4), 1462–1469.

Wang, L., Wu, J., Luo, X., Li, Y., Wang, R., Li, Y., … & Chen, Z. (2018). Dynamic high-pressure microfluidization treatment of rice bran: Effect on Pb (II) ions adsorption in vitro. *Journal of Food Science, 83*(7), 1980–1989.

Wang, T., Raddatz, J., & Chen, G. (2013). Effects of microfluidization on antioxidant properties of wheat bran. *Journal of Cereal Science, 58*(3), 380–386.

Wang, T., Sun, X., Zhou, Z., & Chen, G. (2012). Effects of microfluidization process on physicochemical properties of wheat bran. *Food Research International, 48*(2), 742–747.

Welti-Chanes, J., Morales-de la Peña, M., Jacobo-Velázquez, D. A., & Martín-Belloso, O. (2017). Opportunities and challenges of ultrasound for food processing: An industry point of view. *Ultrasound: Advances for Food Processing and Preservation*, 457–497. https://doi.org/10.1016/B978-0-12-804581-7.00019-1

Xu, J., Fan, X., Xu, X., Deng, D., Yang, L., Song, H., & Liu, H. (2023). Microfluidization improved hempseed yogurt's physicochemical and storage properties. *Journal of the Science of Food and Agriculture, 104* (4), 2252–2261.

Yan, N. J., Liu, G. Q., Chen, L. Y., & Liu, X. Q. (2013). Emulsifying and foaming properties of wheat gluten influenced by high pressure microfluidization. *Advanced Materials Research, 690*, 1327–1330.

Yishak, Y. (2014). Byproducts utilization from wheat milling industries for development of value added products. (Master's thesis, Sc. Addis Ababa University).

Zamaratskaia, G., Gerhardt, K., & Wendin, K. (2021). Biochemical characteristics and potential applications of ancient cereals – An underexploited opportunity for sustainable production and consumption. *Trends in Food Science & Technology, 107*, 114–123.

Zhao, G., Zhang, R., Dong, L., Huang, F., Tang, X., Wei, Z., & Zhang, M. (2018). Particle size of insoluble dietary fiber from rice bran affects its phenolic profile, bioaccessibility and functional properties. *LWT, 87*, 450–456.

Zheng, J., Wang, N., Huang, S., Kan, J., & Zhang, F. (2021). In vitro digestion and structural properties of rice starch modified by high methoxyl pectin and dynamic high-pressure microfluidization. *Carbohydrate Polymers, 274*, 118649.

# 10 Microfluidization of Meat- and Egg-Based Products

*Rajat Suhag and Atul Dhiman*

## 10.1 INTRODUCTION

Homogenizers help overcome interfacial forces by generating high disruptive forces, causing particle breakdown. Using an emulsifier alone reduces the interfacial tension between droplets (Tobin et al. 2015; McClements 2016). A microfluidizer is also a type of high-pressure homogenizer but is advanced in terms of generating high, constant pressure profiles, along with producing smaller particles with a narrow particle size distribution range compared to traditional homogenizers (Mert 2020). Microfluidization has been reported to reduce the particle size to the micron level, increasing the surface area of the droplets and altering physico-chemical and functional properties (swelling capacity, water holding capacity, oil holding capacity, cation exchange capacity, antioxidant capacity, and solubility) of the food matrix along with having preservation potential. Microfluidization induces structural and rheological changes in food. It has also been used to formulate ingredients that provide techno-functional advantages like reduced syneresis, delaying retrogradation, inducing gelatinization, and improving bioavailability (Dhiman and Prabhakar 2021). Microfluidization has been used to modify properties of macromolecules like proteins/enzymes, polysaccharides, dietary fiber, and starch, leading to modulations with improved functional properties (Guo et al. 2020). In comparison to other emerging processing technologies like ultrasonic homogenizer, the microfluidizer generates less temperature in the product, which is beneficial for thermally labile compounds, as reported in a study by Salvia-Trujillo et al. (2014), where the product received from the ultrasonic homogenizer's outlet had twice the temperature than that of microfluidized one. High-temperature generation in ultrasonics may also lead to low yield and cause detrimental physical and chemical effects on the food system by generating free radicals, thereby causing oxidation and forming off-flavors (Bhargava et al. 2021), whereas the microfluidizer has not shown any such limitations. Microfluidization can also serve as a better non-thermal alternative to high-pressure processing (HPP) as it uses much less pressure compared to HPP. In a study by Feijoo et al. (1997), microfluidization was reported to bring spore inactivation up to 68% in ice cream mix at 200 MPa, whereas when using HPP, pressures up to 800 MPa might be required for inactivating bacterial spores (Hogan, Kelly, and Sun 2005). In addition to the high-pressure effect, intense shear, turbulence, and cavitation might have led to biophysical and mechanical damage on the cell membrane, thus inactivating the microbes. Processing egg yolk and egg white with the microfluidizer has brought tremendous changes in their functionality and structural properties, as shown in Figure 10.1.

DOI: 10.1201/9781032632599-10

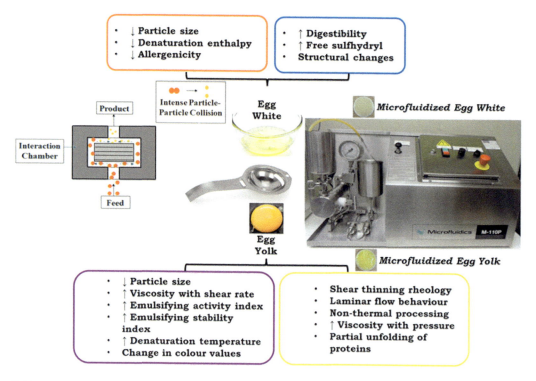

**FIGURE 10.1** Changes brought in egg yolk and egg white after processing with microfluidizer.

## 10.2 MICROFLUIDIZATION OF MEAT AND MEAT-BASED PRODUCTS

### 10.2.1 Effect on Solubility and Turbidity

Myofibrillar protein stands out as the predominant and pivotal protein found in muscle tissue (Wang et al. 2022). This protein plays a crucial role in shaping the characteristics of meat products due to its unique attributes, including solubility, gelling capacity, and emulsifying capabilities, all of which significantly influence the quality of meat-based products. Notably, myofibrillar protein plays a pivotal role in maintaining the homogeneous distribution of fat droplets and preventing unwanted water and oil leakage. These factors collectively impact the texture and yield of meat products (Han et al. 2022; Xiong et al. 2019).

The solubility of myofibrillar protein emerges as a key factor affecting both the viscosity and the consistency of minced meat products, while its wettability significantly influences overall texture and yield. It is crucial to recognize that the solubility and wettability of these proteins are intricately interconnected with various other functional properties. Nevertheless, there remains a need to enhance the inherent properties of natural myofibrillar protein to align with the evolving demands of meat processing. For instance, it is worth noting that myofibrillar protein exhibits poor solubility in water but demonstrates improved solubility in saline solutions, as highlighted by Chen et al. (2018). To address this challenge, numerous studies have concentrated their efforts on enhancing myofibrillar protein properties through a range of physical and chemical methods or cutting-edge technologies, including microfluidization.

Microfluidization treatment has demonstrated significant enhancements in the solubility of both chicken (Han et al. 2023) and pork myofibrillar proteins (Zhang et al. 2022). Untreated chicken myofibrillar proteins exhibited a low solubility of 11.3%, which notably improved with increasing microfluidization pressure (30, 60, 90, and 120 MPa for 3 cycles). The highest solubility, reaching 61.9%, was achieved at a microfluidization pressure of 120 MPa (Han et al. 2023). Similarly,

# Microfluidization of Meat- and Egg-Based Products

untreated pork myofibrillar proteins starting with a solubility of 71.82% increased to 78.59% post-microfluidization treatment at 20 kpsi for 3 cycles. Moreover, when lysine and arginine were added to pork myofibrillar proteins in conjunction with microfluidization, there was a further boost in solubility. This highlights the synergistic effect of microfluidization and enzymatic activity in enhancing the solubility of pork myofibrillar proteins (Zhang et al. 2022). The improvement in the solubility of myofibrillar proteins can be attributed to the disruption of the interaction forces responsible for maintaining the protein's structure. This disruption is achieved through the application of shearing force and cavitation generated by high pressure (Wu et al. 2019). This process exposes the hydrophilic (Liu et al. 2011) and charged groups located within the protein's structure. Furthermore, the microfluidization treatment effectively breaks down large protein aggregates, leading to an increased interaction area between the protein and water molecules (Han et al. 2023).

Similarly, microfluidization had a significant impact on the turbidity of both chicken (Han et al. 2023) and pork myofibrillar proteins (Zhang et al. 2022). In both cases, turbidity decreased after microfluidization treatment, and, for chicken myofibrillar proteins, the turbidity value decreased linearly with increasing microfluidization pressure (Han et al. 2023; Zhang et al. 2022). Furthermore, when lysin and arginine were combined with microfluidization treatment for pork myofibrillar proteins, there was an additional reduction in turbidity, indicating an additive effect (Zhang et al. 2022). Turbidity of the suspension is closely linked to the degree of particle aggregation (Chen et al. 2018). Microfluidization treatment reduces the particle size, thereby improving the accessibility of myofibrillar proteins to the solvent and reducing turbidity.

Contact angle measurements ($\theta$) were employed to assess the wettability of chicken myofibrillar proteins. The ability of an emulsifier to create emulsions is closely linked to the wettability of solid particles (Xiao, Li, and Huang 2016). Untreated chicken myofibrillar proteins exhibited a contact angle ($\theta$) of 59.5°, which notably increased following microfluidization treatment. Specifically, the $\theta$ value for chicken myofibrillar proteins treated at 90 MPa rose to 68.4°, signifying a considerable enhancement in chicken myofibrillar proteins hydrophobicity (Han et al. 2023). According to Kaptay (2006), oil-in-water (O/W) emulsions are stabilized by emulsifier particles with contact angles ranging from 15° to 90°. A higher $\theta$ value is indicative of better emulsifier wettability and, consequently, greater stability in the resulting emulsion. Based on the $\theta$ measurements for chicken myofibrillar proteins, it was determined that the optimal microfluidization condition was 90 MPa. However, it's important to note that as the pressure was increased to 120 MPa, chicken myofibrillar proteins wettability decreased, potentially attributed to the formation of hydrophobic aggregates under excessively high pressure (Han et al. 2023).

## 10.2.2 Effect on Surface Hydrophobicity and Sulfhydryl Groups

Surface hydrophobicity is an important property of proteins that influences their structure and function. Hydrophobic residues tend to occur preferentially in the core of proteins, while hydrophilic residues tend to occur on the surface. Hydrophobicity is thought to be one of the primary forces driving the folding of proteins. The hydrophobicity of protein surfaces can be quantified using different measures, such as liquid chromatography surface hydrophobicity. Proteins tend to bury hydrophobic residues inside their cores during the folding process to provide stability to the protein structure and to prevent aggregation. Nevertheless, proteins do expose some "sticky" hydrophobic residues to the solvent, which can play an important functional role, for example, in protein–protein and membrane interactions. The molecular structure of a protein can be changed or modified to expose more hydrophobic residues on the molecular surface, which may improve the surface properties of proteins (Jiang et al. 2015; Young, Jernigan, and Covell 1994). Upon microfluidization treatment, the surface hydrophobicity of chicken myofibrillar proteins showed a consistent linear increase as the microfluidization pressure increased from 0 to 120 MPa, with a maximum at 120 MPa (Han et al. 2023). This suggests that the protein molecules underwent unfolding as a result of the combined influences of high-pressure conditions, intense shear rates, and cavitation. Consequently, numerous hydrophobic groups originally nestled within the interior of chicken myofibrillar protein molecules gradually became exposed on the protein surface (Hu et al. 2021).

The sulfhydryl group, known for its high oxidation sensitivity, is the most reactive functional group in proteins and is often used to assess protein oxidation levels (Du et al. 2021). The total sulfhydryl group content serves as an indicator of changes in the tertiary structure and intermolecular interactions of proteins (Zhang et al. 2017). Microfluidization has been shown to reduce the total sulfhydryl group content of both chicken (Han et al. 2023) and pork myofibrillar protein (Zhang et al. 2022). In the case of chicken myofibrillar protein, the total sulfhydryl group content decreases with increasing microfluidization pressure up to 90 MPa, with further increase in pressure having no significant effect (Han et al. 2023). Microfluidization treatment disrupts the coarse and fine filament structures of myofibrillar proteins, exposing more sulfhydryl groups that are subsequently oxidized to form disulfide bonds, thereby reducing the overall sulfhydryl content (Zhang et al. 2022). The reduction in sulfhydryl content can have several positive effects on myofibrillar proteins. For instance, it can improve solubility and emulsification properties, as well as enhance resistance to heat and proteolysis. These improvements make myofibrillar proteins more suitable for a variety of food applications, such as meat processing, bakery, and dairy products.

## 10.2.3 Effect on Secondary Structure

The secondary structure of meat myofibrillar proteins refers to the organization of their amino acid backbone. There are three primary types of secondary structure in myofibrillar proteins: α-helices, β-sheets, and random coils. α-Helix are coiled structures where the amino acid backbone wraps around itself, stabilized by hydrogen bonds between the amino acids. They are highly stable and common in proteins like myosin and actin, playing a crucial role in their structure. β-Sheets are flat, pleated structures formed by hydrogen bonds between amino acids in different strands. They are less stable than α-helix and are found in proteins like titin and nebulin, contributing to their specific functions. Random coils lack a regular arrangement, leading to a disordered structure. They are the least stable form and are present in certain myofibrillar proteins like tropomyosin and troponin. The secondary structure of myofibrillar proteins is vital for their functions. For instance, myosin's α-helical structure enables it to bind with actin, facilitating muscle contraction. Titin's β-sheet structure provides elasticity, allowing it to stretch and recoil, a crucial property for muscle flexibility and movement.

Microfluidization pressure significantly affected the secondary structure of chicken myofibrillar proteins. α-Helix content decreased from 52.1% to 25.9% as pressure increased from 0 to 90 MPa, while β-sheet content increased from 13.1% to 26.0%. However, α-helix and β-sheet contents showed the opposite trend at 120 MPa. β-Turn content increased significantly after microfluidization treatment, but random coil content did not change significantly (Han et al. 2023). α-Helix structures are held together by hydrogen bonds within the protein chain (Chen et al. 2018). The shear force, high-speed impact, high-frequency vibration, and cavitation generated during microfluidization break these hydrogen bonds and unfold the protein structure (Sun et al. 2016).

## 10.2.4 Effect on Zeta Potential

The zeta potential, which characterizes the electric charge distribution on the surface of myofibrillar proteins, plays a central role in governing their behavior and functionality. This property holds significant implications for the stability and performance of these proteins in various food systems. A higher zeta potential, for instance, acts as a potent force in repelling fat droplets within emulsions, effectively preventing their unwanted aggregation and ensuring the desired dispersion throughout the system. Additionally, the favorable electrostatic interactions enabled by a high zeta potential promote the binding of water molecules to myofibrillar proteins, facilitating the formation of robust protein–water networks and enhancing the gelation process. This, in turn, contributes to the

# Microfluidization of Meat- and Egg-Based Products 203

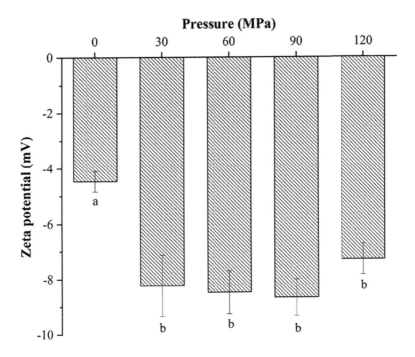

**FIGURE 10.2** Zeta potential of chicken myofibrillar proteins treated at different microfluidization pressures. Different letters on each column indicate significant difference (p < 0.05) among different treatment samples. (Reprinted from Han et al. (2023) with permission from Elsevier.)

development of desirable textural attributes, optimal moisture retention, and improved sensory characteristics in meat products. Thus, the intricate interplay between the zeta potential and the functional properties of myofibrillar proteins underscores its critical significance in shaping the quality and consumer acceptance of meat-based formulations. Microfluidization treatment led to a significant increase in the absolute zeta potential when compared to untreated chicken myofibrillar proteins. Interestingly, no notable distinctions were found among the samples treated at various pressures (Figure 10.2). Furthermore, the zeta potential of chicken myofibrillar proteins–camellia oil emulsion also increased post-microfluidization treatment (Han et al. 2023). The rise in the zeta potential value signifies that the microfluidization treatment strengthened the electrostatic repulsion among chicken myofibrillar proteins–camellia oil emulsion droplets. This increased electrostatic repulsion has the potential to prevent particle aggregation by counteracting the attractive forces between emulsion droplets, including van der Waals forces and hydrophobic interactions (Liu et al. 2021), thereby stabilizing the emulsion.

## 10.2.5 Effect on Emulsifying Properties

EAI, which stands for the emulsifying activity index (EAI), indicates the protein's ability to prevent separation, flocculation, and coalescence of emulsions formed at the oil–water interface (Li et al. 2020). On the other hand, ESI, or Emulsion Stability Index, reflects how well the emulsion maintains stability over a specific duration. The application of microfluidization significantly improved both EAI and ESI of chicken myofibrillar proteins. Specifically, EAI increased significantly as the microfluidization pressure rose from 0 to 90 MPa, plateauing afterward with no significant change beyond 90 MPa. ESI showed an increase after microfluidization treatment, reaching its peak value of 62.5% at 90 MPa, but then decreased to 50.6% at 120 MPa. Furthermore, the optimum condition for preparing chicken myofibrillar proteins–camellia oil emulsion was determined to be a microfluidization pressure of 90

**FIGURE 10.3** Microstructure of chicken myofibrillar proteins treated by different microfluidization pressures. (Reprinted from Han et al. (2023) with permission from Elsevier.)

MPa, based on a thorough assessment of EAI and ESI. However, when the pressure surpassed 90 MPa, the ESI of chicken myofibrillar proteins declined. This decrease occurred due to the exposed hydrophobic groups, which promoted chicken myofibrillar proteins aggregation through hydrophobic interactions and led to a reduction in exposed negative charges (Han et al. 2023). The enhancements observed in both EAI and ESI were a result of the unfolding of chicken myofibrillar proteins during microfluidization treatment, leading to the exposure of more hydrophobic, hydrophilic, and charged groups. This simultaneous exposure improved interactions between chicken myofibrillar proteins and both camellia oil and water. In typical emulsions, the relationship between droplet floatation velocity and droplet size adheres to Stokes' law: smaller droplets result in slower floatation velocities, hindering emulsion droplet flocculation and enhancing emulsion stability (Fennema 1996).

### 10.2.6 Microstructure

Figure 10.3 illustrates the microstructure (magnified 100×) of freeze-dried chicken myofibrillar proteins subjected to varying microfluidization pressures. The untreated chicken myofibrillar proteins displayed substantial clumps with uneven distribution and noticeable gaps between clusters. With increasing pressure, the degree of aggregation and surface roughness of chicken myofibrillar protein samples steadily diminished, resulting in a more uniform and organized structure. High-speed shear force and the sudden pressure drop during microfluidization broke down the massive aggregates into particles. At 120 MPa, significant portions of chicken myofibrillar protein aggregates vanished, leading to an even dispersion of the remaining aggregates (Han et al. 2023).

## 10.3 MICROFLUIDIZATION OF EGG AND EGG-BASED PRODUCTS

In the past decade, food scientists from India and China have worked on processing egg yolk and egg white using microfluidizer (Chi, Li, and Zhao 2017; Suhag et al. 2021, 2022; Tu et al. 2009). These scientists have worked on microfluidization which has been found to bring tremendous changes in physico-chemical, techno-functional, structural, thermal, and rheological properties of egg. These changes have been linked to be caused by various disruptive forces generated inside the interaction chamber which led to reduction of particle size, protein denaturation, and rearrangement of molecules (Kumar et al. 2022). Changes brought by microfluidization increase the potential of egg yolk and egg white as a potential food ingredient discussed in later sections.

### 10.3.1 Effect on Physico-chemical Properties

Microfluidization processing significantly decreased the particle size of egg white protein within a pressure range of 20 to 160 MPa (Tu et al. 2009). Similarly, egg yolk particle size decreased from 85 to 37 μm with an increasing pressure from 103 to 172 MPa. In contrast, further increasing the pressure to 207 MPa leads to an increase in the particle size to 87 μm (Suhag et al. 2021). Beyond a critical point in pressure and number of passes, termed as "over processing," it has been observed that emulsion particle size increases instead of decreasing, despite the high energy input. This phenomenon is attributed to increased Brownian motion, resulting in more collisions and re-coalescence at higher energy inputs (Kumar et al. 2022).

In terms of color characteristics, compared to the control sample, the L* value of the microfluidized egg yolk was considerably higher, rising from 55.29 (control) to 62.05. Reduced particle size enhances light scattering and changes in color values. Microfluidization also lowered the pH and refractive index values of egg yolk, though the reduction in these values was up to such an extent that it did not affect the quality of egg yolk. Variation in physico-chemical characteristics can also be attributed to molecular rearrangement brought post-microfluidization, which was also observed in the FTIR spectra by Suhag et al. (2021). With an increase in pressure from 103 to 207 MPa, the intensity of peaks increased, as shown in Figure 10.4.

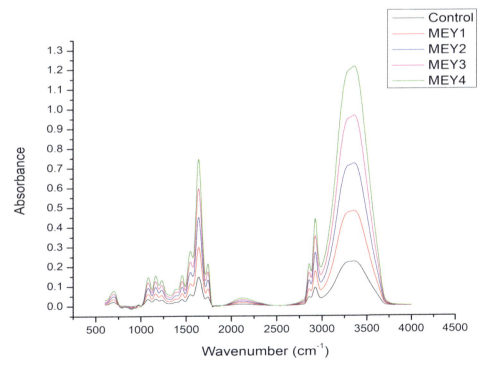

**FIGURE 10.4** FTIR spectrum of control and microfluidized egg yolk, where MEY1 is the microfluidized egg yolk at 103 MPa, MEY2 is the microfluidized egg yolk at 138 MPa, MEY3 is the microfluidized egg yolk at 172 MPa, and MEY4 is the microfluidized egg yolk at 207 MPa. (Reprinted from Suhag et al. (2021) with permission from Elsevier.)

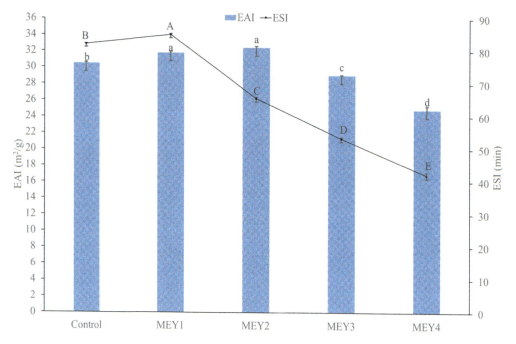

**FIGURE 10.5** Impact of microfluidization on emulsifying activity index (EAI) and emulsifying stability index (ESI), where MEY1 is the microfluidized egg yolk at 103 MPa, MEY2 is the microfluidized egg yolk at 138 MPa, MEY3 is the microfluidized egg yolk at 172 MPa, and MEY4 is the microfluidized egg yolk at 207 MPa. (Reprinted from Suhag et al. (2021) with permission from Elsevier.)

### 10.3.2 Effect on Functional Properties

Microfluidization has also been reported to reduce the denaturation enthalpy on increasing the pressure from 20 to 100 MPa and 120 to 160 MPa. Microfluidization also led to changes in solubility and increase in surface hydrophobicity with a maximum value at 100 MPa (Tu et al. 2009). Microfluidization has been found to decrease the allergenicity and increase the digestibility of egg white proteins. In a study performed by Chi, Li, and Zhao (2017), egg white protein was microfluidized in a range of 40–200 MPa. Pressure ranging from 80 to 200 MPa was found to significantly reduce the allergenicity of egg white protein which was measured using ELISA assay. The IgE-binding was reduced up to 64%. An increase in the free sulfhydryl group along with surface hydrophobicity was also observed. In vitro digestion analysis of these samples revealed an increase in the values of digestibility and degree of hydrolysis. Changes in secondary and tertiary structures of protein including disruption of α-helices and their further conversion into random coils and β-sheets were attributed to these changes (Chi, Li, and Zhao 2017). Microfluidization was also effective in increasing the emulsifying activity and stability from 30.49 m²/g to 32.41 m²/g and 82.21 to 85.06 min, respectively, as shown in Figure 10.5. These changes were attributed to mild protein denaturation caused by microfluidization (Suhag et al. 2021).

### 10.3.3 Effect on Thermal Properties

The studies of gelatinization, crystallization, melting, cross-linking, oxidation, and the detection of water loss from food matrix are all subjects of thermal analysis (Farah et al. 2018). In egg yolk samples, the denaturation temperature rose noticeably as the microfluidization pressure rose. Microfluidization sample at 207 MPa showed the highest value of 91.55 °C, whereas the control

Microfluidization of Meat- and Egg-Based Products

sample showed a minimum denaturation temperature of 84.85 °C, which was close to the transition temperatures of both granule (85.9 °C) and plasma (86.1 °C) proteins. The higher denaturation temperature observed in microfluidized egg yolk samples is a result of the method's ability to disrupt protein aggregates and expose more protein molecules to the surrounding environment. This increased protein exposure enhances their susceptibility to denaturation when subjected to heat, leading to a noticeable rise in denaturation temperature compared to untreated samples (Suhag et al. 2021). These changes can be induced because of the significant shear, cavitation, and turbulence produced inside the interaction chamber, thereby leading to denaturation of proteins (Dhiman et al. 2022).

### 10.3.4 Effect on Rheological Properties

Microfluidization has been reported to bring significant changes in the rheological properties of egg yolk (Suhag et al. 2021, 2022). In comparison to raw egg yolk sample which possessed the Newtonian nature, the microfluidized egg yolk had shown shear-thinning or pseudoplastic nature. Microfluidized egg yolk's viscosity dropped as the shear rate rose. The viscosity of the egg yolk tends to decrease with shear rate as a result of degradation of interactions. The viscosity of egg yolk samples increased with an increase in pressure. It was concluded that microfluidization-induced rheological changes will offer production advantages as well as forming emulsions easily (Suhag et al. 2021). Further investigation by Suhag et al. (2022) revealed that the steady-state rheological data was well explained by the Power law, Casson, Bingham, and Herschel-Bulkley models. Time-dependent rheology revealed the microfluidized egg yolk's thixotropic and anti-thixotropic behavior. Shear stress readings for egg yolk microfluidized at lower pressures were greater than the control, according to time-dependent rheology. When the microfluidization pressure was raised to 172 MPa, shear stress almost doubled and was almost four times higher at 207 MPa. This indicates that egg yolk proteins partially unfold at low microfluidization pressures, whereas fully unfolding occurs at high microfluidization pressures. Control and microfluidized egg yolk at 103 and 138 MPa showed solid-like behavior in terms of dynamic viscoelastic properties, but egg yolk microfluidized at higher pressure (172 and 207 MPa) demonstrated liquid-like behavior. Microfluidized egg yolk had laminar flow that was calculated using Reynolds number (Suhag et al. 2022).

## 10.4 CHALLENGES AND POTENTIAL SOLUTIONS TO MICROFLUIDIZATION OF MEAT- AND EGG-BASED PRODUCTS

The process of microfluidizing meat- and egg-based products is intricate, involving the reduction of particle size and homogenization through high-pressure fluid dynamics. A significant hurdle in microfluidizing meat-based products is their natural viscosity, often making it challenging to achieve uniform particle size reduction. This viscosity leads to uneven processing, impacting the final product's texture and consistency. Researchers and food technologists tackle this challenge by adjusting various parameters such as pressure and flow rate. Preliminary treatments, like enzymatic breakdown, are also utilized to reduce viscosity before microfluidization, ensuring a more uniform and efficient process.

Temperature sensitivity poses another obstacle in microfluidizing these products, particularly meat. Temperature fluctuations can adversely affect texture, flavor, and overall quality. Generated heat during microfluidization can denature proteins, altering the product's characteristics. Precise temperature control is vital, and specialized microfluidization equipment with temperature regulation capabilities is used. Cooling methods, like cryogenic fluids or incorporating cooling stages, help mitigate temperature impact on the final product.

Fat content in these products presents a challenge, influencing the emulsion's homogeneity and stability. Achieving a uniform distribution of fat particles is crucial for texture and mouthfeel.

Emulsification techniques and suitable emulsifiers are essential to stabilize fat droplets in the micro-fluidized product. Understanding fat's phase behavior during processing is critical. Manufacturers overcome this challenge by selecting appropriate emulsifiers and optimizing the emulsification process.

Protein denaturation is a concern, especially with egg-based products, as high-pressure processing can negatively impact texture and quality. Researchers focus on optimizing pressure conditions and processing time to minimize protein denaturation. Specialized equipment designed to handle delicate proteins and innovative methods such as pulsed electric fields help mitigate protein denaturation during microfluidization.

Microfluidization often leads to product losses due to adhesion to equipment surfaces caused by high pressure and shear forces. Engineers employ strategies like designing equipment with smooth surfaces and anti-stick coatings, implementing scraping mechanisms, and recirculation systems. Meticulous cleaning procedures prevent cross-contamination, optimizing yield.

Meeting regulatory standards is crucial. Compliance involves adhering to guidelines related to food safety, labeling, and overall product quality. Extensive documentation, quality control, testing for contaminants, validation of processing methods, and accurate ingredient labeling are fundamental. Collaboration with food safety experts and legal advisors ensures that products meet consumer preferences and adhere to legal requirements, fostering trust among consumers and regulatory bodies.

Overall, addressing microfluidization challenges demands a multidisciplinary approach, combining expertise in food science, engineering, and technology. By adjusting parameters, controlling temperature, stabilizing fats, and minimizing protein denaturation, researchers can develop effective strategies for high-quality microfluidized meat and egg-based products meeting consumer expectations.

## 10.5 CONCLUSIONS AND PERSPECTIVES

In the past, microfluidization has been reported to reduce the cholesterol content of cow ghee (Dhiman et al. 2022), cow milk, and buffalo milk (Kumar et al. 2019) by 40%, 42%, and 46%, respectively. Egg yolk does possess a high level of cholesterol that is around 1085 mg/100 g (Réhault-Godbert, Guyot, and Nys 2019), which is even higher than that of cow ghee (250 mg/100 g) (Bhatia et al. 2019) and milk (22 mg/100 mL) (Kumar et al. 2019). Future studies can target investigating the effect of microfluidization on egg cholesterol. This can lead toward exploring and further proposing microfluidization as a mechanical technology for reducing cholesterol in food products. In addition to this, since high pressure and mechanical forces are involved in this process, along with heat generation at higher pressure, investigation related to its effect on microbial load especially pathogens can be explored further. Also, past studies carried out on microfluidization of egg yolk were only on single pass, future studies can investigate the effects of a number of passes on overall quality of egg.

In summary, this chapter elucidates the significant impact of microfluidization technology on the physico-chemical, functional, and structural aspects of meat and egg-based food products. While promising, the implementation of microfluidization faces challenges requiring meticulous control over parameters and consideration of sensory attributes. Achieving consistent product quality demands precision in process parameters, necessitating advanced instrumentation and optimization techniques. Balancing enhanced functionality with sensory appeal is vital for consumer acceptance. Scalability and cost-effectiveness pose hurdles for industrial applications, demanding innovative approaches. Encouragingly, ongoing advancements in computational modeling and material science offer potential solutions. Through interdisciplinary collaborations and continuous research, addressing these challenges can unlock microfluidization's transformative potential, ushering in a new era of innovative, high-quality food products.

# REFERENCES

Bhargava, Nitya, Rahul S. Mor, Kshitiz Kumar, and Vijay Singh Sharanagat. 2021. "Ultrasonics – Sonochemistry Advances in Application of Ultrasound in Food Processing: A Review." *Ultrasonics – Sonochemistry* 70 (June 2020): 105293. https://doi.org/10.1016/j.ultsonch.2020.105293

Bhatia, Piyush, Vivek Sharma, Sumit Arora, and Priyanka Singh Rao. 2019. "Effect of Cholesterol Removal on Compositional and the Physicochemical Characteristics of Anhydrous Cow Milk Fat (Cow Ghee)." *International Journal of Food Properties* 22 (1): 1–8. https://doi.org/10.1080/10942912.2018.1564762

Chen, Xing, Xinglian Xu, Dongmei Liu, Guanghong Zhou, Minyi Han, and Peng Wang. 2018. "Rheological Behavior, Conformational Changes and Interactions of Water-Soluble Myofibrillar Protein during Heating." *Food Hydrocolloids.* https://doi.org/10.1016/j.foodhyd.2017.10.030

Chi, Yujie, Yinnan Li, and Ying Zhao. 2017. "Effect of Dynamic High-Pressure Microfluidization on Egg White Protein Allergenicity and Digestibility." *Nongye Jixie Xuebao/Transactions of the Chinese Society for Agricultural Machinery* https://doi.org/10.6041/j.issn.1000-1298.2017.06.041

Dhiman, Atul, and Pramod K Prabhakar. 2021. "Micronization in Food Processing: A Comprehensive Review of Mechanistic Approach, Physicochemical, Functional Properties and Self-Stability of Micronized Food Materials." *Journal of Food Engineering* 292: 110248. https://doi.org/10.1016/j.jfoodeng.2020.110248

Dhiman, Atul, Rajat Suhag, Kiran Verma, Dhruv Thakur, Anit Kumar, Ashutosh Upadhyay, and Anurag Singh. 2022. "Influence of Microfluidization on Physico-Chemical, Rheological, Thermal Properties and Cholesterol Level of Cow Ghee." *LWT.* https://doi.org/10.1016/j.lwt.2022.113281

Du, Peng Cheng, Zong Cai Tu, Hui Wang, Yue Ming Hu, Jing Jing Zhang, and Bi Zhen Zhong. 2021. "Investigation of the Effect of Oxidation on the Structure of β-Lactoglobulin by High Resolution Mass Spectrometry." *Food Chemistry.* https://doi.org/10.1016/j.foodchem.2020.127939

Farah, Juliana S., Marcia C. Silva, Adriano G. Cruz, and Verônica Calado. 2018. "Differential Calorimetry Scanning: Current Background and Application in Authenticity of Dairy Products." *Current Opinion in Food Science* 22: 88–94. https://doi.org/10.1016/j.cofs.2018.02.006

Feijoo, S.C., W.W. Hayes, C.E. Watson, and J.H. Martin. 1997. "Effects of Microfluidizer® Technology on Bacillus Licheniformis Spores in Ice Cream Mix." *Journal of Dairy Science* 80: 2184–2187. https://doi.org/10.3168/jds.s0022-0302(97)76166-6

Fennema, Owen R. 1996. *Food Chemistry* 3rd Edition. Marcel Dekker Inc.

Guo, Xiaojuan, Mingshun Chen, Yuting Li, Taotao Dai, Xixiang Shuai, Jun Chen, and Chengmei Liu. 2020. "Modification of Food Macromolecules Using Dynamic High Pressure Microfluidization: A Review." *Trends in Food Science & Technology* 100 (April): 223–234. https://doi.org/10.1016/j.tifs.2020.04.004

Han, Keying, Xiao Feng, Yuling Yang, Xiaozhi Tang, and Chengcheng Gao. 2023. "Changes in the Physicochemical, Structural and Emulsifying Properties of Chicken Myofibrillar Protein via Microfluidization." *Innovative Food Science and Emerging Technologies.* https://doi.org/10.1016/j.ifset.2022.103236

Han, Keying, Shanshan Li, Yuling Yang, Xiao Feng, Xiaozhi Tang, and Yumin Chen. 2022. "Mechanisms of Inulin Addition Affecting the Properties of Chicken Myofibrillar Protein Gel." *Food Hydrocolloids.* https://doi.org/10.1016/j.foodhyd.2022.107843

Hogan, E, A. L. Kelly, and D W Sun. 2005. "High Pressure Processing of Foods: An Overview." In *Emerging Technologies for Food Processing, Emerging Technologies for Food Processing*, edited by D. W. Sun, 1–27. Elsevier.

Hu, Xiao, William Kwame Amakye, Peiying He, Min Wang, and Jiaoyan Ren. 2021. "Effects of Microfluidization and Transglutaminase Cross-Linking on the Conformations and Functional Properties of Arachin and Conarachin in Peanut." *LWT.* https://doi.org/10.1016/j.lwt.2021.111438

Jiang, Lianzhou, Zhongjiang Wang, Yang Li, Xianghe Meng, Xiaonan Sui, Baokun Qi, and Linyi Zhou. 2015. "Relationship between Surface Hydrophobicity and Structure of Soy Protein Isolate Subjected to Different Ionic Strength." *International Journal of Food Properties.* https://doi.org/10.1080/10942912.2013.865057

Kaptay, G. 2006. "On the Equation of the Maximum Capillary Pressure Induced by Solid Particles to Stabilize Emulsions and Foams and on the Emulsion Stability Diagrams." *Colloids and Surfaces A: Physicochemical and Engineering Aspects.* https://doi.org/10.1016/j.colsurfa.2005.12.021

Kumar, Anit, Prarabdh C. Badgujar, Vijendra Mishra, Rachna Sehrawat, Onkar A. Babar, and Ashutosh Upadhyay. 2019. "Effect of Microfluidization on Cholesterol, Thermal Properties and in Vitro and in Vivo Protein Digestibility of Milk." *LWT* 116 (December): 108523. https://doi.org/10.1016/j.lwt.2019.108523

Kumar, Anit, Atul Dhiman, Rajat Suhag, Rachna Sehrawat, Ashutosh Upadhyay, and David Julian McClements. 2022. "Comprehensive Review on Potential Applications of Microfluidization in Food Processing." *Food Science and Biotechnology* 31 (1): 17–36. https://doi.org/10.1007/s10068-021-01010-x

Li, Ke, Lei Fu, Ying-Ying Zhao, Si-Wen Xue, Peng Wang, Xing-Lian Xu, and Yan-Hong Bai. 2020. "Use of High-Intensity Ultrasound to Improve Emulsifying Properties of Chicken Myofibrillar Protein and Enhance the Rheological Properties and Stability of the Emulsion." *Food Hydrocolloids* 98 (January): 105275. https://doi.org/10.1016/j.foodhyd.2019.105275

Liu, Haotian, Jingnan Zhang, Hui Wang, Qian Chen, and Baohua Kong. 2021. "High-Intensity Ultrasound Improves the Physical Stability of Myofibrillar Protein Emulsion at Low Ionic Strength by Destroying and Suppressing Myosin Molecular Assembly." *Ultrasonics Sonochemistry*. https://doi.org/10.1016/j.ultsonch.2021.105554

Liu, Ru, Si-Ming Zhao, Bi-Jun Xie, and Shan-Bai Xiong. 2011. "Contribution of Protein Conformation and Intermolecular Bonds to Fish and Pork Gelation Properties." *Food Hydrocolloids* 25 (5): 898–906. https://doi.org/10.1016/j.foodhyd.2010.08.016

Mcclements, David Julian. 2016. "Emulsion Formation." In *Food Emulsions Principles, Practices, and Techniques*, edited by David Julian Mcclements, 245–284. CRC Press Taylor & Francis Group.

Mert, Ilkem Demirkesen. 2020. "The Applications of Microfluidization in Cereals and Cereal-Based Products: An Overview." *Critical Reviews in Food Science and Nutrition*. https://doi.org/10.1080/10408398.2018.1555134

Réhault-Godbert, Sophie, Nicolas Guyot, and Yves Nys. 2019. "The Golden Egg: Nutritional Value, Bioactivities, and Emerging Benefits for Human Health." *Nutrients*. https://doi.org/10.3390/nu11030684

Salvia-Trujillo, Laura, M. Alejandra Rojas-Graü, Robert Soliva-Fortuny, and Olga Martín-Belloso. 2014. "Impact of Microfluidization or Ultrasound Processing on the Antimicrobial Activity against Escherichia Coli of Lemongrass Oil-Loaded Nanoemulsions." *Food Control*. https://doi.org/10.1016/j.foodcont.2013.09.015

Suhag, Rajat, Atul Dhiman, Pramod K. Prabhakar, Arun Sharma, Anurag Singh, and Ashutosh Upadhyay. 2022. "Microfluidization of Liquid Egg Yolk: Modelling of Rheological Characteristics and Interpretation of Flow Behavior under a Pipe Flow." *Innovative Food Science and Emerging Technologies*. https://doi.org/10.1016/j.ifset.2022.103119

Suhag, Rajat, Atul Dhiman, Dhruv Thakur, Anit Kumar, and Ashutosh Upadhyay. 2021. "Physico-Chemical and Functional Properties of Microfluidized Egg Yolk." *Journal of Food Engineering* 294 (April): 110416. https://doi.org/10.1016/j.jfoodeng.2020.110416

Sun, Cuixia, Lei Dai, Fuguo Liu, and Yanxiang Gao. 2016. "Dynamic High Pressure Microfluidization Treatment of Zein in Aqueous Ethanol Solution." *Food Chemistry*. https://doi.org/10.1016/j.foodchem.2016.04.138

Tobin, John, Sinead P. Heffernan, Daniel M. Mulvihill, Thom Huppertz, and Alan L. Kelly. 2015. "Applications of High-Pressure Homogenization and Microfluidization for Milk and Dairy Products." In *Emerging Dairy Processing Technologies: Opportunities for the Dairy Industry*, edited by Nivedita Datta and Peggy M. Tomasula, First, 18:41. https://doi.org/10.1002/9781118560471.ch4

Tu, Zongcai, H. Wang, C.-M. Liu, G.-X. Liu, G. Chen, and Roger Ruan. 2009. "Dynamic High Pressure Micro-Fluidization Effects on Structure and Physico-Chemical Properties of Egg White Protein." *Chemical Research in Chinese Universities* 25 (3): 302–305.

Wang, Ke, Yan Li, Yimin Zhang, Ming Huang, Xinglian Xu, Harvey Ho, He Huang, and Jingxin Sun. 2022. "Improving Physicochemical Properties of Myofibrillar Proteins from Wooden Breast of Broiler by Diverse Glycation Strategies." *Food Chemistry*. https://doi.org/10.1016/j.foodchem.2022.132328

Wu, Fan, Xiaojie Shi, Henan Zou, Tingyu Zhang, Xinran Dong, Rui Zhu, and Cuiping Yu. 2019. "Effects of High-Pressure Homogenization on Physicochemical, Rheological and Emulsifying Properties of Myofibrillar Protein." *Journal of Food Engineering*. https://doi.org/10.1016/j.jfoodeng.2019.07.009

Xiao, Jie, Yunqi Li, and Qingrong Huang. 2016. "Recent Advances on Food-Grade Particles Stabilized Pickering Emulsions: Fabrication, Characterization and Research Trends." *Trends in Food Science & Technology* 55 (September): 48–60. https://doi.org/10.1016/j.tifs.2016.05.010

Xiong, Yao, Qianru Li, Song Miao, Yi Zhang, Baodong Zheng, and Longtao Zhang. 2019. "Effect of Ultrasound on Physicochemical Properties of Emulsion Stabilized by Fish Myofibrillar Protein and Xanthan Gum." *Innovative Food Science and Emerging Technologies*. https://doi.org/10.1016/j.ifset.2019.04.013

Young, L., R.L. Jernigan, and D.G. Covell. 1994. "A Role for Surface Hydrophobicity in Protein-Protein Recognition." *Protein Science* 3 (5): 717–729. https://doi.org/10.1002/pro.5560030501

Zhang, Dong, Zhicheng Wu, Jinggang Ruan, Yizhi Wang, Xueyi Li, Min Xu, Jie Zhao, et al. 2022. "Effects of Lysine and Arginine Addition Combined with High-Pressure Microfluidization Treatment on the Structure, Solubility, and Stability of Pork Myofibrillar Proteins." *LWT*. https://doi.org/10.1016/j.lwt.2022.114190

Zhang, Ziye, Yuling Yang, Peng Zhou, Xing Zhang, and Jingyu Wang. 2017. "Effects of High Pressure Modification on Conformation and Gelation Properties of Myofibrillar Protein." *Food Chemistry*. https://doi.org/10.1016/j.foodchem.2016.09.040

# 11 Microfluidization-Assisted Extraction of Bioactive Compounds from Biological Resources

*Meemansha Sharma, Rakesh Karwa, S. Ilavarasan, Mamta Meena, Manju Gari, Ranjna Sirohi, Anshuk Sharma, and Thakur Uttam Singh*

## 11.1 INTRODUCTION

The extraction of bioactive constituents has gained significant importance because of its numerous health-promoting attributes, including its antibacterial, antioxidant, and anti-carcinogenic qualities. These substances are a great source of molecules that can be used to make food additives, nutraceuticals, and other flavoring chemicals (Khan et al., 2019). One of the major issues that pharmaceutical and food industries are dealing with is the safe and efficient extraction of these bioactive substances. The extraction process has a significant impact on the active components' yield, purity, chemical makeup, and biological efficacy. There are a variety of extraction techniques available today, which can be categorized as traditional and non-traditional (Salam et al. 2019). A few examples of traditional techniques are maceration, Soxhlet, hot water extraction, alkaline extraction, and acid extraction (Zhang et al. 2018). These are well-established industrial techniques that benefit from simple operations and inexpensive installation. Nonetheless, the primary obstacles linked to them include extended extraction durations, reduced extraction specificity, elevated temperatures, substantial usage of organic solvents, potential extraction loss as a result of extra cleaning procedures, and so on. Numerous non-traditional technologies have been suggested to address these drawbacks, such as enzyme-assisted extraction, ultrasonic extraction, microwave extraction, infrared extraction, negative pressure cavitation, and dynamic high-pressure microfluidization (DHPM) (Azmir et al. 2013). These techniques are safer for the environment since they require less industrial and organic chemicals, which results in greater output, quicker operation times, and faster operating times. DHPM is a technique that shows promise for boosting the concentration of bioactive substances while preserving their bioactivities and benefiting from low-operating temperatures and its eco-friendly characteristics (Huang et al. 2023). It is a high-pressure homogenization process that combines forces including strong shear, impact force, and cavitation via a small opening to generate liquid materials at extremely high pressures (Jing et al. 2016). DHPM can increase the release of contents from cells and the synthesis of active ingredients. Apart from the extraction process, it is quite effective in encapsulating bioactive substances as well as creating cutting-edge delivery systems (Kavinila et al. 2023). It is able to structurally alter biological macromolecules for better techno-functional qualities because of many active mechanical forces (Kavinila et al. 2023).

212

DOI: 10.1201/9781032632599-11

The two primary focuses of modern microfluidization research are the morphological and physicochemical alterations of treated materials with this technique and the breaking down of cell structures to extract intracellular substances from plant cells, such as flavones, polysaccharides, and polyphenols. Nevertheless, there hasn't been much information published about the use of microfluidization in actual industrial food production. It is worthwhile to investigate in depth the discrepancy between theoretical studies and practical utilization of microfluidization in food processing. With the expanding popularity of microfluidization in mind, this chapter provides detailed explanations of the technique's basic structure and operating principle, along with some unique applications pertaining to the extraction of bioactive contents from plants and food sources.

## 11.2  WORKING OF A MICROFLUIDIZER

Microfluidization is a unique method for creating homogenous products, different from traditional techniques like homogenization and ultrasonication. Microfluidization is a combination of water jet technology, impinging stream technology, and high-pressure homogenization (Li et al. 2022). A liquid feed is divided into two or more microstreams and made to clash with one another inside a precisely designed interaction chamber during the microfluidization process. This process produces a special mix of high pressure, high-speed vibrations, shear rate, and cavitation caused by hydrodynamics (Ozturk & Turasan 2021). The breakdown and homogenization occur mainly in the interaction chamber due to inertial forces in turbulent flow and cavitation.

The standard components of a microfluidizer include an inlet reservoir, an interaction chamber, an intensifier pressure pump, a heat exchanger, a cooling coil, and a collecting reservoir (Bai & McClements, 2016). Core microfluidizer components that directly impact processing efficiency are the intensifier pump and the interaction chamber. The intensifier pump speeds up the liquid during the microfluidization process up to 400 m/s velocity with high pressure at 30,000 psi (Huang et al. 2023). This intense pressure and speed cause shearing between the liquid and the channel walls and subsequently between liquid droplets. The high-speed liquid is then pumped into the interaction chamber, where there is a strong Y- or Z-shaped impact. Interaction chambers are available in two versions that include Y-type and Z-type (Leyva-Daniel et al. 2020). The high-pressure liquid stream in a Y-type chamber is split into two sub-streams, each of which collides with the microchannel walls and one another at a speed of up to 400 m/s (Leyva-Daniel et al. 2020). In contrast, the liquid stream in a Z-type chamber is driven via zigzag microchannels where it collides (Mert 2020). Z-type chambers are typically utilized for the dispersion of solid materials in suspensions and the breakdown of cell structures, whereas Y-type chambers are mostly employed for liquid–liquid dispersions (Mert 2020).

## 11.3  EXTRACTION PROCEDURE OF BIOACTIVE COMPOUND WITH DHPM

The extraction of bioactive chemicals using DHPM necessitates three steps: sample preparation, DHPM treatment, and extraction (Figure 11.1). The initial phase of sample preparation includes the removal of moisture from the plant parts. This is achieved through a controlled drying process. Maintain the drying temperature at either 50 or 60 °C to prevent thermal degradation, as suggested by various studies (Sun et al. 2013; Huang et al. 2012; Jing et al. 2016). Once dried, the plant sample is crushed and sieved through a mesh screen to obtain a uniform particle size. To maintain sample purity, immerse the crushed and sieved plant material in ethyl ether or petroleum ether, along with other appropriate solvents. Further, utilize extraction procedures such as Soxhlet or reflux to effectively remove lipids, pigments, and fat-soluble components from the immersed sample (Huang et al. 2023).

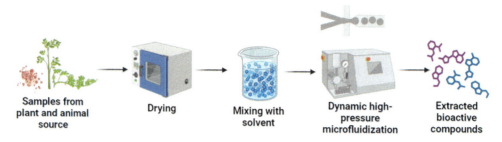

**FIGURE 11.1** Process of extraction of bioactive compounds using DHPM.

To prepare the specimen for DHPM treatment, the next step process entails first soaking it in extraction solvents and then prehomogenizing it (Huang et al. 2023). Due to the small size of the DHPM interaction chamber, a valve homogenizer is used initially to prevent clogging of the microfluidizer (Guo et al. 2020). The homogenizer employs hydraulic shear action, cavitation, and impact action to finely crush the material. Subsequently, the treated sample is processed in the microfluidizer to further improve extraction efficiency (Putri et al. 2022). For extraction, the treated sample is kept in a container, and a solvent is added before opening the reflux system (Qin et al. 2019; Huang et al. 2023). Following extraction, the resulting extract is filtered and centrifuged at 500–4000 rpm for 10–15 minutes (Qin et al. 2019; Huang et al. 2023). The supernatant is collected, evaporated, and stored for subsequent analysis.

## 11.4 FACTORS AFFECTING DHPM-TREATED EXTRACTION OF BIOACTIVE COMPOUNDS

The effectiveness of DHPM-treated extraction is influenced by some significant factors including the selection of solvent, solid–liquid ratio, temperature, and pressure. The selection of solvent in extraction processes is governed by various factors, including the solubility of the target component, the solvent's ability to penetrate and interact with the sample matrix, and the dielectric constant of the solvent (Ameer et al. 2017). For some extractions, an aqueous solution of a specific organic solvent is preferred, as water can enhance mass transfer by increasing the solvent's relative polarity, improving permeability in the sample matrix, and increasing the surface area for solute–solvent interaction (Mandal & Mandal 2010). Organic solvents like ethanol, methanol, and acetone are also effective in extractions, with differences observed in extraction yields and antioxidant properties (Anwar & Przybylski 2012). Once the solvent is selected, attention must be given to the solid–liquid ratio, as it significantly influences extraction yield. A low solvent proportion in the solid may concentrate the active compound in a specific area, limiting its movement from the cell matrix to the solvent and hindering mass transfer (Khaw et al. 2017).

In addition, temperature and pressure are other variables that affect the bioactive chemical extraction yield during DHPM treatment. Controlling pressure during microfluidization led to better particle size reduction and even dispersion (Li et al. 2022). Higher pressure resulted in more severe cell rupture, enhancing the dissolution of active components in cells. DHPM treatment facilitated quicker and more even dispersion of sample particles in the solvent, yielding higher extraction rates (Tu et al. 2010). Additionally, continuous pressure rise might cause slight decreases in yields, possibly due to high-frequency oscillations or changes in dissolved oxygen promoting oxidation and altering compound structures (Huang et al. 2023). The stability of the active ingredient and the intended extraction yield influence the extraction temperature selection (Chuo et al. 2022). Excessive heat risks thermal degradation, oxidation, protein denaturation, and increased solution viscosity lead to lower solubility and yield (Huang et al. 2023). Generally, elevated temperatures promote the mass transfer rate of molecules, highlighting the importance of temperature control in extraction processes.

## 11.5 FLAVONOIDS BY DHPM TREATMENT

Flavonoids are commonly found in medicinal plants and have vital pharmacological qualities (Karak 2019). Flavonoids contain numerous pharmacological properties, including anti-oxidation, anticancer, antimutation, anti-inflammatory, antibacterial, and anti-aging; hence, they can produce positive effects on human and animal health (Wan 2019). Moreover, they are also utilized in the food and cosmetics sectors as pigments and biopreservatives (Tzanova et al. 2020). DHPM significantly boosted the amount of overall flavonoids extracted from *C. esculentus* leaves (Jing et al. 2016). DHPM pretreatment not only increased the total flavonoid yield but also strengthened the antioxidant activity of these flavonoids. Huang et al. (2013) have reported the higher content of flavonoids in leaves obtained from sweet potato by microfluidization. For the extraction of flavonoids, the leaf extract of sweet potato was treated with DHPM homogenizer at 100 MPa twice. Similarly, in another study, the yield of flavonoids and antioxidant activity has been improved with DHPM treatment in *Ipomoea batatas* leaves (Zhang et al. 2017). Further, Gali et al. (2020) have shown the highest recovered amount of some low water-soluble flavonoids such as rutin and quercetin from *R. chalepensis* plant extract by high-pressure homogenization at 100 MPa and in 10 passes. Nayak and co-workers (2023) reported significant yield of extracted flavonoids like hesperidin (284.07 mg/L, 1.7-fold higher) and naringin (1.8-fold higher) when the juice from granulated dancy-tangerine (citrus fruit) was subjected to microfluidization at 124 MPa three times in comparison with non-microfluidized juice. Another study demonstrated the increased extraction yield of bioactive flavonoids from Peruvian purple corn cob extract by dynamic high pressure at 60 MPa and 45 °C. According to this study, the flavonoids concentration was 108% higher in comparison with different conventional extraction processes (Guillén Sánchez et al. 2023). Sun et al. (2013) showed that DHPM treatment resulted in the highest recovery of flavonoid content (3.49%) from the leaves of *Radix tetrastigmae* as compared to the old reflux method. Additionally, the higher extracted concentration of flavonoid reported from DHPM-treated leaves obtained from *Gynura procumbens* at 80 MPa compared to the reflux method (Tu et al. 2018). All research findings related to DHPM-treated extraction of flavonoids mentioned in this section are summarized in Table 11.1.

## TABLE 11.1
## Application of DHPM in Flavonoid Extraction

| Bioactive Compounds | Source | Processing Conditions | Key Findings | Reference |
|---|---|---|---|---|
| Flavonoid | *Cyperus esculentus* leaves | 120 MPa pressure<br>2 passes cycles<br>M-110EH Microfluidizer (USA) | Increased yield and free radical scavenging activity | Jing et al. (2016) |
| Flavonoid | *Ipomoea batatas* leaves | 100 MPa pressure<br>2 passes cycles<br>M-110EH Microfluidizer (USA) | Higher flavonoid content<br>Increased free radical scavenging | Huang et al. (2013) |
| Flavonoid | *Ruta chalepensis* plant extract | 100 MPa pressure<br>2 passes cycles | Enhanced extraction yield of flavonoids | Gali et al. (2020) |
| Flavonoid | Dancy tangerine citrus fruit juice | 124 MPa pressure<br>3 passes cycles<br>M-110P Microfluidics (USA) | High flavonoids yield<br>Reduced enzymatic activity<br>Good organoleptic properties | Nayak et al. (2023) |
| Flavonoid | Peruvian purple corn cob extract | 60 MPa pressure and 40 °C temperature in one step | Increased extraction amount of flavonoids<br>Increased antioxidant activities | Guillén Sánchez et al. (2023) |
| Flavonoid | *Radix tetrastigmae* leaves | 120 MPa pressure<br>Extraction temperature 70 °C<br>90 minutes extraction time | Highest extraction of flavonoid following DHPM | Sun et al. (2013) |
| Flavonoid | *Gynura procumbens* leaves | 80 MPa pressure<br>65 °C temperature<br>90-minute extraction time | Increased flavonoid content (1.95%) compared to reflux method | Tu et al. (2018) |

## 11.6 TOTAL PHENOLICS BY DHPM TREATMENT

An essential process in the field of food processing industry is extraction (Jha & Sit 2022). The output, pureness, and bioactivity of the isolated component determine the success of bioactive compound extraction. Microfluidization serves as a great disruptive technique for retrieving bioactive constituents because of its special combination of several mechanical forces (Kumar et al. 2021). The cellular structure of plants is disrupted by these mechanical stresses, increasing yield as a result. Apart from plants, these bioactive constituents are also found in abundance in vegetable and fruit products; however, excessive processing, especially heat treatment, causes an extensive degradation of these molecules (Sorrenti et al. 2023). Microfluidization is a non-thermal method that has been demonstrated to have a good impact on these bioactive substances. Zhang et al. (2015) demonstrated that using microfluidization-assisted technology enhanced the extracted amount of total phenolics (16.35 mg/g) and free radical scavenging activity from orange-fleshed sweet potato leaves. One other study suggested that the microfluidization technique significantly increased the quantity of total phenolic and total antioxidant capacity of strawberry (*F. ananasaa*) juice through homogenization at high pressure of up to 100 MPa (Karacam et al. 2015). A higher amount of total phenolic contents (56.79–65.28 mg/100 ml) has been obtained through microfluidization processing from ginger rhizome juice at 103.42 MPa and three passes cycles (Suhag et al. 2023). Further, the significant extracted amount of sulforaphane has been noted by microfluidization from Broccoli seeds (Xing et al. 2019). Tarafdar et al. (2019) have evaluated the total phenolic content and total flavonoids in sugarcane juice with microfluidization at 120 MPa and found enhanced free radical scavenging activity. In addition, the impact of DHPM on polyphenol levels and antioxidant qualities in peach juice was evaluated by Wang et al. (2019). They found that the antioxidative activity of peach juice has been preserved despite the lower concentration of total phenolic components. Furthermore, cereal grains are the good resource of several phenolic acids, such as ferulic acid, and numerous kinds of other bioactive substances (Kasote et al. 2021). The increased yield of bioactive phenolic acids such as p-coumarin acid (51.1%) and ferulic acid (45.1%) has been noted in corn bran when microfluidized through 87 μm interaction chamber for 5 passes (Wang et al. 2014). Table 11.2

## TABLE 11.2
### Application of DHPM in extraction of Total Phenolics

| Bioactive Substances | Sources | Processing Conditions | Key Findings | Reference |
|---|---|---|---|---|
| Total phenolics | Orange flashed sweet potato Leaves | Microfluidization M-110EH Microfluidizer (USA) 100 MPa pressure 2 passes cycles | Increased content of total phenolics and flavonoids Better free radical scavenging activity | Zhang et al. (2015) |
| Total phenolics | Strawberry (*F. ananassa*) juice | 60 and 100 MPa pressure 2–5 passes cycles Nanodisperser NLM 100 (South Korea) | Improved total phenolic content and antioxidant activities | Karacam et al. (2015) |
| Total phenolics | Ginger rhizome (Zingiber officinale roscoe) | 103.42 MPa pressure 3 passes cycles | Improved free radical scavenging activity Microbial reduction Increased yield of total phenolics | Suhag et al. (2023) |

*(Continued)*

# TABLE 11.2 (CONTINUED)
## Application of DHPM in extraction of Total Phenolics

| Bioactive Substances | Sources | Processing Conditions | Key Findings | Reference |
|---|---|---|---|---|
| Sulforane | Seeds of broccoli | 5000 Psi pressure 5 passes cycle | Increased amount of sulforane at 5 passes and 5000 psi pressure | Xing et al. (2019) |
| Total phenolic and total flavonoids | Sugarcane juice | 120 MPa pressure 1–2 passes cycles M-110P Microfluidizer (USA) | Increased content of total phenolics and flavonoids Enhanced free radical scavenging activity | Tarafdar et al. (2019) |
| Total phenolic content | Peach juice | 20–160 MPa pressure 1 or 3 passes cycles | Decreased phenolic contents Antioxidant activity stable | Wang et al. (2019) |
| Total phenolic | Corn Bran | 124–172 MPa pressure 1–5 passes cycles Room temperature for inlet temperature | Increased yield of bioactive phenolic acids such as p-coumarin acid (51.1%) and ferulic acid (45.1%) in corn bran | Wang et al. (2014) |

provides an overview of all research findings pertaining to the DHPM-treated total phenolics extraction discussed in this section.

## 11.7 POLYSACCHARIDES BY DHPM TREATMENT

Leaching and boiling are the traditional methods employed to harvest water-soluble polysaccharides; nevertheless, these methods utilize higher temperatures and lengthy processing times, which can cause deterioration of polysaccharides and a decrease in pharmacological activity. Over the past 10 years, DHPM has been utilized to extract polysaccharides from plants, resulting in increased extraction rate, antioxidant activity, and bioaccessibility (Ke et al. 2023). With the use of microfluidization technique, Zhang and co-workers (2015) reported the highest extraction yield of polysaccharides (6.31%) in shorter time (50 min) from lotus leaves as compared with 2.95% yield obtained by leaching (90 min). The most suitable microfluidization configurations have been identified at 180 MPa pressure for two passes for polysaccharide extraction (Zhang et al. 2015). Additionally, the microfluidization procedure may alter the molecular weight, biochemical structure, and free radical scavenging activity of the plant extract. Huang et al. (2018) found that the enhanced content of total sugar and uronic acid of polysaccharide which was treated following high-pressure microfluidization treatment and obtained from *Mesona chinensis benth*. In addition, DHPM-treated polysaccharide samples showed higher free radical scavenging activity in comparison with non-DHPM-treated samples. An additional study utilized DHPM to extract lentinans from *Lentinus edodes* at a temperature of 83 °C and a pressure of 147 MPa with two passes. This purified lentinan from microfluidization has shown higher free radical scavenging activity in comparison to hot water extraction (Huang et al. 2012). Qin et al. (2019) showed the increased extraction yield of polysaccharides obtained from *Auricularia auricula* plant following DHPM treatment at pressure 140 MPa than that of microwave, hot water, and ultrasonic extraction. Moreover, DHPM treatment increased the level of polysaccharides from maize pollen grains compared to dry extraction (Tu et al. 2010).

**TABLE 11.3**

**Application of DHPM in Polysaccharides Extraction**

| Bioactive Substances | Sources | Processing Conditions | Key Findings | Reference |
|---|---|---|---|---|
| Polysaccharides | Lotus leaves | 180 MPa pressure and 2 passes cycles | Increased polysaccharides content and antioxidant activities | Zhang et al. (2015) |
| Polysaccharides | *Mesona chinensis benth* | 120 MPa pressure and 6 passes cycles | Enhanced free radical scavenging activity Molecular weight and composition of monosaccharide changed with DHPM treatment | Huang et al. (2018) |
| Lentinans | *Lentinus edodes* | 147 MPa pressure with two passes | Increased extraction yield Improved antioxidant activity | Huang et al. (2012) |
| Polysaccharides | *Auricularia auricula* | 140 MPa pressure 90 °C temperature 120-minute extraction time | Highest yield of polysaccharide content compared to microwave hot water and ultrasonic extraction | Qin et al. (2019) |
| Polysaccharides | Maize pollen | 120 MPa pressure 70 °C extraction temperature 150 minutes extraction time | Increased polysaccharides concentration than the dry extraction | Tu et al. (2010) |

An overview of all the study results related to the extraction of polysaccharides treated with DHPM that are covered in this part is given in Table 11.3.

## 11.8 BIOACTIVE PIGMENTS BY DHPM TREATMENT

Microfluidization may be utilized to optimize the efficiency of pigment extraction through its unique fluid dynamics from biological sources (Ke et al. 2023). Pigments are often recovered from plant-derived products, algae, or other organisms for various applications such as food coloring, cosmetics, and pharmaceuticals (Di Salvo et al. 2023). Ke et al. (2023) reported the application of novel industry-scale microfluidization (ISM) for better extraction of pigments and other biomolecules from *Chlorella pyrenoidosa* microalgae. The higher extraction yield of chlorophyll b and lipid content has been found in microalgae. Lycopene is part of a pigment family called carotenoids, which includes carotenes (Guerra et al. 2021). Koley and co-workers (2020) reported the significantly enhanced extraction yield of beta-carotene and lutein bioactive constituents in orange carrot juice by microfluidization with an increasing number of passes and pressure up to 68.95 MPa. The higher lycopene content and improved physical properties were reported in tomato ketchup mixes using the microfluidization technique at treatment pressure range between 200 and 1200 bar (Mert 2012). Further, Cha et al. (2012) noted that microfluidization improved 10 times more bioaccessibility of zeaxanthine and beta-carotene from *C. ellipsoidea* during the digestion process at 20,000 Psi. Table 11.4 provides a summary of all research findings for the DHPM-treated pigment extraction discussed in this section.

# Microfluidization-Assisted Extraction of Bioactive Compounds

**TABLE 11.4**
**Application of DHPM in Pigment Extraction**

| Bioactive Substances | Sources | Processing Conditions | Key Findings | Reference |
|---|---|---|---|---|
| Chlorophyll b | *Chlorella pyrenoidosa* (microalgae) | 120 MPa pressure<br>One pass | Increased chlorophyll b content and lipid yield | Ke et al. (2023) |
| β-Carotene and lutein | Orange carrot juice | Pressure up to 68.95 MPa<br>Increase number of passes 1–3<br>M-110P Microfluidizer (USA) | Enhanced yield of lutein and β-carotene | Koley et al. (2020) |
| Lycopene | Ketchup mixes | Pressure range 200–2000 bar<br>M-110Y Microfluidics (USA) | Increased lycopene content in ketchup mixes | Mert (2012) |
| Zeaxanthine and beta-carotene | *Chlorella ellipsoidea* | Pressure 5000, 10,000, 20,000 Psi<br>One pass<br>M-110EH Microfluidizer (USA) | Improved zeaxanthine and beta-carotene bioaccessibility | Cha et al. (2012) |

## 11.9 PROS AND CONS OF DHPM-TREATED EXTRACTION

DHPM-assisted removal is a dependable technique for extracting active compounds from plants, according to numerous research studies on their uses and effectiveness. The advantages of DHPM-assisted extraction include good homogenization effects, shortened extraction time, preservation of compound structure, enhanced dissolution of antioxidant compounds, room temperature operation, closed system for sample protection, and easy control of working parameters. However, DHPM also has limitations, such as a lack of advantage in extraction time compared to emerging methods, a small interaction chamber that cannot handle large particles, low processing capacity, and relatively high equipment costs.

## 11.10 CONCLUSIONS AND PERSPECTIVES

In this chapter, many studies on using the microfluidization technique for the isolation of bioactive compounds such as flavonoids, total phenolic, polysaccharides, and pigments are included. Various findings demonstrated that the DHPM treatment significantly improved the extraction yield, antioxidants, and physiochemical properties of compounds that are biologically active. The outstanding processing efficacy and ongoing operational efficiency of the microfluidization technique indicate its promising potential use in commercial processing of foods. It has the ability to structurally change biological macromolecules for improved techno-functional qualities due to the action of several mechanical forces. Despite investigations that have focused on various uses and benefits of the microfluidization technique, ramping up the technology remains a challenge. Furthermore, effort should be needed for designing and developing industry-scale microfluidizers, along with reducing operational and instrument costs, for successful deployment of the technology at an industrial scale.

Microfluidization has been identified as a feasible technology for isolating bioactive constituents from numerous sources. Ongoing research may concentrate on optimizing microfluidization variables including adjusting pressures, flow rates, and other process conditions to boost the extraction efficacy of biologically active substances. Researchers and industries may work on ramping up microfluidization processes for commercial production. This involves adapting the technology to larger volume and ensuring cost-effectiveness for widespread utilization in pharmaceuticals, cosmetics, and food industries. Furthermore, microfluidization may be integrated with other developing technologies, such as ultrasound-assisted extraction or nanotechnology, to create synergistic effects and improve the entire process of extraction. Continued efforts may be directed toward making microfluidization more environmentally friendly. This could entail developing sustainable and biodegradable solvents or optimizing processes to minimize waste generation. With advancements in understanding the individual variations in bioactive compound metabolism, microfluidization could be useful in developing personalized extraction processes for tailored healthcare solutions. Future research may delve into a better comprehension of the mechanisms involved in microfluidization-based extraction, including the impact on cell structures, release kinetics, and the effect of various factors on the entire process.

## REFERENCES

Ameer, K., Shahbaz, H. M., & Kwon, J. H. (2017). Green extraction methods for polyphenols from plant matrices and their byproducts: A review. *Comprehensive Reviews in Food Science and Food Safety, 16*(2), 295–315.

Anwar, F., & Przybylski, R. (2012). Effect of solvents extraction on total phenolics and antioxidant activity of extracts from flaxseed (*Linum usitatissimum* L.). *ACTA Scientiarum Polonorum Technologia Alimentaria, 11*(3), 293–302.

Azmir, J., Zaidul, I. S. M., Rahman, M. M., Sharif, K. M., Mohamed, A., Sahena, F., … & Omar, A. K. M. (2013). Techniques for extraction of bioactive compounds from plant materials: A review. *Journal of Food Engineering, 117*(4), 426–436.

Bai, L., & McClements, D. J. (2016). Development of microfluidization methods for efficient production of concentrated nanoemulsions; comparison of single- and dual-channel microfluidizers. *Journal of Colloid and Interface Science, 466*, 206–212. https://doi.org/10.1016/j.jcis.2015.12.039

Cha, K. H., Koo, S. Y., Song, D. G., & Pan, C. H. (2012 Sep 19). Effect of microfluidization on bioaccessibility of carotenoids from Chlorella ellipsoidea during simulated digestion. *Journal of Agricultural and Food Chemistry, 60*(37), 9437–9442. https://doi.org/10.1021/jf303207x. Epub 2012 Sep 10. PMID: 22946699.

Chuo, S. C., Nasir, H. M., Mohd-Setapar, S. H., Mohamed, S. F., Ahmad, A., Wani, W. A., … & Alarifi, A. (2022). A glimpse into the extraction methods of active compounds from plants. *Critical Reviews in Analytical Chemistry, 52*(4), 667–696.

Di Salvo, E., Lo Vecchio, G., De Pasquale, R., De Maria, L., Tardugno, R., Vadalà, R., & Cicero, N. (2023). Natural pigments production and their application in food, health and other industries. *Nutrients, 15*(8), 1923. https://doi.org/10.3390/nu15081923

Gali, L., Bedjou, F., Velikov, K. P., Ferrari, G., & Donsì, F. (2020). High-pressure homogenization-assisted extraction of bioactive compounds from Ruta chalepensis. *Journal of Food Measurement and Characterization, 14*, 2800–2809.

Guillén Sánchez, J. S., Betim Cazarin, C. B., Canesin, M. R., Reyes Reyes, F., Hoshi Iglesias, A., & Cristianini, M. (2023). Extraction of bioactive compounds from Peruvian purple corn cob (Zea Mays L.) by dynamic high pressure. *Scientia Agropecuaria, 14*(3), 367–373.

Guerra, A. S., Hoyos, C. G., Molina-Ramírez, C., Velásquez-Cock, J., Vélez, L., Gañán, P., … & Zuluaga, R. (2021). Extraction and preservation of lycopene: A review of the advancements offered by the value chain of nanotechnology. *Trends in Food Science & Technology, 116*, 1120–1140.

Guo, X., Chen, M., Li, Y., Dai, T., Shuai, X., Chen, J., & Liu, C. (2020). Modification of food macromolecules using dynamic high pressure microfluidization: A review. *Trends in Food Science & Technology, 100*, 223–234.

Huang, L., Shen, M., Zhang, X., Jiang, L., Song, Q., & Xie, J. (2018). Effect of high-pressure microfluidization treatment on the physicochemical properties and antioxidant activities of polysaccharide from Mesona chinensis Benth. *Carbohydrate Polymers, 200*, 191–199.

Huang, X., Li, C., & Xi, J. (2023). Dynamic high pressure microfluidization-assisted extraction of plant active ingredients: A novel approach. *Critical Reviews in Food Science and Nutrition, 63*(33), 12413–12421. https://doi.org/10.1080/10408398.2022.2101427

Huang, X., Tu, Z., Jiang, Y., Xiao, H., Zhang, Q., & Wang, H. (2012). Dynamic high pressure microfluidization-assisted extraction and antioxidant activities of lentinan. *International Journal of Biological Macromolecules, 51*(5), 926–932. https://doi.org/10.1016/j.ijbiomac.2012.07.018

Huang, X., Tu, Z., Xiao, H., Li, Z., Zhang, Q., Wang, H., … & Zhang, L. (2013). Dynamic high pressure microfluidization-assisted extraction and antioxidant activities of sweet potato (Ipomoea batatas L.) leaves flavonoid. *Food and Bioproducts Processing, 91*(1), 1–6.

Jha, A. K., and N. Sit. 2022. Extraction of bioactive compounds from plant materials using combination of various novel methods: A review. *Trends in Food Science & Technology, 119*, 579–591. https://doi.org/10.1016/j.tifs.2021.11.019

Jing, S., Wang, S., Li, Q., Zheng, L., Yue, L., Fan, S., & Tao, G. (2016). Dynamic high pressure microfluidization-assisted extraction and bioactivities of Cyperus esculentus (C. esculentus L.) leaves flavonoids. *Food Chemistry*, *192*, 319–327. https://doi.org/10.1016/j.foodchem.2015.06.097

Karacam, C. H., Sahin, S., & Oztop, M. H. (2015). Effect of high-pressure homogenization (microfluidization) on the quality of Ottoman Strawberry (F. Ananassa) juice. *LWT – Food Science and Technology*, *64*(2), 932–937.

Karak, P. (2019). Biological activities of flavonoids: An overview. *International Journal of Pharmaceutical Sciences and Research*, *10*(4), 1567–1574.

Kasote, D., Tiozon, R. N., Jr, Sartagoda, K. J. D., Itagi, H., Roy, P., Kohli, A., Regina, A., & Sreenivasulu, N. (2021). Food processing technologies to develop functional foods with enriched bioactive phenolic compounds in cereals. *Frontiers in plant science*, *12*, 771276. https://doi.org/10.3389/fpls.2021.771276

Kavinila, S., Nimbkar, S., Moses, J. A., & Anandharamakrishnan, C. (2023). Emerging applications of microfluidization in the food industry. *Journal of Agriculture and Food Research*, *12*, 100537.

Ke, Y., Chen, J., Dai, T., Liang, R., Liu, W., Liu, C., & Deng, L. (2023). Developing industry-scale microfluidization for cell disruption, biomolecules release and bioaccessibility improvement of Chlorella pyrenoidosa. *Bioresource Technology*, *387*, 129649. https://doi.org/10.1016/j.biortech.2023.129649

Khan, S. A., Aslam, R., & Makroo, H. A. (2019). High pressure extraction and its application the extraction of bio-active compounds: A review. *Journal of Food Process Engineering*, *42*(1), e12896.

Khaw, K. Y., Parat, M. O., Shaw, P. N., & Falconer, J. R. (2017). Solvent supercritical fluid technologies to extract bioactive compounds from natural sources: A review. *Molecules*, *22*(7), 1186.

Koley, T. K., Nishad, J., Kaur, C., Su, Y., Sethi, S., Saha, S., Sen, S., & Bhatt, B. P. (2020). Effect of high-pressure microfluidization on nutritional quality of carrot (*Daucus carota* L.) juice. *Journal of Food Science and Technology*, *57*(6), 2159–2168. https://doi.org/10.1007/s13197-020-04251-6

Kumar, A., Dhiman, A., Suhag, R., Sehrawat, R., Upadhyay, A., & McClements, D. J. (2021). Comprehensive review on potential applications of microfluidization in food processing. *Food Science and Biotechnology*, *31*(1), 17–36. https://doi.org/10.1007/s10068-021-01010-x

Leyva-Daniel, D. E., Alamilla-Beltrán, L., Villalobos-Castillejos, F., Monroy-Villagrana, A., Jiménez-Guzmán, J., & Welti-Chanes, J. (2020). Microfluidization as a honey processing proposal to improve its functional quality. *Journal of Food Engineering*, *274*, 109831.

Li, Y., Deng, L., Dai, T., Li, Y., Chen, J., Liu, W., & Liu, C. (2022). Microfluidization: A promising food processing technology and its challenges in industrial application. *Food Control*, *137*, 108794.

Mandal, V., & Mandal, S. C. (2010). Design and performance evaluation of a microwave based low carbon yielding extraction technique for naturally occurring bioactive triterpenoid: Oleanolic acid. *Biochemical Engineering Journal*, *50*(1–2), 63–70.

Mert, B. (2012). Using high pressure microfluidization to improve physical properties and lycopene content of ketchup type products. *Journal of Food Engineering*, *109*(3), 579–587.

Mert, I. D. (2020). The applications of microfluidization in cereals and cereal-based products: An overview. *Critical Reviews in Food Science and Nutrition*, *60*(6), 1007–1024.

Nayak, S. L., Sethi, S., Saha, S., Dubey, A. K., & Bhowmik, A. (2023). Microfluidization of juice extracted from partially granulated citrus fruits: Effect on physical attributes, functional quality and enzymatic activity. *Food Chemistry Advances*, *2*, 100331.

Ozturk, O. K., & Turasan, H. (2021). Applications of microfluidization in emulsion-based systems, nanoparticle formation, and beverages. *Trends in Food Science & Technology*, *116*, 609–625.

Putri, N. I., Celus, M., Van Audenhove, J., Nanseera, R. P., Van Loey, A., & Hendrickx, M. (2022). Functionalization of pectin-depleted residue from different citrus by-products by high pressure homogenization. *Food Hydrocolloids*, *129*, 107638.

Qin, L., Zhou, J., Cui, S., Luo, S., & Gao, Y. (2019). Study on extraction of polysaccharides from Auricularia auricular by dynamic high pressure micro-fluidization technology. *Food Research and Development*, *40*(19), 155–159

Salam, A. M., Lyles, J. T., & Quave, C. L. (2019). Methods in the extraction and chemical analysis of medicinal plants. In *Methods and Techniques in Ethnobiology and Ethnoecology*, 257–283. Berlin, Germany: Springer.

Sorrenti, V., Burò, I., Consoli, V., & Vanella, L. (2023). Recent advances in health benefits of bioactive compounds from food wastes and by-products: Biochemical aspects. *International Journal of Molecular Sciences*, *24*(3), 2019. https://doi.org/10.3390/ijms24032019\

Sun, Y., H. Y. Li, X. R. Liu, and Z. Y. Deng. (2013). Application of the dynamic high-pressure microfluidization technology in Radix Tetrastigmae leaves flavonoids extraction. *Science and Technology of Food Industry (China)*, *34* (2), 273–276. doi: 10.13386/j. issn1002-0306.2013.02.022

Suhag, R., Singh, S., Kumar, Y., Prabhakar, P. K., & Meghwal, M. (2023). Microfluidization of ginger rhizome (zingiber officinale roscoe) juice: Impact of pressure and cycles on physicochemical attributes, antioxidant, microbial, and enzymatic activity. *Food and Bioprocess Technology*, *17*(4), 1045–1058.

Tarafdar, A., Nair, S. G., & Pal Kaur, B. (2019). Identification of microfluidization processing conditions for quality retention of sugarcane juice using genetic algorithm. *Food and Bioprocess Technology*, *12*, 1874–1886.

Tu, Z. C., Wang, Y. M., Liu, C. M., Zhang, X. C., & Liu, W. (2010). Applications of dynamic high pressure micro-fluidization technology in maize pollen polysaccharides extraction. *Science and Technology of Food Industry (China)*, *31*(6), 212–217.

Tu, Z. X., X. Xie, J. W. Hu, and B. H. Huang. (2018). Application of the dynamic high-pressure microfluidization technology in Gynura procumbens leaves flavonoids extraction. *Biological Chemical Engineering (China) 4* (4), 6–9.

Tzanova, M., Atanasov, V., Yaneva, Z., Ivanova, D., & Dinev, T. (2020). Selectivity of current extraction techniques for flavonoids from plant materials. *Processes*, *8*(10), 1222.

Wan, X. H. (2019). Applications of new methods in extraction of flavonoids from Chinese materia medica. *Chinese Traditional and Herbal Drugs*, *24*, 3691–3699.

Wang, T., Zhu, Y., Sun, X., Raddatz, J., Zhou, Z., & Chen, G. (2014). Effect of microfluidisation on antioxidant properties of corn bran. *Food Chemistry*, *152*, 37–45.

Wang, X., Wang, S., Wang, W., Ge, Z., Zhang, L., Li, C., … & Zong, W. (2019). Comparison of the effects of dynamic high-pressure microfluidization and conventional homogenization on the quality of peach juice. *Journal of the Science of Food and Agriculture*, *99*(13), 5994–6000.

Xing, J. J., Cheng, Y. L., Chen, P., Shan, L., Ruan, R., Li, D., & Wang, L. J. (2019). Effect of high-pressure homogenization on the extraction of sulforaphane from broccoli (Brassica oleracea) seeds. *Powder Technology*, *358*, 103–109.

Zhang, L., Lu, Y., Tu, Z., Xie, X., Sha, X., Wang, H., & Fu, Z. (2017). Mechanism of dynamic high pressure microfluidization assisted-extraction on the effect of antioxidant activities of polyphenols from Ipomoea batatas leaves. *Food and Fermentation Industries*, *43*(6), 169–174.

Zhang, L., Tu, Z. C., Wang, H., Fu, Z. F., Wen, Q. H., Chang, H. X., & Huang, X. Q. (2015). Comparison of different methods for extracting polyphenols from Ipomoea batatas leaves, and identification of antioxidant constituents by HPLC-QTOF-MS2. *Food Research International*, *70*, 101–109.

Zhang, Q. W., Lin, L. G., & Ye, W. C. (2018). Techniques for extraction and isolation of natural products: A comprehensive review. *Chinese Medicine*, *13*, 1–26.

# 12 Microfluidization of Nut-Based Proteins

## Modulation of Structural, Physicochemical, and Functional Properties

*Geetarani Loushigam, S. Prithya, T. P. Sari, and Prarabdh C. Badgujar*

## 12.1 INTRODUCTION

The rise in the world population has resulted in a significant need for food production to ensure sufficient nutrition (Munialo et al., 2022). With an estimated global population of around 9.7 billion by 2050, there is a crucial requirement to sustainably address nutritional security without compromising environmental sustainability for future generations (United Nations, 2023). The protein ingredients market worldwide, with a worth of USD 77.69 billion in 2022, is expected to attain a 5.8% compound annual growth rate (CAGR) from 2023 to 2030, with plant proteins experiencing a notable 9.1% CAGR due to rising consumer preferences for the plant-based diet (Grand View Research, 2023). Plant-derived foods have gained significant attention in the food and pharmaceutical sectors due to their eco-friendly practices, reduced harm to forests, alleviation of climate change impacts, decreased chance of infection and pollution, attraction to veggie consumers, versatility, and cost-effectiveness promoting sustainable energy (Poore & Nemecek, 2018). However, plant proteins are often criticized for their lower functionality, including poor emulsifying, foaming, solubility, and gelling capabilities (Loushigam & Shanmugam, 2023). Different protein extraction and modification methods have been explored to overcome these limitations (Sim et al., 2021). These include pulsed-electric field extraction (Liang et al., 2018), enzymatic extraction (Sari et al., 2024a), gamma irradiation (Malik & Saini, 2017), high-pressure homogenization (Saricaoglu, 2020), ultrasound-assisted extraction (Loushigam & Shanmugam, 2023), ultrafiltration (Eckert et al., 2019), and cold atmospheric plasma (Ji et al., 2018).

Nuts, characterized as dry fruits with a hard shell and an edible seed, are particularly rich in proteins (Gonçalves et al., 2023). In 2022–23, the worldwide output of tree nuts, such as almonds, walnuts, pistachios, cashews, and hazelnuts, exceeded 5.3 million metric tons (MMT). Almonds held the largest production at 27%, subsequently followed by walnuts, cashews, pistachios, and hazelnuts, which are 22%, 20%, 14%, and 11%, respectively (International Nut & Dried Fruit Council Foundation, 2023). An emerging trend in recent decades involves extracting oil from these nuts through cold pressing, with applications ranging from functional foods to cosmetic products (Barreira et al., 2019). Following oil extraction, the by-products, such as cakes or residues/meals,

DOI: 10.1201/9781032632599-12

are rich in organic matter like carbohydrates, proteins, and vital nutrients (Sari et al., 2022). Traditionally, these residues are used as animal feed or disposed of in landfills. However, considering the global protein deficiency, there is potential for better utilization of these residues for human consumption. The protein content of these residues varies significantly. For example, the protein percentage in walnut oil-extraction residue is around 55.96%, whereas the protein percentage of defatted cashew flour rises from 20.2% to 40.74% (Zhao et al., 2022). Similarly, protein levels in defatted hazelnut powder following cold-press oil extraction range from 35% to 41%, and defatted almond flours are rich in protein, ranging from 44.46% to 55.88% (Ceylan et al., 2022; Roncero et al., 2021). Defatted peanut meal usually contains around 50% proteins (Tu & Wu, 2019). Nut proteins extracted from by-products of the nut oil extraction process have the potential to supplement the increasing global demand for proteins. However, these nut proteins lack the functional properties necessary for utilization in the food industries (Zhao et al., 2022). Therefore, there is a need for protein modification to enhance these properties. By enhancing the structural and functional attributes of these proteins, they can be more effectively exploited in a range of food commodities, contributing to sustainable resource utilization and addressing protein deficiencies on a global scale. Nut protein functional properties have been improved through a range of modification methods, including extrusion (Chen et al., 2018), cold atmospheric plasma (Ji et al., 2018), electron beam irradiation (Zhao et al., 2017), pulsed electric field (Liang et al., 2018), high-pressure homogenization (Saricaoglu et al., 2018), sonication (Zhu et al., 2018), enzymatic processes (Zhang et al., 2020), and fermentation (Yang et al., 2016).

Microfluidization (MF) is an advanced homogenization method that uses high pressure (up to 200 MPa) to process fluids through microchannels in an interaction chamber called a microfluidizer (Guo et al., 2020). The technology involves various forces like high-velocity impact, shear rate, pressure drop, cavitation, and vibration (Ozturk & Turasan, 2022). The interaction chamber, which can be Y- or Z-shaped, facilitates turbulent mixing and energy dissipation, ensuring a consistent pressure profile and leading to a narrow particle size distribution. In the Y-type, the fluid is divided into two streams that collide, causing deformation and rupture. However, in the Z-type, the fluid flows through zigzag channels, changing direction and causing collisions which enhances the impact force (Kavinila et al., 2023). Compared to the other methods, MF stands out for its benefits, including no need for additional chemicals, minimal nutrients loss, continuous operation (case by case basis), low-processing temperature, and a substantially quick processing time (Hu et al., 2011).

MF has emerged as a successful technique for modifying various proteins, with notable applications on the soy proteins (Shen & Tang, 2012), peanut protein isolate (PPI) (X. Hu et al., 2011), zein (Sun et al., 2016), and pea albumins (Djemaoune et al., 2019). Recently, there has been an increasing attention to the modification of nut proteins. This chapter delves into the MF-driven alterations in the nut proteins, providing insights into the physicochemical, functional, structural, and biological changes induced by this process. Additionally, the chapter addresses the challenges encountered in the commercialization of microfluidized nut proteins. By examining both the applications and challenges, this chapter contributes to a comprehensive understanding of the potential and limitations of the MF in the context of nut protein modification.

## 12.2 ROLE OF MF IN NUT PROTEIN MODIFICATION

In food industries, the usage of protein mainly depends on its functional characteristics such as viscosity, emulsifying activity, solubility, foaming properties, binding capacity, colloidal stability, and storage stability. Likewise, here we are discussing the modification of nut protein due to MF treatment is a high-pressure non-thermal processing, which plays a major role in the structural modification of components and also alters various functional properties in positive and negative ways.

## 12.2.1 SOLUBILITY

The application of MF in protein treatment leads to significant modifications in protein structure, notably through partial denaturation and matrix disintegration. This process transforms β-sheet structures into α-helices and creates a three-dimensional (3D) network that enhances the emulsifying activity of proteins (Hu et al., 2011). During MF, non-covalent protein bonds are disrupted, leading to depolymerization and the prevention of insoluble aggregate formation. Furthermore, protein structures unfold, resulting in the production of low molecular weight components with increased solubility. The particle size, indicated by parameters like D43 (average volume diameter) and D32 (average surface area diameter) of microfluidized protein emulsions, exhibits higher emulsifying activity compared to unprocessed protein emulsions, attributed to improved solubility, reduced particle size, and enhanced physical stability.

Saricaoglu et al. (2018) reported that the impact of high-pressure homogenization at 0–100 MPa pressure on hazelnut meal proteins showed increased solubility by unfolding hazelnut protein structure with an improved protein–solvent interaction. The increase in the pressure from 100 to 150 MPa reduces protein solubility by which may be attributed to additional protein unfolding, production of -SH residues, and an increase in the hydrophobic groups. This may lead to protein aggregation leading to decreased protein solubility.

The MF treatment of cross-linked groundnut proteins (arachin and conarachin) under 160 MPa pressure resulted in an effective improvement in functional attributes like solubility and emulsifying properties of groundnut protein isolate (Hu et al., 2022). The research emphasizes the importance of solubility in influencing other functional properties. Specifically, the MF-treated product exhibited a significant increase in protein solubility (from 35.7% to 45.9% for arachin; from 63.8% to 68.2% for conarachin), along with enhancements in water holding capacity, gelation, thermal stability, and surface hydrophobicity, showcasing the broad impact of MF treatment on protein functionality. Sari et al. (2024b) reported the solubility of microfluidized almond protein isolate (API) in different pressures (0, 40, 80, 120, and 160 MPa/1 pass). They noted a rise in the solubility of the API as the pressure increased from 0 MPa (75.13%) to 120 MPa (88.89%), followed by a slight decrease at 160 MPa (78.06%). This enhanced solubility was attributed to the shear stress induced by MF, which caused protein unwinding and the cleavage of hydrophobic and hydrogen bonds. Consequently, this process released more hydrophobic and hydrophilic amino acid residues and reduced particle size, resulting in higher surface charge density and specific surface area, ultimately leading to improved protein hydration and increased solubility.

## 12.2.2 COLLOIDAL STABILITY

Microfluidized hazelnut yogurt, developed by Demirkesen et al. (2018) using a pressure of 135 MPa for 2 passes and subsequent pasteurization at 85 °C/30 min as pretreatment, achieved a stable creamy consistency in the product. The MF-treated formulation exhibited significantly increased colloidal stability, likely attributed to structural modifications in hazelnut proteins. Typically, hazelnut yogurt has a liquid gel-like structure due to its storage proteins, soluble and insoluble fibers, leading to suspension and product instability. However, the MF-treated yogurt formulation showed substantial disintegration of storage proteins and fiber matrix, resulting in a 3D gel structure with random accumulation of proteins and fibers during fermentation with lactic acid bacteria. This led to increased colloidal stability, reduced particle size (D43 – from 121.21 to 42.15 μm; D32 – from 29.15 to 13.32 μm), improved water holding capacity, and enhanced rheological, textural properties, and firmness of the product. Sari et al. (2024b) also noted a significant increase in the zeta potential of MF-treated APIs at various pressures (40–160 MPa), indicating enhanced electrostatic repulsion between colloidal particles and improved protein isolate stabilization through MF as a high-pressure treatment.

### 12.2.3 Physical and Storage Stability

The MF effect on physical parameters such as creaming index, zeta potential, particle size, rheological properties, color, texture, and storage stability of protein emulsions is significant. Ling et al. (2023) developed a microfluidized walnut protein emulsion and beverage at 20 MPa pressure, resulting in a notable decrease in particle size from 14.17 μm (untreated) to 10.03 μm with an even dispersal. This reduction was attributed to the dispersion and separation of protein molecules under high pressure, shear, and impact force during MF. To enhance MF efficacy, a pretreatment process of homogenizing at 12,000 rpm/5 min was employed, ensuring continuous dispersion and stable emulsion with reduced droplet size. The increased zeta potential of the emulsion from −22 (untreated) to −28 (microfluidized) indicated greater surface charge density, promoting electrostatic repulsion and emulsion stability. The rheological properties of the emulsion improved due to the strong interaction of protein molecules under rotational motion, high-frequency dissipation, and winding movement during MF. This resulted in a shear-thinning fluid behavior, as evidenced by a decrease in viscosity from 743 to 147 MPa (Ling et al., 2023). The developed walnut protein emulsion exhibited increased shelf life and freshness retention for 175 days at 4 °C, attributed to particle disintegration, increased shear stress, and attractive forces between molecules.

In an industry-scale MF system, whole peanut milk prepared at different pressures (0–120 MPa) showed improved storage stability without stratification for 67 h at 4 °C (Dai et al., 2022). The decreased sedimentation rate and improved stability were linked to reduced particle size (from 48.9 to 25.15 μm with uniform distribution) and changes in apparent viscosity, transforming polydisperse particles to monodisperse.

Sari et al. (2024b) processed APIs using MF at varying pressures (40–160 MPa/1 pass), noting a significant decrease in the particle size (from 395.91 to 225.59 nm) with intensifying pressure (40 to 120 MPa). The disruption of protein structures by MF-induced cavitation resulted in reduced particle size for all applied pressures, although at 160 MPa, there was a slight increase in particle size due to high shear turbulent forces breaking down complex protein structures.

### 12.2.4 Enhancing Functional Properties

Many previous studies have reported that MF enhances and stabilizes the functional properties of food products, including emulsifying activity, antioxidant activity, and the production of low-molecular-weight peptides. Ceylan et al. (2022) investigated hazelnut protein hydrolysates through enzymatic hydrolysis using alcalase and neutrase enzymes. The synergistic effect of enzymatic hydrolysis and MF treatment resulted in protein hydrolysates with increased radical scavenging activities (96.63% DPPH and 98.31% ABTS), attributed to the unfolding of protein β-sheet structures, protection of hydrophobic amino acids, disintegration into α-helix structures, and formation of random 3D coil structures, enhancing the degree of protein hydrolysis and the formation of low-molecular-weight peptides.

MF also improved the emulsifying activity index of protein hydrolysates from 84.32 m²/g (untreated) to 88.04 m²/g (treated). Previous studies have noted that processing proteins at ≤120 MPa pressure causes disintegration and disaggregation of molecules, leading to the formation of many smaller hydrophobic peptides with high antihypertensive bioactivity (97.30%). However, processing at >120 MPa can lead to re-aggregation of particles due to modifications in intermolecular hydrogen bonds, van der Waals forces, and electrostatic repulsion of protein molecules, resulting in the formation of larger peptides with a lower degree of hydrolysis (Gong et al., 2017).

The microfluidized APIs studied by Sari et al. (2024b) exhibited increased radical scavenging activity with increasing pressure (from 40 to 120 MPa), with subsequent reductions observed at 160 MPa. The shear and impact forces generated by MF caused the unfolding of protein structures and the revelation of hydrophobic amino acid residues. This led to increased free sulfhydryl content and repositioning of amino acid sequences in the peptide chain, enhancing the antioxidant activity of proteins at 120 MPa (17.17 and 15.35 µM TE/5 mg protein) for ABTS and DPPH, respectively.

## 12.3 IMPACT OF MF ON PHYSICOCHEMICAL PROPERTIES

Nut-derived proteins have become gradually vital due to environmental, economic, and therapeutic reasons. The extraction, modification, and enhancement of the functionalities of nut proteins have attained global focus owing to their numerous applications in food industries. Nut proteins are considered as "promising alternative to animal proteins" owing to their excellent nutritional profile and high gastrointestinal bioavailability (Cui et al., 2023). However, the inherent drawbacks like immune reactivity, complex secondary structures, resistance to proteolysis, and poor solubility often limit their application in food systems. MF has been deemed a "green modification technique" for food macromolecules which facilitates reduction in antinutritional components and enhancement in the functional attributes of proteins from various sources. In recent years, several studies have reported MF as an effective method to functionalize major macromolecules of food while maintaining their nutritional value. MF often imparts favorable structural and functional characteristics to the major nutritional components of food, especially proteins, polysaccharides, and dietary fiber (Guo et al., 2020). Proteins, the major body-building nutrient of foods, often encounter the drawbacks of allergenicity and poor solubility (Zhu et al., 2022). Functionalization through MF involves breaking the non-covalent bonds of proteins, thereby changing the secondary and quaternary structures and often enhancing their functionality. Even though data are scarce on the effect of MF on nut proteins, some studies have reported its beneficial effects, especially on almond proteins (Sari et al., 2024b), walnut proteins (Ling et al., 2023), peanut proteins (Hu et al., 2011), and hazelnut proteins (Saricaoglu et al., 2018).

### 12.3.1 IMPACT OF MF ON STRUCTURAL CHARACTERISTICS OF PROTEINS

#### 12.3.1.1 Effect of MF on Distribution of Molecular Weight

Sodium dodecyl-sulfate polyacrylamide gel electrophoresis (SDS-PAGE), a frequently used electrophoretic procedure to analyze the distribution of molecular weight of protein shows variations in the molecular mass of proteins treated with high-pressure MF. A downward movement and fading of protein bands under high-intensity MF have often been reported by researchers (Hu et al., 2021). Saricaoglu and co-workers observed a reduction in the bandwidth when the pressure increased beyond 150 MPa on hazelnut meal proteins which could be attributed to the decreased solubility of proteins (Saricaoglu et al., 2018). An applied MF pressure beyond 90 MPa induced complete disappearance of bands on almond proteins as evaluated by SDS-PAGE (Sari et al., 2024b). The complete disappearance of protein bands, especially those with low molecular weight, further strengthens the hypothesis of structural modification of proteins by MF (Adjei-Fremah et al., 2019). Further, the probable breakdown of peptide chains into low-molecular-weight segments often enhances the functionalities of proteins since the hydrophobic inner core is exposed. Apart from inducing molecular weight reduction, changes in the electrophoretic bands are also possible due to the establishment of protein–polyphenol multiplexes under high-shear stress. Studies report MF as an effective technique to enhance the stability of protein–polysaccharide complexes, resulting in a wide range of applications in the food and biomedical fields (Hu et al., 2022). However, it is observed by many researchers that beyond certain pressure levels, the implementation of intense shear forces induces protein–protein aggregates which could unfavorably influence their attributes (He et al., 2021).

#### 12.3.1.2 Effect of MF on Microstructure and Particle Size

The unrivalled shear and intense pressure drop associated with high-frequency vibrations often induce structural changes in the proteins (Guo et al., 2020). The variations in the size and morphology of

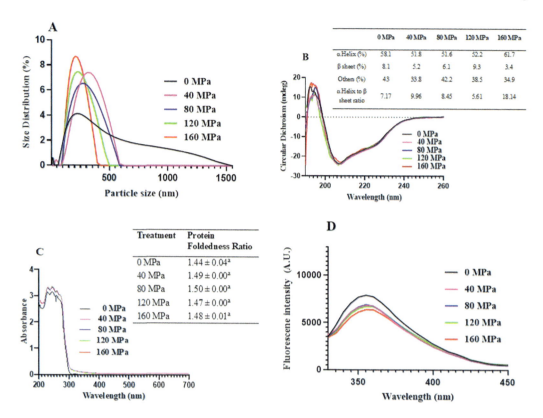

**FIGURE 12.1** Effect of MF on particle size distribution (a), CD spectra (b), ultraviolet absorption spectra (c), and intrinsic fluorescence spectra (d) of almond protein isolate. (Adapted from Sari et al., 2024b.)

protein molecules following MF could be observed by analyzing the particle size and electron microscopy imaging. Optimal MF pressures are often associated with a reduction in the particle size, thereby enhancing functional attributes of proteins (Ling et al., 2023). Reduction in particle size facilitates the formation of more uniformly sized particles with enhanced stability as reported in almond proteins (Figure 12.1a) (Sari et al., 2024b) and hazelnut nut proteins (Saricaoglu et al., 2018). Moreover, a narrow particle size distribution could benefit the digestibility of proteins by enhancing the accessibility of enzymes and thereby enhancing the nutritive value of proteins. High-pressure homogenization induced a significant reduction in the particle size when compared with high-intensity ultrasound conditions studied on peanut protein emulsions. Additionally, high pressure of 500 bar induced improved gel strength when compared to the ultrasound-treated peanut emulsions (Jiang et al., 2021). When compared to the conventional homogenization, MF induced reduced particle size and improved stability to walnut protein emulsion and beverage produced from walnut meal as reported by Ling et al. (2023).

The microstructure of proteins as observed by scanning electron microscope (SEM) following MF often exhibits smaller and more uniform fragments in comparison with the untreated parent protein (Zhang et al., 2017). Saricaoglu et al. (2018) reported a decrease in the particle size and increment in the zeta potential of hazelnut meal proteins with increasing MF pressures from 0 to 150 MPa. It can be concluded that the disintegration of complex protein structures results in particle size reduction which is predominantly associated with an enhancement in MF pressures. However, beyond certain pressures, the disintegrated protein molecules aggregate to form larger structures, leading to an enhancement in the particle size (Hu et al., 2021). Therefore, analyzing the microstructure and size of particles allows researchers to optimize MF conditions for proteins to achieve maximum functional properties.

### 12.3.1.3 Effect of MF on Secondary Structures of Protein

The application of MF has shown considerable influence on the secondary and tertiary structures of protein. Distinct changes in the main secondary structural components of proteins, especially β-sheets and α-helix that are stabilized by intermolecular and intramolecular hydrogen bonds, respectively, were described to be influenced by the pressure processing (Takano et al., 2022). The intermolecular hydrogen bonds are mostly affected by the MF facilitating the development of new structures and connections among the protein molecules, especially α-helix to β-sheet transition (Hu et al., 2011). However, a simultaneous reduction in the β-sheets and α-helix has also been reported in microfluidized APIs (Figure 12.1b) under MF, facilitating increment in other structural protein forms (turns, loops, and random coils) (Sari et al., 2024b). A decrease/increase pattern of the major secondary structural forms indicates dissociation effect of cavitation under high-shear stress (Hu et al., 2021). However, variations in the structural integrity and amino acid sequences of different proteins lead to variations in their response under high-shear stress. A change in α-helix to β-sheets ratio also indicates unfolding of complex secondary protein structures. It can be concluded that MF imparts unfolding of the complex secondary structures of proteins, and these structural rearrangements lead to differences in their functional and physicochemical attributes.

### 12.3.1.4 Effect of MF on the Tertiary Structure of Proteins

The tertiary structure of proteins is the 3D arrangement of amino acid chains, which are stabilized by the interactions of hydrophilic hydrogen and ionic bonds on the surface and hydrophobic interactions at the core (Rehman et al., 2021). The variances in the exposure of specific hydrophobic amino acid residues, particularly tryptophan, tyrosine, and phenylalanine, due to MF are typically detected under UV light within the range of 260–280 nm. The absorption levels vary depending on the extent of exposure to these amino acids, which are normally buried within the protein's core in its native state. An increase in the protein foldedness ratio, which serves as a quantifiable indicator of protein unfolding, also signifies the exposure of the hydrophobic amino acid core (Biter et al., 2019). An increment in the protein foldedness ratio (Figure 12.1c), a quantifiable indicator of the unwinding of highly structured almond proteins following MF, indicates the exposure of tryptophan residues (Sari et al., 2024b). Tertiary structural rearrangements of protein are also analyzed through tryptophan fluorescence, also known as "natural protein fluorescence" (Ladokhin, 2000). Changes in the maximum fluorescence intensity and wavelength shift at maximum fluorescence indicate the exposure of buried groups. Sari et al. (2024b) also observed and reported variations in the fluorescence intensity of microfluidized API as compared to the control API sample, as depicted in Figure 12.1d, evincing exposure of buried tryptophan residues. MF causes protein molecules to partially scatter, exposing hydrophobic amino acid chains which have been confirmed by researchers on several other plant-based proteins (Hu et al., 2021; He et al., 2021; Zhao et al., 2021). This loosening of the protein structure further explains why MF-treated proteins exhibit improved functional attributes like solubility, emulsifying properties, and protein bioaccessibility.

### 12.3.1.5 Effect of MF on Other Structural Properties of Proteins

Hydrophobicity ($H_0$) and free sulfhydryl content (SH) are often studied to determine the effect of MF on the surface properties and stability of the proteins, respectively (Zhao et al., 2021). In comparison to the native state of protein, the $H_0$ value increases in a pressure-dependent manner under MF, as described by several studies (Cheng et al., 2022; He et al., 2021). The dissociation of the secondary structures and rearrangement of the tertiary structures often expose hydrophobic amino acids to the surface, thereby enhancing the $H_0$ value. Similarly, an increment in the SH content indicates the conformational rearrangement following high shear stress. Moreover, the disulfide (SS) content increases to a maximum value following the exposure of SH groups under MF. However, the extent of pressure has a significant effect on the $H_0$, SH, and SS values, since beyond a certain pressure level, the disruption of SS bonds causes a reduction in the SH and SS values (He et al., 2021).

Breaking of SS bonds usually promotes the exposure of hydrophobic as well as hydrophilic sides, thereby enhancing the solubility and emulsifying activities of the protein simultaneously (He et al., 2021) due to improvement in amphiphilic properties of proteins. However, optimization of the MF pressure is critical to achieving optimum structural changes leading to maximum functional benefits of nut proteins, as reported by several studies. An optimum pressure of 80 MPa was found suitable for almond proteins (Sari et al., 2024b), whereas peanut proteins exhibited better functional properties at 120 MPa (Hu et al., 2011).

## 12.3.2 Effect of MF on Functional Characteristics of Proteins

The modifications in the protein functional properties following MF have been linked to the variations in the protein structure, conformation, and aggregation state. Furthermore, reduction in the protein particle size contributes to the enhancement of functional attributes (Guo et al., 2020). Modifications brought about by MF treatment for different nut proteins are summarized in Table 12.1.

### 12.3.2.1 Effect of MF on Solubility of Proteins

Solubility is crucial for protein functionality as it directly influences denaturation and aggregation (Huang et al., 2019) and most importantly its application in the food/other domain. Enhancing solubility through different techniques/treatments can result in improved functional features of proteins, including enhanced surface activity, rheological properties, hydrodynamics, emulsification, and gelation (Yang et al., 2018). There are hardly any reports about MF of tree-nut proteins. In a study involving PPI samples, the solubility exhibited a noteworthy increase of 11.4% as pressure surged from 40 to 120 MPa. However, a subsequent rise in the pressure from 120 to 160 MPa resulted in a slight 2.1% reduction in the solubility (Hu et al., 2011).

In another study focusing on API processed using dynamic high-pressure MF, various pressure levels (0, 40, 80, 120, and 160 MPa) were applied in a single pass (Sari et al., 2024b). The solubility of the API was increased by 13.76% from 0 to 120 MPa, but then slightly decreased by 10.83% at 160 MPa compared to 120 MPa. The authors inferred that the increased solubility under high-pressure MF can be attributed to two key factors. Firstly, pressure-induced shear stress unfolded proteins, exposing hydrophilic and hydrophobic amino acid residues, enhancing their interaction with the solvent, and boosting solubility. Secondly, the pressure also reduced API particle size, increasing surface charge density and specific surface area. This amplified interaction with the solvent led to stronger protein hydration and overall improved solubility (Sari et al., 2024b).

A considerable enhancement in the solubility of arachin (16.34%) and conarachin (6.75%), the two main protein fractions in the peanut flour, following MF treatment at 120 MPa has been reported by Hu et al. (2021). This increase was due to the disruptive effects of high-velocity impact, cavitation, high-frequency vibration, and high pressure leading to alteration of insoluble aggregates into soluble forms. The MF treatment not only enhanced the solubility of PPI but also raised the solubility of transglutaminase-cross-linked arachin (28.5%) and conarachin (6.8%) (Hu et al., 2022). This suggests that MF effectively depolymerized and disrupted the insoluble aggregates formed by transglutaminase cross-linking, leading to the unfolding of their protein structures and an overall improvement in the solubility.

### 12.3.2.2 Effect of MF on Emulsification of Proteins

The emulsification properties of proteins are influenced by factors like hydrophobicity–hydrophilicity ratio and folding-unfolding ease (Aryee et al., 2017). Among different MF pressure levels used (compared to 40, 80, 120, and 160 MPa), 120 MPa has been reported to be more effective in enhancing the emulsification of PPI (Hu et al., 2011). Combining MF with transglutaminase (TGase) treatment proved to be a successful strategy for improving emulsification performance in peanut protein fractions. MF induces a shift to a bimodal particle size distribution with an increase in small particle volume, while TGase treatment shifts the distribution towards larger particles. The synergistic effect

# TABLE 12.1
## Modifications to Nut Proteins by MF

| Type of Nut | Processing Conditions | Modification/Change in Properties | Reference |
|---|---|---|---|
| Peanut protein isolate | 40, 80, 120, or 160 MPa/1 | • Unfolding of proteins, shifting secondary structures, and exposing buried hydrophobic groups, leading to increased dispersibility, solubility, and reduced particle size distribution.<br>• Re-aggregation under higher pressure decreases sulfhydryl content. | Hu et al. (2011) |
| Hazelnut slurry | 135 MPa/1 | Enhanced strength and colloidal stability in yogurt. | Demirkesen et al. (2018) |
| Hazelnut protein | 25, 50, 75, 100, or 150 MPa/1 | • Increases protein unfolding, hydrophobic groups, and -SH residue production.<br>• Decreases particle size, improves water solubility, and enhances emulsifying and foaming properties.<br>• Increases surface area and charged sites, thereby increasing zeta potential of hazelnut meal protein suspension. | Saricaoglu et al. (2018) |
| Peanut protein isolate | 30, 60, 90, 120, 150, 180, or 210 MPa | • Improves hydrophobicity on the surface and modifies molecular interactions.<br>• Modifies electrostatic repulsion, causing proteins to become metastable.<br>• Causes dynamic processes in aqueous dispersion that involve disaggregation and reaggregation.<br>• Produces fractions of peanut peptide with potent antihypertensive properties. | Gong et al. (2017) |
| Peanut protein isolate | 30, 60, 90, 120, 150, 180, or 210 MPa | • Hydrophobicity on the surface is favored by disintegration up to 120 MPa.<br>• Substantial aggregation under intense pressure. | Gong et al. (2019) |
| Peanut protein | 120 MPa | • A rise in exposed sulfhydryl groups.<br>• Enhanced hydrophobicity on the surface. | Hu et al. (2021) |
| Peanut milk | 30, 60, 90, or 120 MPa /1 | • Oil droplets are wrapped by deformed proteins, which create a three-dimensional network.<br>• Increased stability. | Dai et al. (2022) |
| Walnut protein emulsion and beverage | 20 MPa/2 | • Reduction in droplet size and enhanced stability in emulsion<br>• Efficient protein aggregate disruption and uniform distribution.<br>• Enhanced the viscosity, color, zeta potential, and centrifugal precipitation rate of beverages.<br>• Extended the shelf life of oil-in-water emulsions. | Ling et al. (2023) |
| Almond protein isolate | 40, 80, 120, 160 MPa/1 | • Reduced particle size and increased zeta potential.<br>• Significant improvements in antioxidant activity, protein solubility, and emulsifying properties.<br>• Structural analysis found dissociation up to 80 MPa, followed by reformation.<br>• API at 80 MPa demonstrated improved protein digestibility (1.16-fold increase) and PDCAAS (1.15-fold increase). | Sari et al. (2024b) |

of MF and TGase treatment results in a bimodal particle size distribution, showcasing its effectiveness in enhancing the emulsion properties of PPI. In a related study on the PPIs by the same group of authors, MF has been said to enhance the emulsification potential of arachin and conarachin proteins (Hu et al., 2021). An improvement in the emulsification potential observed after undergoing MF can be attributed to the enhanced surface hydrophobicity and increased solubility of the proteins. Conarachin outperformed arachin in emulsifying activity and stability, and the application of MF contributes to reduce particle size values, thereby improving overall emulsion properties. TGase treatment initially increased particle size values but later restrained growth during storage, indicating a trade-off between declining emulsifying activity and enhanced stability (Hu et al., 2021).

The microfluidized API samples showed significantly higher EAI values compared to the control API, with the largest increase observed at 80 MPa, corresponding to a percentage increase of about 23.86% (441.34 ± 8.57 mg$^2$/g) (Sari et al., 2024b). Subsequently, there was a gradual decrease in EAI values at 120 MPa (a decrease of 8.55% from the peak) and 160 MPa (a decrease of 12.7% from the peak) following the 80 MPa treatment. These variations in EAI can be attributed to the exposure of hydrophilic and hydrophobic amino acid residues during pressure treatment, thereby enhancing the proteins' amphipathic characteristics (Sari et al., 2024b). Likewise, in both walnut protein emulsion and walnut protein beverage, the application of MF demonstrated a positive impact on the emulsifying properties by reducing the particle size (Ling et al., 2023).

### 12.3.3 Effect of MF on Bioactivity of Proteins/Peptides

#### 12.3.3.1 Degree of Hydrolysis

Peptide bond cleavage can be detected using a widely used measure of degree of hydrolysis (DH). Regardless of pressure, the peanut protein hydrolysate (PPH) made from PPI that underwent MF treatment showed noticeably greater DH (maximum 36.60 ± 1.85% at 120 MPa) compared to untreated sample (Gong et al., 2017). Higher pressures (120 MPa) led to more efficient hydrolysis than lower pressures (30 or 60 MPa), even if re-aggregation did occur. Re-aggregation produced decreased hydrolysis, but disaggregation produced higher hydrolysis. This has been better explained by the researchers who microfluidized whey proteins and suggested that the disrupted fractions, rather than initial protein particles, are mostly involved in re-aggregation at higher pressures, allowing easier hydrolysis (Oboroceanu et al., 2011).

#### 12.3.3.2 Oligopeptide Profiles

Gong et al. (2017) investigated the oligopeptide profiles of PPH after MF pretreatment of PPI under different pressures. Based on the MF treatment conditions, the results revealed substantial differences: 39 oligopeptides were detected in the PPH treated at 120 MPa, compared to 29 and 35 in the control and 210 MPa treatments, respectively. This suggests that disaggregation, possibly promoting exposure of hindered cleavage sites in native proteins, influences the oligopeptide composition (Jin et al., 2016). The presence of aromatic and branched-chain amino acids in the oligopeptide composition is noteworthy, as these have been associated with strong inhibitory activities against renin and ACE (Aluko, 2015b). MF pretreatment of PPI is an effective alternative method for producing oligopeptides with antihypertensive inhibitory activity, suggesting that MF treatment, particularly at 120 MPa, enhances the oligopeptide composition of PPH, contributing to its potential as an effective peptide source with antihypertensive activity (Gong et al., 2017).

#### 12.3.3.3 Antihypertensive Properties

Protein hydrolysate fractions with a high concentration of low-molecular-weight peptides (usually <3 kDa) are reported to be more effective than larger peptides at inhibiting renin and angiotensin-converting enzyme (ACE) (Aluko, 2015a). Gong et al. (2017) examined the antihypertensive activity of PPH after MF pretreatment of PPI under different pressures. The results showed that PPH (with maximum number of small-size peptides) produced from the PPI treated with 120 MPa MF treatment demonstrated the highest effective inhibitory activity against renin and ACE. Renin inhibition was significantly higher at 120 MPa, indicating a twofold increase in the inhibitory activity. Conversely, the maximum ACE inhibition of PPH was 68.03 ± 1.84% at 120 MPa, slightly surpassing the control. The differences between renin and ACE inhibition in the PPH after MF treatment might be explained by the protein isolate's surface hydrophobicity and peptide composition (Aluko et al., 2015). The study suggests that the antihypertensive activity of PPH can be significantly influenced by the pressure conditions during MF treatment, with 120 MPa emerging as a favorable pressure for achieving enhanced inhibitory effects on renin and ACE (Gong et al., 2017).

### 12.3.3.4 Antioxidant Properties

Sari et al. (2024b) examined the antioxidant properties of microfluidized (40–160 MPa) and untreated API using DPPH and ABTS free radical scavenging assays. The results showed a notable percentage increase in radical scavenging capacity at 120 MPa compared to the control: approximately 12.44% for DPPH and about 13.57% for ABTS. Conversely, at 160 MPa, there was a slight decrease (approximately 1.79%) in ABTS value compared to the control, although this difference was not statistically significant (Figure 12.2a). The enhancement in the antioxidant activity at 120 MPa can be attributed to several factors. Firstly, the high-pressure treatment induced protein unfolding, particularly affecting tertiary and quaternary structures, which exposed buried hydrophobic amino acid residues (Landim et al., 2021). This exposure, along with an increase in free SH content, is known to play a vital role in enhancing free radical scavenging activity. Additionally, the shear stress during high-pressure treatment contributed further to the improved antioxidant capacity of API by exposing hydrophobic amino acids (Feng et al., 2022). These findings shed light on the mechanisms underlying the enhanced antioxidant properties of microfluidized API under specific pressure conditions.

### 12.3.3.5 Protein Digestibility

Sari et al. (2024b) also delved into the protein digestibility of microfluidized and untreated API. Their findings revealed a notable increase in the protein digestibility across all microfluidized samples (Figure 12.2b) compared to the untreated samples (77.17 ± 2.83%), although statistical significance was not reached for samples subjected to high pressures of 160 MPa (84.24 ± 3.09%) and 120 MPa (85.91 ± 2.97%). The most substantial improvement in the protein digestibility was observed at 40 MPa (90.66 ± 3.68%) and 80 MPa (89.85 ± 1.60%). This notable enhancement is due to the partial unwinding of secondary and tertiary protein structures under high-shear stress, which exposed buried amino acid residues to digestive enzymes. Furthermore, reducing the particle size and relaxation

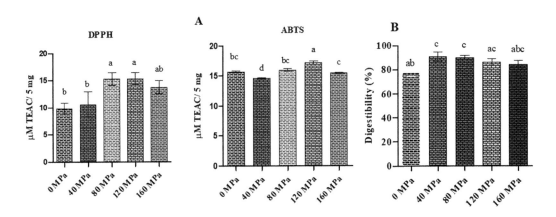

**FIGURE 12.2** Effect of MF on antioxidant property (a) and protein digestibility (b) of almond protein isolate treated at different pressure. (Adapted from Sari et al., 2024b.)

of complex protein microstructures contributed significantly to the improved protein digestibility observed (Kumar et al., 2021). However, contrasting results were noted in studies by Chatur et al. (2022), where very high pressure led to protein re-aggregation, potentially hindering the accessibility of digestive enzymes to active protein sites. Despite these variations, the favorable effects of pressure treatment on protein bioaccessibility underscore the probable utility of microfluidized API in enhancing the nutritive value of API-fortified supplements.

### 12.3.3.6 Amino Acid and Protein Digestibility Corrected Amino Acid Score (PDCAAS)

Comparing the amino acid profiles of microfluidized API at 80 MPa and untreated API revealed that essential amino acids (EAAs) constituted 53.56% and 52.51% of all amino acids, respectively, signifying the high nutritive value of the API. In both samples, leucine was identified as the first limiting amino acid, with EAA ratios of 0.68 and 0.67 for the untreated and 80 MPa-treated samples, respectively. The higher proportion of EAA and improved digestibility of protein resulted in a 1.15-fold increase in PDCAAS for the 80 MPa-treated API. The unwinding of the dense protein structure under high pressure probably improved the availability of proteolytic enzymes, thus improving protein digestibility and contributing to the observed increase in PDCAAS for the 80 MPa-treated API.

## 12.4 CURRENT APPLICATIONS OF MICROFLUIDIZED NUT PROTEINS

Homogenization is a crucial unit process in the food, chemical, and pharmaceutical industries, which is achieved through blending, agitation, and mixing. The small particle size of food ingredients is crucial for preserving the emulsion stability of food beverages (Karacam et al., 2015). MF is a useful tool for reducing particle size of liquid components, improving their stability and sensory qualities (Kavinila et al., 2023). The structural elements of plant cells are damaged by MF, which can result in a decrease in the size of the particle in beverages (Wang et al., 2019). With increase in the number of passes and pressure, the efficiency of the reduction in the particle size is enhanced. However, since too much pressure and too many treatment cycles result in larger particle sizes, adjustment of the pressure and number of passes is required (Karacam et al., 2015).

Fermented yogurt-type product was produced from the hazelnut slurry by applying pressure at 135 MPa for two passes, without the need for any additional ingredients like milk powder or hydrocolloids (Demirkesen et al., 2018). Pressure reduced the particle size and enhanced the stability which improved the hazelnut yogurt's stiffness and colloidal stability. It resulted in high concentration of vital amino acids and high-quality protein in the yogurt. Additionally, the shredded fibrous material of hazelnut skin was altered by MF, giving the cake a gluten-like elasticity and strength that improved the firmness and springiness ratings (Cikrikci et al., 2016). Cake with microfluidized hazelnut skin exhibited reduced retrogradation and staling propensity because of their increased water-holding capacity and lower starch content. Samples with microfluidized hazelnut skin exhibited a darker color, which can be preferred for some baked goods (Cikrikci et al., 2016). The hazelnut skin that has been microfluidized was also used to formulate cookies (Yildiz et al., 2016). The creation of a finely divided fibrous structure with a larger surface area was the result of two passes of MF at 150 MPa, which enhanced the ability of hazelnut skin samples to bind well with the water. It increased hardness, produced cookies with a finer crumb structure, were darker in color, and had a longer shelf life when added to cookie formulas. Furthermore, at similar moisture concentrations, the resulting microfluidized hazelnut skin containing cookies displayed less water activity. The study's findings indicated that fibrous hazelnut skin samples with better physical qualities may be produced using MF as a novel technique. More intriguingly, they can be used to increase the water content of baked goods without causing them to have excessively high water activity. In a different investigation, the MF treatment increased the physical stability and flavor profile of whole peanut milk using an industry-scale MF system. Reduced particle size, higher apparent viscosity, and the breakdown of proteins, which created 3D network structures to encircle oil droplets, were probably the factors that might have contributed to the stability (Dai et al., 2022). In order to provide a more stable emulsion,

Microfluidization of Nut-Based Proteins

walnut protein emulsion, and beverage processed by MF homogenization may guarantee effective protein aggregation disruption and tiny droplet size (Ling et al., 2023). This occurred because of higher pressures and a greater shear effect on the emulsion during MF, which caused the macromolecules to disperse and divide. The discoveries of this study may serve as a guide for MF homogenization, which can greatly increase beverage shelf life, even if there haven't been many research on the impact of MF on the shelf life of other protein beverages. The most important factor while applying food components is to use pressure and passes as optimally as possible. Even though these studies have been conducted, more in-depth information regarding the shelf life, bioavailability, and bioaccessibility of proteins should be investigated using various food matrices.

## 12.5 CONCLUSIONS AND PERSPECTIVES

MF presents a promising technology for enhancing the structural, physical, and chemical attributes of materials in the food industry. Nevertheless, there are a number of obstacles that prevent its effective deployment. The small size of the interaction chamber (10–100 mm) limits operating capacity, necessitating additional equipment for size reduction and posing challenges in the cleaning due to product residue accumulation, potentially causing microbial contamination (Ozturk & Turasan, 2022). Current studies often focus on individual components, overlooking the complexity of real food formulations, particularly those containing biological macromolecules. Scaling up MF to an industrial level is impeded by its high cost, requirement of additional equipment like homogenizers and millers, and, most importantly, its design (very narrow channel), which limits the production capacity to 1–10 L/min. A high-pressure homogenizer (valve) costs 20–30 times less than the existing microfluidizer with the same processing capacity (Li et al., 2022). Moreover, the non-thermal processing nature of MF may result in temperature increases affecting material properties, a factor not extensively studied at larger scales (Ozturk & Turasan, 2022).

MF stands out due to its advantages, including the absence of exogenous chemicals, minimal loss of nutritional components, low temperature, and less processing time (Guo et al., 2020). While it excels in developing emulsions with even droplet size and enhanced stability, a pre-emulsification step may be required, which dual-channel microfluidizers can address by feeding two phases separately (Bai & McClements, 2016). However, MF is prone to over-processing, resulting in increased droplet size and recoalescence, emphasizing the need for proper optimization and the use of surfactants (Bai & McClements, 2016). Effective emulsification requires careful process optimization to minimize these issues and optimize energy consumption. Integrating MF with different physical, chemical, or enzymatic processes can enhance its functionality, particularly in modifying biological molecules. However, even in commercial systems with multiple-slotted chambers, the technology is limited by the size of the chamber, which is less than 300 μm (Kavinila et al., 2023).

Addressing these challenges, an innovative solution has emerged in the form of an industrial-scale MF system (ISMS). ISMS integrates crushing, homogenization, coarse grinding, and emulsification into one unit, overcoming many limitations of conventional MF equipment. It boasts a processing capability of 83.3 L/min, meeting industrial production needs (Li et al., 2022). Successfully applied in manufacturing whole corn and soybean milk slurry, ISMS demonstrates superior performance compared to the traditional methods, offering a glimpse into the potential of MF technology for industrial applications. In conclusion, while MF holds a great promise in the food industry, ongoing research and innovative solutions such as ISMS are crucial to overcoming current limitations and fully realizing its potential. Further research should focus on successful scale-up across different products, utilizing computational modelling and simulation tools for pilot and industrial-scale studies while also aiming to increase the capacity and reduce costs without compromising product quality (Kavinila et al., 2023). Designing of MF channel for improving the processing capacity and efficiency according to individual proteins is also worth exploring. Further investigations are needed to study the behavior of MF-modified ingredients in the actual/final complex food products.

The lack of studies on MF of tree nut proteins represents a significant research gap. This technique offers promising potential for valorizing and improving various tree nut proteins extracted from meals or deoiled cakes. Nut proteins, such as those from almonds, walnuts, hazelnuts, and others, possess unique nutritional profiles and functional properties that could benefit from MF. By subjecting these proteins to MF, their solubility, emulsifying capacity, gelation properties, and overall functionality can be enhanced. Further, effect of MF on allergenicity of nut proteins could be another promising area to research on considering large number of people experience allergies towards nut proteins. Research focusing on MF of the tree nut proteins holds promise for developing innovative and value-added products that can cater to the increasing necessity for plant-derived protein sources in the food sector.

In conclusion, MF, an emerging homogenization technology, holds a great field of research owing to a non-thermal green modification technique that is scalable for commercialization. MF has been proven to be a promising technique and has already found its way into industries. Numerous studies concluded the potential effect of MF on enhancing the functionalities of proteins. Since the beneficial effect of MF depends on the processing parameters, careful optimization is required to achieve the desired functionality. The future is full of great possibilities for MF-based modification of major macromolecules of tree nuts, especially proteins that have a variety of applications in the food industry. Numerous results indicated that MF could serve as a helpful modification method to increase the number of potential uses for proteins as it aids in reducing the anti-nutritional components and enhancing the key functional properties including protein solubility, bioaccessibility, emulsifying properties, and antioxidant activities. Still, there are numerous interesting research subjects to delve into, especially focusing on MF as a modification technique to reduce the immunoreactivity/allergenicity of nut proteins. Intensive research to determine how pressure levels, number of passes, and temperature affect nut protein immunoreactivity can address the prevalence of nut allergy worldwide. Since data on combination of MF and other non-thermal techniques on the functionality and physico-chemical attributes of nut proteins is limited, further research is needed to explore their combined impact. Pre-treatment with MF to enhance enzymatic hydrolysis so as to yield low molecular weight bioactive peptides could also be a promising area of research.

## REFERENCES

Adjei-Fremah, S., Worku, M., De Erive, M. O., He, F., Wang, T., & Chen, G. (2019). Effect of microfluidization on microstructure, protein profile and physicochemical properties of whole cowpea flours. *Innovative Food Science and Emerging Technologies*, *57*. https://doi.org/10.1016/j.ifset.2019.102207

Aluko, R. E. (2015a). Antihypertensive peptides from food proteins. *Annual Review of Food Science and Technology*, *6*, 235–262. https://doi.org/10.1146/ANNUREV-FOOD-022814-015520

Aluko, R. E. (2015b). Structure and function of plant protein-derived antihypertensive peptides. In *Current Opinion in Food Science* (Vol. 4). https://doi.org/10.1016/j.cofs.2015.05.002

Aluko, R. E., Girgih, A. T., He, R., Malomo, S., Li, H., Offengenden, M., & Wu, J. (2015). Structural and functional characterization of yellow field pea seed (Pisum sativum L.) protein-derived antihypertensive peptides. *Food Research International*, *77*. https://doi.org/10.1016/j.foodres.2015.03.029

Aryee, A. N. A., Agyei, D., & Udenigwe, C. C. (2017). Impact of processing on the chemistry and functionality of food proteins. In *Proteins in Food Processing*, 2nd Edition. https://doi.org/10.1016/B978-0-08-10072 2-8.00003-6

Bai, L., & McClements, D. J. (2016). Development of microfluidization methods for efficient production of concentrated nanoemulsions: Comparison of single- and dual-channel microfluidizers. *Journal of Colloid and Interface Science*, *466*. https://doi.org/10.1016/j.jcis.2015.12.039

Barreira, J. C. M., Nunes, M. A., da Silva, B. V., Pimentel, F. B., Costa, A. S. G., Alvarez-Ortí, M., Pardo, J. E., & Oliveira, M. B. P. P. (2019). Almond cold-pressed oil by-product as ingredient for cookies with potential health benefits: Chemical and sensory evaluation. *Food Science and Human Wellness*, *8*(3). https://doi.org/10.1016/j.fshw.2019.07.002

Biter, A. B., Pollet, J., Chen, W. H., Strych, U., Hotez, P. J., & Bottazzi, M. E. (2019). A method to probe protein structure from UV absorbance spectra. *Analytical Biochemistry*, *587*. https://doi.org/10.1016/j.ab.2019.113450

Ceylan, F. D., Adrar, N., Günal-Köroğlu, D., Gültekin Subaşl, B., & Capanoglu, E. (2022). Combined neutrase-alcalase protein hydrolysates from hazelnut meal, a potential functional food ingredient. *ACS Omega*. https://doi.org/10.1021/acsomega.2c07157

Chatur, P., Johnson, S., Coorey, R., Bhattarai, R. R., & Bennett, S. J. (2022). The effect of high pressure processing on textural, bioactive and digestibility properties of cooked Kimberley large Kabuli chickpeas. *Frontiers in Nutrition*, *9*. https://doi.org/10.3389/fnut.2022.847877

Chen, L., Chen, J., Yu, L., Wu, K., & Zhao, M. (2018). Emulsification performance and interfacial properties of enzymically hydrolyzed peanut protein isolate pretreated by extrusion cooking. *Food Hydrocolloids*, *77*. https://doi.org/10.1016/j.foodhyd.2017.11.002

Cheng, M., Li, Y., Luo, X., Chen, Z., Wang, R., Wang, T., Feng, W., Zhang, H., He, J., & Li, C. (2022). Effect of dynamic high-pressure microfluidization on physicochemical, structural, and functional properties of oat protein isolate. *Innovative Food Science and Emerging Technologies*, *82*. https://doi.org/10.1016/j.ifset.2022.103204

Cikrikci, S., Demirkesen, I., & Mert, B. (2016). Production of hazelnut skin fibres and utilisation in a model bakery product. *Quality Assurance and Safety of Crops and Foods*, *8*(2). https://doi.org/10.3920/QAS2015.0587

Cui, S., McClements, D. J., Xu, X., Jiao, B., Zhou, L., Zhou, H., Xiong, L., Wang, Q., Sun, Q., & Dai, L. (2023). Peanut proteins: Extraction, modifications, and applications: A comprehensive review. *Grain and Oil Science and Technology*, *6*(3). https://doi.org/10.1016/j.gaost.2023.07.001

Dai, T., Shuai, X., Chen, J., Li, C., Wang, J., Liu, W., Liu, C., & Wang, R. (2022). Whole peanut milk prepared by an industry-scale microfluidization system: Physical stability, microstructure, and flavor properties. *LWT*, *171*. https://doi.org/10.1016/j.lwt.2022.114140

Demirkesen, I., Vilgis, T. A., & Mert, B. (2018). Effect of microfluidization on the microstructure and physical properties of a novel yoghurt formulation. *Journal of Food Engineering*, *237*. https://doi.org/10.1016/j.jfoodeng.2018.05.025

Djemaoune, Y., Cases, E., & Saurel, R. (2019). The effect of high-pressure microfluidization treatment on the foaming properties of pea albumin aggregates. *Journal of Food Science*, *84*(8). https://doi.org/10.1111/1750-3841.14734

Eckert, E., Han, J., Swallow, K., Tian, Z., Jarpa-Parra, M., & Chen, L. (2019). Effects of enzymatic hydrolysis and ultrafiltration on physicochemical and functional properties of faba bean protein. *Cereal Chemistry*, *96*(4). https://doi.org/10.1002/cche.10169

Feng, Y., Yuan, D., Kong, B., Sun, F., Wang, M., Wang, H., & Liu, Q. (2022). Structural changes and exposed amino acids of ethanol-modified whey proteins isolates promote its antioxidant potential. *Current Research in Food Science*, *5*. https://doi.org/10.1016/j.crfs.2022.08.012

Gonçalves, B., Pinto, T., Aires, A., Morais, M. C., Bacelar, E., Anjos, R., Ferreira-Cardoso, J., Oliveira, I., Vilela, A., & Cosme, F. (2023). Composition of nuts and their potential health benefits – An overview. *Foods*, *12*(5). https://doi.org/10.3390/foods12050942

Gong, K., Chen, L., Xia, H., Dai, H., Li, X., Sun, L., Kong, W., & Liu, K. (2019). Driving forces of disaggregation and reaggregation of peanut protein isolates in aqueous dispersion induced by high-pressure microfluidization. *International Journal of Biological Macromolecules*, *130*. https://doi.org/10.1016/j.ijbiomac.2019.02.123

Gong, K., Deng, L., Shi, A., Liu, H., Liu, L., Hu, H., Adhikari, B., & Wang, Q. (2017). High-pressure microfluidisation pretreatment disaggregate peanut protein isolates to prepare antihypertensive peptide fractions. *International Journal of Food Science and Technology*, *52*(8). https://doi.org/10.1111/ijfs.13449

Grand View Research. (2023). Protein ingredients market size, share & trends analysis report by product (plant proteins, animal/dairy proteins, microbe-based proteins, insect proteins), by application, by region, and segment forecasts, 2021–2028. *Grandview Research*. https://www.grandviewresearch.com/industry-analysis/protein-ingredients-market

Guo, X., Chen, M., Li, Y., Dai, T., Shuai, X., Chen, J., & Liu, C. (2020). Modification of food macromolecules using dynamic high pressure microfluidization: A review. *Trends in Food Science and Technology*, *100*. https://doi.org/10.1016/j.tifs.2020.04.004

He, X., Chen, J., He, X., Feng, Z., Li, C., Liu, W., Dai, T., & Liu, C. (2021). Industry-scale microfluidization as a potential technique to improve solubility and modify structure of pea protein. *Innovative Food Science and Emerging Technologies*, *67*. https://doi.org/10.1016/j.ifset.2020.102582

Hu, X., Amakye, W. K., He, P., Wang, M., & Ren, J. (2021). Effects of microfluidization and transglutaminase cross-linking on the conformations and functional properties of arachin and conarachin in peanut. *LWT, 146.* https://doi.org/10.1016/j.lwt.2021.111438

Hu, X., He, Z., He, P., & Wang, M. (2022). Microfluidization treatment improve the functional and physico-chemical properties of transglutaminase cross-linked groundnut arachin and conarachin. *Food Hydrocolloids, 130.* https://doi.org/10.1016/j.foodhyd.2022.107723

Hu, X., Zhao, M., Sun, W., Zhao, G., & Ren, J. (2011). Effects of microfluidization treatment and transglutaminase cross-linking on physicochemical, functional, and conformational properties of peanut protein isolate. *Journal of Agricultural and Food Chemistry, 59*(16), 8886–8894. https://doi.org/10.1021/jf201781z

Huang, L., Ding, X., Li, Y., & Ma, H. (2019). The aggregation, structures and emulsifying properties of soybean protein isolate induced by ultrasound and acid. *Food Chemistry, 279.* https://doi.org/10.1016/j.foodchem.2018.11.147

International Nut & Dried Fruit Council Foundation. (2023). Nuts & dried fruits-statistical yearbook 2022–2023. In *Statistical Yearbook,* 6–7.

Ji, H., Dong, S., Han, F., Li, Y., Chen, G., Li, L., & Chen, Y. (2018). Effects of dielectric barrier discharge (DBD) cold plasma treatment on physicochemical and functional properties of peanut protein. *Food and Bioprocess Technology, 11,* 344–354. https://doi.org/10.1007/s11947-017-2015-z

Jiang, Y. S., Zhang, S. B., Zhang, S. Y., & Peng, Y. X. (2021). Comparative study of high-intensity ultrasound and high-pressure homogenization on physicochemical properties of peanut protein-stabilized emulsions and emulsion gels. *Journal of Food Process Engineering, 44*(6). https://doi.org/10.1111/jfpe.13710

Jin, J., Ma, H., Wang, B., Yagoub, A. E. G. A., Wang, K., He, R., & Zhou, C. (2016). Effects and mechanism of dual-frequency power ultrasound on the molecular weight distribution of corn gluten meal hydrolysates. *Ultrasonics Sonochemistry, 30.* https://doi.org/10.1016/j.ultsonch.2015.11.021

Karacam, C. H., Sahin, S., & Oztop, M. H. (2015). Effect of high pressure homogenization (microfluidization) on the quality of Ottoman Strawberry (*F. Ananassa*) juice. *LWT, 64*(2). https://doi.org/10.1016/j.lwt.2015.06.064

Kavinila, S., Nimbkar, S., Moses, J. A., & Anandharamakrishnan, C. (2023). Emerging applications of micro-fluidization in the food industry. *Journal of Agriculture and Food Research, 12.* https://doi.org/10.1016/j.jafr.2023.100537

Kumar, D., Mishra, A., Tarafdar, A., Kumar, Y., Verma, K., Aluko, R., Trajkovska, B., & Badgujar, P. C. (2021). In vitro bioaccessibility and characterisation of spent hen meat hydrolysate powder prepared by spray and freeze-drying techniques. *Process Biochemistry, 105.* https://doi.org/10.1016/j.procbio.2021.03.029

Ladokhin, A. S. (2000). Fluorescence spectroscopy in peptide and protein analysis. *Encyclopedia of Analytical Chemistry.* https://doi.org/10.1002/9780470027318.a1611

Landim, A. P. M., Chávez, D. W. H., da Rosa, J. S., Mellinger-Silva, C., & Rosenthal, A. (2021). Effect of high hydrostatic pressure on the antioxidant capacity and peptic hydrolysis of whey proteins. *Ciência Rural, 51*(4). https://doi.org/10.1590/0103-8478cr20200560

Li, Y., Deng, L., Dai, T., Li, Y., Chen, J., Liu, W., & Liu, C. (2022). Microfluidization: A promising food processing technology and its challenges in industrial application. *Food Control, 137.* https://doi.org/10.1016/j.foodcont.2021.108794

Liang, R., Cheng, S., & Wang, X. (2018). Secondary structure changes induced by pulsed electric field affect antioxidant activity of pentapeptides from pine nut (*Pinus koraiensis*) protein. *Food Chemistry, 254.* https://doi.org/10.1016/j.foodchem.2018.01.090

Ling, Y., Cheng, L., Bai, X., Li, Z., Dai, J., & Ren, D. (2023). Effects of microfluidization on the physical and storage stability of walnut protein emulsion and beverages. *Plant Foods for Human Nutrition, 78*(2). https://doi.org/10.1007/s11130-023-01073-7

Loushigam, G., & Shanmugam, A. (2023). Modifications to functional and biological properties of proteins of cowpea pulse crop by ultrasound-assisted extraction. *Ultrasonics Sonochemistry, 106448.* https://doi.org/10.1016/J.ULTSONCH.2023.106448

Malik, M. A., & Saini, C. S. (2017). Gamma irradiation of alkali extracted protein isolate from dephenolized sunflower meal. *LWT, 84,* 204–211. https://doi.org/10.1016/J.LWT.2017.05.067

Munialo, C. D., Stewart, D., Campbell, L., & Euston, S. R. (2022). Extraction, characterisation and functional applications of sustainable alternative protein sources for future foods: A review. *Future Foods, 6.* https://doi.org/10.1016/j.fufo.2022.100152

Oboroceanu, D., Wang, L., Kroes-Nijboer, A., Brodkorb, A., Venema, P., Magner, E., & Auty, M. A. E. (2011). The effect of high pressure microfluidization on the structure and length distribution of whey protein fibrils. *International Dairy Journal, 21*(10). https://doi.org/10.1016/j.idairyj.2011.03.015

Ozturk, O. K., & Turasan, H. (2022). Latest developments in the applications of microfluidization to modify the structure of macromolecules leading to improved physicochemical and functional properties. *Critical Reviews in Food Science and Nutrition, 62*(16). https://doi.org/10.1080/10408398.2021.1875981

Poore, J., & Nemecek, T. (2018). Reducing food's environmental impacts through producers and consumers. *Science, 360*(6392), 987–992. https://doi.org/10.1126/SCIENCE.AAQ0216

Rehman, I., Kerndt, C. C., & Botelho, S. (2021). Biochemistry, tertiary protein structure. *StatPearls [Internet],* 1–3. https://www.ncbi.nlm.nih.gov/books/NBK470269/

Roncero, J. M., Álvarez-Ortí, M., Pardo-Giménez, A., Rabadán, A., & Pardo, J. E. (2021). Influence of pressure extraction systems on the performance, quality and composition of virgin almond oil and defatted flours. *Foods, 10*(5). https://doi.org/10.3390/foods10051049

Sari, T. P., Dhamane, A. H., Pawar, K., Bajaj, M., Badgujar, P. C., Tarafdar, A., Bodana, V., & Pareek, S. (2024b). High-pressure microfluidisation positively impacts structural properties and improves functional characteristics of almond proteins obtained from almond meal. *Food Chemistry,* 139084. https://doi.org/10.1016/j.foodchem.2024.139084

Sari, T. P., Sirohi, R., Krishania, M., Bhoj, S., Samtiya, M., Duggal, M., Kumar, D., & Badgujar, P. C. (2022). Critical overview of biorefinery approaches for valorization of protein rich tree nut oil industry by-product. *Bioresource Technology, 362.* https://doi.org/10.1016/j.biortech.2022.127775

Sari, T. P., Sirohi, R., Tyagi, P., Tiwari, G., Pal, J., Kunadia, N. N., Verma, K., Badgujar, P. C., & Pareek, S. (2024a). Protein hydrolysates prepared by Alcalase using ultrasound and microwave pretreated almond meal and their characterization. *Journal of Food Science and Technology,* 1–8. https://doi.org/10.1007/s13197-024-05945-x

Saricaoglu, F. T. (2020). Application of high-pressure homogenization (HPH) to modify functional, structural and rheological properties of lentil (Lens culinaris) proteins. *International Journal of Biological Macromolecules, 144.* https://doi.org/10.1016/j.ijbiomac.2019.11.034

Saricaoglu, F. T., Gul, O., Besir, A., & Atalar, I. (2018). Effect of high pressure homogenization (HPH) on functional and rheological properties of hazelnut meal proteins obtained from hazelnut oil industry by-products. *Journal of Food Engineering, 233.* https://doi.org/10.1016/j.jfoodeng.2018.04.003

Shen, L., & Tang, C. H. (2012). Microfluidization as a potential technique to modify surface properties of soy protein isolate. *Food Research International, 48*(1). https://doi.org/10.1016/j.foodres.2012.03.006

Sim, S. Y. J., Srv, A., Chiang, J. H., & Henry, C. J. (2021). Plant proteins for future foods: A roadmap. *Foods, 10*(8). https://doi.org/10.3390/FOODS10081967/TITLE/PLANT_PROTEINS_FOR_FUTURE_FOODS_A_ROADMAP

Sun, C., Dai, L., Liu, F., & Gao, Y. (2016). Dynamic high pressure microfluidization treatment of zein in aqueous ethanol solution. *Food Chemistry, 210.* https://doi.org/10.1016/j.foodchem.2016.04.138

Takano, Y., Kondo, H. X., & Nakamura, H. (2022). Quantum chemical studies on hydrogen bonds in helical secondary structures. *Biophysical Reviews, 14*(6). https://doi.org/10.1007/s12551-022-01034-5

Tu, J., & Wu, W. (2019). Critical functional properties of defatted peanut meal produced by aqueous extraction and conventional methods. *Journal of Food Science and Technology, 56*(10). https://doi.org/10.1007/s13197-019-03922-3

United Nations. (2023). *The World in 2100.* https://www.un.org/en/global-issues/population#:~:text=Theworldpopulationisprojected,and10.4billionby2100. Last assessed on 05-August-2024.

Wang, X., Wang, S., Wang, W., Ge, Z., Zhang, L., Li, C., Zhang, B., & Zong, W. (2019). Comparison of the effects of dynamic high-pressure microfluidization and conventional homogenization on the quality of peach juice. *Journal of the Science of Food and Agriculture, 99*(13). https://doi.org/10.1002/jsfa.9874

Yang, F., Liu, X., Ren, X., Huang, Y., Huang, C., & Zhang, K. (2018). Swirling cavitation improves the emulsifying properties of commercial soy protein isolate. *Ultrasonics Sonochemistry, 42,* 471–481. https://doi.org/10.1016/j.ultsonch.2017.12.014

Yang, X., Teng, D., Wang, X., Guan, Q., Mao, R., Hao, Y., & Wang, J. (2016). Enhancement of nutritional and antioxidant properties of peanut meal by bio-modification with Bacillus licheniformis. *Applied Biochemistry and Biotechnology, 180*(6). https://doi.org/10.1007/s12010-016-2163-z

Yildiz, E., Demirkesen, I., & Mert, B. (2016). High pressure microfluidization of agro by-product to functionalized dietary fiber and evaluation as a novel bakery ingredient. *Journal of Food Quality, 39*(6). https://doi.org/10.1111/jfq.12246

Zhang, S. B., Wang, X. H., Li, X., & Yan, D. Q. (2020). Effects of tween 20 and transglutaminase modifications on the functional properties of peanut proteins. *JAOCS, Journal of the American Oil Chemists' Society, 97*(1). https://doi.org/10.1002/aocs.12309

Zhang, Z., Yang, Y., Zhou, P., Zhang, X., & Wang, J. (2017). Effects of high pressure modification on conformation and gelation properties of myofibrillar protein. *Food Chemistry, 217.* https://doi.org/10.1016/j.foodchem.2016.09.040

Zhao, F., Liu, C., Bordoni, L., Petracci, I., Wu, D., Fang, L., Wang, J., Wang, X., Gabbianelli, R., & Min, W. (2022). Advances on the antioxidant peptides from nuts: A narrow review. *Antioxidants*, 11(10). https://doi.org/10.3390/antiox11102020

Zhao, Q., Yan, W., Liu, Y., & Li, J. (2021). Modulation of the structural and functional properties of perilla protein isolate from oilseed residues by dynamic high-pressure microfluidization. *Food Chemistry*, *365*. https://doi.org/10.1016/j.foodchem.2021.130497

Zhao, Y., Sun, N., Li, Y., Cheng, S., Jiang, C., & Lin, S. (2017). Effects of electron beam irradiation (EBI) on structure characteristics and thermal properties of walnut protein flour. *Food Research International*, *100*. https://doi.org/10.1016/j.foodres.2017.08.004

Zhu, Q., Tang, X., Lu, M., & Chen, J. (2022). Structure and immunoreactivity of purified Siberian apricot (Prunus sibirica L.) kernel allergen under high hydrostatic pressure treatment. *Food Bioscience*, *48*. https://doi.org/10.1016/j.fbio.2022.101727

Zhu, Z., Zhu, W., Yi, J., Liu, N., Cao, Y., Lu, J., Decker, E. A., & McClements, D. J. (2018). Effects of sonication on the physicochemical and functional properties of walnut protein isolate. *Food Research International*, *106*. https://doi.org/10.1016/j.foodres.2018.01.060

# 13 Microfluidization of Seed Storage Proteins

*Neeraj Ghanghas, Yogesh Kumar, and Rajat Suhag*

## 13.1 INTRODUCTION

As the global population is anticipated to escalate to 10.4 billion individuals by the year 2100 (United Nations Department of Economic and Social Affairs, 2022), the challenge of producing and distributing adequate food supplies to the entire human population has become increasingly formidable. This challenge is about meeting basic caloric needs and addressing the essential requirement for proteins, which rank as the second most crucial macronutrient for human health and survival. The concern regarding protein production is particularly pressing, given that conventional animal protein sources rely heavily on extensive land use and significant resource input (Calicioglu et al., 2019; Ghanghas et al., 2024). In this context, plant proteins emerge as a promising alternative. Their advantages stem from a well-established history of cultivation, comparatively lower production costs, and their widespread availability in many parts of the globe.

Furthermore, the environmental sustainability of plant-based proteins adds to their appeal as a preferable option for addressing the protein needs of a growing global population (Ghanghas, 2023; Willett et al., 2019). Given the higher cost of animal proteins, populations in developing countries predominantly rely on seed protein to fulfill their entire protein needs. However, in contrast to animal-derived proteins such as casein and egg albumin, which offer a more balanced profile of all essential amino acids, plant proteins (seed proteins) often lack one or more of these crucial nutrients. The utilization of seed proteins in various applications is intricately linked to their functional properties, which are, in turn, fundamentally dependent on their structural characteristics. The structural integrity of these proteins determines their behavior and interaction in different environments, directly influencing their functional attributes such as solubility, emulsification capacity, gelation, foaming properties, and water and oil binding capacities (OBCs).

Seed proteins are generally categorized into two main types: housekeeping proteins, which play a crucial role in sustaining regular cell metabolism, and storage proteins. Recent advances in the classification further divided these proteins into storage, structural, and biologically active types. Notably, biologically active proteins, including lectins, enzymes, and enzyme inhibitors, are present in smaller quantities but may present a more balanced amino acid profile than storage proteins. Conversely, storage proteins are non-enzymatic and primarily provide essential proteins (sources of nitrogen and sulfur) necessary for germinating and developing new plants (Mandal & Mandal, 2000). Historically, the classification of seed proteins was conducted by Osborne (1924), who categorized them based on their solubility. This classification includes albumins (soluble in water), globulins (soluble in dilute salt solutions), prolamins (soluble in aqueous alcohol), and glutelins (soluble in weakly acidic or alkaline solutions).

A diverse array of seed crops, encompassing cereals, pulses, and legumes, are employed in extracting proteins for many applications, spanning both food and non-food sectors. Key examples include soy, wheat, peas, canola, corn, rice, hemp, millets, lentils, fenugreek, and

DOI: 10.1201/9781032632599-13

sunflower, each of which can be utilized to produce an extensive range of food products. These encompass plant-based dairy alternatives as well as extruded products such as protein-enriched snack foods and meat substitutes (Ghanghas et al., 2024; Jeske et al., 2018; Luo et al., 2020; Nasrabadi et al., 2021; Singh et al., 2023; Zahari et al., 2020). However, it is crucial to note that the seed proteins derived from these various sources exhibit substantial differences in their chemical composition, molecular structure, and functional properties. As a result, these proteins' specific application and suitability vary markedly depending on their inherent characteristics (Navaf et al., 2023).

This variation in the protein profile is a critical factor in determining the end-use of these plant-derived proteins. For instance, the protein content, solubility, emulsifying properties, and textural attributes play a significant role in their application in food products (Ghanghas et al., 2019). In non-food uses, these properties influence their utility in areas such as biodegradable materials, pharmaceuticals, and cosmetics (Ghanghas et al., 2019; Yashwant et al., 2023). Understanding and harnessing the unique properties of these diverse seed proteins are essential in optimizing their use in various industrial applications. This includes protein modifications to meet the demands of different food textures, nutritional profiles, and functional requirements and exploiting their unique characteristics for innovative non-food uses.

Tailoring the properties of seed proteins to meet specific demands is a critical step in optimizing their use across a range of applications. This customization can be effectively achieved through various methods, such as physical, chemical, and biological (enzymatic and fermentation) techniques. Physical methods include high-pressure processing or ultrasonication, which modify the protein's structure without altering its chemical composition. Chemical methods may involve pH adjustments or the incorporation of additives to change protein solubility and functionality. Enzymatic approaches use specific enzymes to hydrolyze proteins, enhancing their functional properties (Nasrabadi et al., 2021; Navare et al., 2023; Sharma et al., 2023). Various biological and chemical techniques have been employed for protein modifications, including enzymatic digestion, glycation, and deamidation. These methods aim to enhance its solubility, emulsification, and foaming characteristics, thereby expanding its range of uses (Jiang et al., 2015; Nivala et al., 2017; Zhang et al., 2015). However, due to various health and safety issues, such modifications often prove unsuitable for food-related applications. Recently, a growing interest has been in modifying proteins through physical methods such as microfluidization and ultrasonication. These methods are gaining popularity due to their use of non-toxic substances and their operation under gentle conditions (Cheng et al., 2022; Nikbakht Nasrabadi et al., 2021; Sharma et al., 2023).

Among these, non-thermal techniques, especially microfluidization, stand out for their efficacy in protein modification. Microfluidization involves processing fluids through microchannels at high velocities, generating intense shear forces that can significantly alter the structure and characteristics of proteins. This technique is particularly beneficial for modifying seed proteins, improving solubility, emulsifying properties, and texture without affecting their nutritional value. Such enhancements are pivotal for developing plant-based food products where texture and flavor are key considerations (Ghanghas et al., 2021; Han et al., 2023).

Furthermore, microfluidization is an environmentally friendly and energy-efficient approach, aligning well with the sustainable attributes of plant-based protein sources. The ability of microfluidization to modify protein structure and functionality at the molecular level presents new opportunities for developing nutritious, sustainable, and consumer-appealing plant-based food products. Its potential extends beyond food applications, offering innovative solutions in non-food industries such as biodegradable materials, pharmaceuticals, and personal care products. Therefore, the application of microfluidization in seed protein modification represents a significant advancement in capitalizing on the full potential of these versatile and sustainable protein sources.

## 13.2 MICROFLUIDIZATION TECHNOLOGY OVERVIEW

### 13.2.1 Principles of Microfluidization

Microfluidization is one of the emerging processing technologies that has brought a tremendous and desirable change in the food matrix. By generating high cavitation, shear, velocity impact, and turbulent forces, microfluidizers brought structural modifications in food, which led to significant improvements in the physicochemical, functional, nutritional, rheological, and sensory properties of food products without affecting their natural flavor (Kumar et al., 2022). The microfluidization operating principle is different from conventional high-energy homogenization techniques such as homogenization and ultrasonication, which confers a decisive advantage due to its synergistic amalgamation of water jet technology, impinging stream technology, and high-pressure homogenization (Kavinila et al., 2023). The process entails directing a liquid feed through an interaction chamber, where it undergoes division into two or more microstreams, subsequently colliding with each other (Figure 13.1). This intricate sequence engenders a combination of high pressure, elevated velocity, vibrations, pressure differentials, shear rates, and hydrodynamic cavitation (I. D. Mert, 2020). The breakdown and homogenization processes primarily occur within the interaction chambers, as illustrated in Figure 13.1. These phenomena are governed by inertial forces in turbulent flow and cavitation. The presence of turbulent flow can be confirmed by the Reynolds number, as given by Equation (13.1).

$$Re = \frac{\rho v d}{\mu} \tag{13.1}$$

where $\rho$ is the density of the liquid, $v$ is the velocity of the liquid, $d$ is the diameter of microchannel, and $\mu$ is the dynamic viscosity.

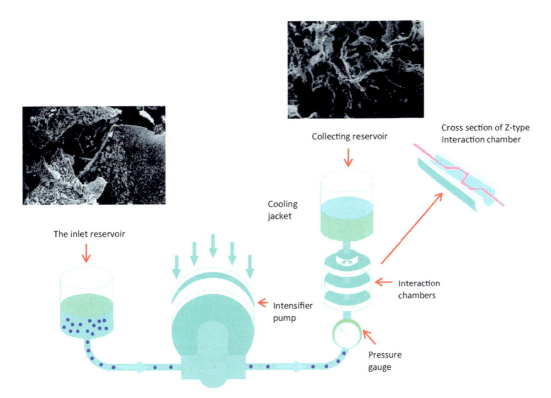

**FIGURE 13.1** Microfluidizer processor. (Source: Mert, 2020.)

## 13.2.2 Description of Microfluidization Equipment and Process Parameters and Their Impact

The microfluidizer comprises an inlet reservoir, an intensifier pump, a pressure gauge, an interaction chamber, a cooling jacket, and a collecting reservoir (Figure 13.1). Furthermore, the configuration can be adjusted according to specific purposes and requirements. Broadly categorized based on the number of inlet reservoirs and processing steps, microfluidizers can be classified as either single- or dual-channel systems. The intensifier pump amplifies the pressure delivered by the hydraulic power system to achieve an optimal operating level within the interaction chamber. The pump oscillates through two strokes, encompassing suction and compression, thereby pressurizing the air to approximately 276 MPa (Kumar et al., 2022; Ozturk & Turasan, 2022). The interaction chamber is a continuous microchannel responsible for turbulent mixing and energy dissipation. It serves as a stabilized geometry that generates a homogeneous pressure profile, facilitating the attainment of a narrow size and distribution of particles (Ocampo-Salinas et al., 2016). The interaction chamber exists in Y-type and Z-type (Figure 13.2). In a Y-type interaction

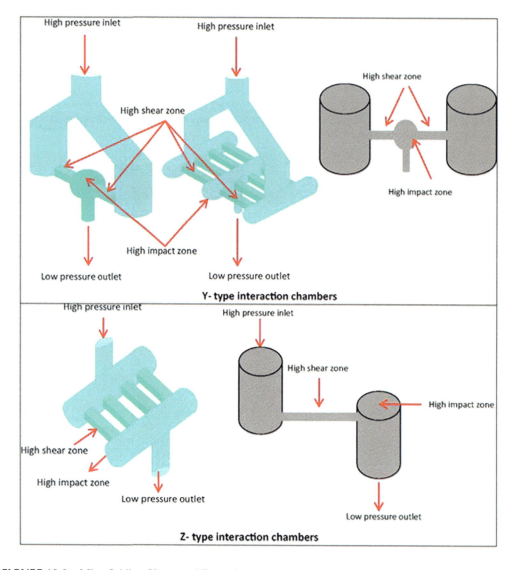

**FIGURE 13.2** Microfluidizer Y-type and Z-type interaction chambers. (Source: Mert, 2020.)

Microfluidization of Seed Storage Proteins  245

chamber, a high-speed liquid stream undergoes bifurcation into two microstreams upon entering the chamber, where the sudden reduction in pipe diameter significantly increases fluid velocity (B. Mert, 2012). Subsequently, these two microstreams collide at remarkably high velocities, reaching up to 400 m/s, generating a shear rate of up to $10^7$ s$^{-1}$ (Villalobos-Castillejos et al., 2018). Y-type chambers are used for processing liquid–liquid dispersions, encompassing the production of liposomes, emulsions, and polymer-based encapsulation. The dimensions of these chambers typically fall within the range of 75–125 μm (Suhag et al., 2023). The Z-type interaction chamber subjects a feed to high pressure, compelling it through intricately designed zig-zag microchannels. Within these channels, the feed undergoes significant particle–particle and particle–wall collisions, leading to the breakdown of droplets. This unique configuration is particularly well-suited for applications in solid–liquid dispersion, water-in-oil emulsion formation, cellular disruption, extraction processes, and deagglomeration. The dimensions of Z-type chambers typically span the range of 87–1100 μm (Dhiman & Prabhakar, 2021; Kumar et al., 2022; I. D. Mert, 2020).

Moreover, a microfluidizer's performance can be enhanced by incorporating an auxiliary processing module (APM). In Y-type chambers, the APM is strategically positioned downstream of the interaction chamber, acting as a flow stabilization unit bridging the transition from high pressure to atmospheric pressure. In Z-type interaction chambers, the APM is located above the interaction chamber and serves the crucial role of premixing or pre-processing. The cooling jacket is used to reduce temperature elevation resulting from energy dissipation during the process (Huang et al., 2023).

## 13.3 MICROFLUIDIZATION FOR PROTEIN STRUCTURAL MODIFICATION

### 13.3.1 MICROFLUIDIZATION-INDUCED MORPHOLOGICAL CHANGES IN SEED PROTEINS

Microfluidization, a process employing high-pressure treatment, has been identified as a significant influencer in altering the morphology of seed proteins. This advanced technique, as evidenced through comprehensive scanning electron microscopy studies, profoundly affects protein structures, leading to notable alterations in size, shape, surface characteristics, and aggregation behavior. These changes were observed in a variety of seed proteins such as oat protein isolate (OPI) (Cheng et al., 2022), fenugreek protein concentrate (FPC) (Ghanghas et al., 2021), and rapeseed protein isolate (RPI) (N. Zhang et al., 2022), which are vital to understanding the impact of microfluidization on the functionality and application of seed proteins.

#### 13.3.1.1 Morphological Transformations Across Seed Proteins

When subjected to microfluidization, seed proteins exhibit dramatic changes in their physical structure. For example, OPI, under a treatment pressure of 120 MPa, transitions from its original state of large, irregular flakes with rough surfaces to much smaller, smoother fragments. This drastic reduction in particle size and smoothening of surfaces is indicative of the protein's structural disaggregation, a phenomenon similarly observed in other seed proteins like fenugreek and rapeseed protein (Cheng et al., 2022; Ghanghas et al., 2021; N. Zhang et al., 2022).

In the case of fenugreek protein, microfluidization leads to a fragmentation of the protein structure, resulting in an assortment of smaller particles with irregular shapes and various surface deformities. The extent and nature of these morphological changes, such as bent sheet-like formations and surface roughness, are vividly captured in microscopic images (Ghanghas et al., 2021).

An intriguing aspect of microfluidization is its pressure-dependent impact on protein morphology. While lower pressures generally result in disaggregation and smoother surfaces, higher pressures, as seen at 150 MPa for OPI and 80 MPa for RPI, can lead to the re-aggregation of protein particles. This aggregation is thought to be a consequence of the exposure and interaction of functional groups within the proteins, such as hydrophobic groups, which drive the formation of new protein networks and aggregates.

The influence of microfluidization on seed proteins is further evidenced by shifts in particle size distribution. For MOPI, a notable shift toward smaller particle sizes is observed when compared to native OPI. The average particle size significantly reduces from $1322.67 \pm 97.5$ nm to $369.13 \pm 10.3$ nm as treatment pressure increases, indicating the mechanical force-driven disaggregation of protein particles. This phenomenon, however, is not linear; at higher pressures, the trend reverses, leading to an increase in particle size, possibly due to hydrophobic interactions (Cheng et al., 2022).

Similar trends are seen in RPI, where microfluidization at 40 MPa results in a more uniform particle size distribution, decreasing median particle size (D50) from 239.2 nm to approximately 170 nm. However, at 80 MPa, the particle distribution becomes uneven, and new peaks of protein aggregation emerge, highlighting the complex interplay between processing pressure and protein structure (N. Zhang et al., 2022). The morphological changes induced by microfluidization have significant implications for the functional properties of seed proteins. The reduction in particle size and the creation of smoother surfaces enhance solubility and improve surface hydrophobicity, which is beneficial for applications like foam formation and emulsion stabilization. The altered morphology, therefore, has the potential to expand the application scope of these proteins in various industries, particularly in food science and technology.

In summary, microfluidization profoundly influences the morphology of seed proteins, impacting their size, shape, surface characteristics, and aggregation behavior in complex and varied ways. These changes, as illustrated in OPI, fenugreek protein, and RPI, underscore the importance of this processing technique in modulating the functional properties of seed proteins. The detailed insights provided here emphasize the significance of understanding these morphological transformations for optimizing the use of seed proteins in diverse applications, paving the way for innovative product development and enhanced utilization in the food industry and beyond the impact of microfluidization on seed storage protein.

### 13.3.2 MICROFLUIDIZATION'S ROLE IN MODIFYING PROTEIN MOLECULAR SIZE: A CONTINUATION FROM MORPHOLOGICAL TRANSFORMATIONS

Following the detailed exploration of how microfluidization alters the morphology of seed proteins, it is equally crucial to examine its impact on the molecular size distribution of these proteins. This aspect, analyzed through advanced methods such as sodium dodecyl sulfate polyacrylamide gel electrophoresis (SDS-PAGE) and size exclusion chromatography by high-performance liquid chromatography (SEC-HPLC), builds upon the previously discussed morphological changes, offering a more complete picture of microfluidization's influence on protein structure.

The morphological transformations observed in seed proteins such as OPI, FPC, and RPI through microfluidization, including alterations in size, shape, and aggregation behavior, are closely related to changes in molecular size distribution. These changes in molecular size are a direct consequence of the structural reorganization at the molecular level, as microfluidization's high-pressure treatment affects both the physical appearance and the molecular structure of proteins.

Microfluidization, a sophisticated process employing ultra-high pressures, significantly influences the molecular size distribution of seed storage proteins. This advanced processing technique's effects on various proteins are comprehensively analyzed using SDS-PAGE and SEC-HPLC. These analytical methods provide a detailed understanding of the changes in protein molecular size and structure post-microfluidization. The findings from these analyses offer profound insights into the nuanced impacts of microfluidization on different seed proteins, elucidating the complex interplay between processing conditions and protein structure.

In examining the effects of microfluidization on RPI, SDS-PAGE analysis provides a clear picture post-microfluidization. The analysis reveals no significant deviations in the protein band patterns compared to control samples. This observation is crucial as it indicates that the primary molecular components of RPI, including napin and cruciferin polypeptides, retain their molecular size even when subjected to high-pressure processing. Maintaining molecular size in RPI is essential for preserving its native functional properties, which are critical in various applications. The stability of these molecular components under the rigorous conditions of microfluidization speaks about the robustness of RPI's structure and its suitability for processing methods that require maintaining protein integrity (N. Zhang et al., 2022).

Contrasting with RPI, hemp protein isolate (HPI) shows notable alterations under non-reducing conditions in SDS-PAGE analysis. The disappearance of the edestin band in microfluidized HPI is a significant finding, suggesting a substantial change in the protein's molecular structure. This change is likely due to the disruption of disulphide bonds within the protein, a transformation that could lead to a reduction in molecular size or modifications in the protein's tertiary and quaternary structures. Such structural modifications have essential implications for HPI's solubility and functional properties, pivotal in its application in food products and other uses (N. Zhang et al., 2022; Z. hui Zhang et al., 2024). The ability of microfluidization to induce these changes highlights its potential as a tool for modifying protein structures to achieve desired functional properties.

Additionally, examining common bean and lentil proteins through SEC-HPLC sheds light on subtle shifts in molecular weight distribution following microfluidization. The overall protein profiles of these beans remain essentially like control samples, affirming the preservation of major protein structures. However, the slight adjustments observed in the balance of protein fractions indicate minor molecular size and aggregation state changes. These shifts reflect microfluidization's capability to induce structural reorganization at the molecular level, subtly altering protein properties without causing extensive disruption (Lopes et al., 2023).

The comparative analysis of microfluidization with other processing methods, such as ultrasound and heat treatments, further underscores its unique impact on seed proteins. While ultrasound and heat treatments often lead to more pronounced structural alterations in proteins, microfluidization maintains the native molecular weight distribution (Lopes et al., 2023), with any modifications occurring more subtly at the molecular level. This distinction is particularly significant in food science and technology, where preserving the native structure of proteins is crucial for maintaining their functional qualities.

In conclusion, the combined use of SDS-PAGE and SEC-HPLC to analyze the impact of microfluidization on seed storage proteins provides an extensive understanding of how this processing technique affects protein molecular size and structure. The analyses reveal that while microfluidization can maintain the protein's molecular size in cases like RPI, it can also induce significant structural changes in proteins such as HPI (Cheng et al., 2022; Ghanghas et al., 2021; Lopes et al., 2023; N. Zhang et al., 2022; Z. hui Zhang et al., 2024). These findings highlight the diverse effects of microfluidization on different seed proteins and underscore its potential for tailoring protein properties in a controlled and specific manner. This versatility makes microfluidization a valuable tool in a wide range of applications, particularly in the food industry, where manipulating protein structures is often vital to developing new and functional food products.

## 13.4 FUNCTIONAL, RHEOLOGICAL, AND THERMAL PROPERTIES MODIFICATIONS

Microfluidization improves properties such as protein solubility, foaming, and emulsion properties and seed protein's water- and oil-holding capacities. Besides, it alters the rheological behavior and thermal properties of seed storage proteins. The modified seed protein exhibits enhanced functional properties, making it a valuable ingredient in formulating various products, including cakes, meringues, whipped creams, sausages, and meat analogs (Ghanghas et al., 2020). This enhanced protein variant demonstrates superior functional characteristics, contributing to the improved quality of the end products. Table 13.1 shows the effect of microfluidization on the functional, rheological, and thermal properties of seed storage protein. Ghanghas et al. (2021) studied the effect of microfluidization on the functional, rheological, and thermal properties of fenugreek seed proteins. The results indicated that water binding capacity (WBC) and foaming capacity decreased, while

**TABLE 13.1**

**Effect of Microfluidization on the Functional, Rheological, and Thermal Properties of Seed Storage Protein**

| Product | Interaction Chamber Type | Processing Condition | Key Findings ($S_c \rightarrow S_m$) | References |
|---|---|---|---|---|
| Ajmer fenugreek seed protein | Y-type | P: 68.9 MPa, passed through microfluidizer twice (double pass) | • Reduction in the water-binding capacity (3.67 → 2.37 mL/g)<br>• Oil-binding capacity increased (9.99 → 12.92 mL/g)<br>• No significant change in emulsion activity (62.2%)<br>• Emulsion stability increased (76.3 → 90.5%)<br>• Foaming capacity decreased (57.9 → 46.9%)<br>• Higher mid-temperature (Tm) during DSC<br>• Solid-like behavior | Ghanghas et al. (2021) |
| Eucommia ulmoides Oliv. seed meal proteins | Y-type | P: 40, 80, 120, and 160 MPa, passed through microfluidizer once (one pass) | • The solubility increased within the 40–80 MPa pressure range and slightly decreased when pressure exceeded 80 MPa<br>• Water-holding capacity decreased with increased pressure in pH range 7–11 and increase in pH range 1–6<br>• Foaming capacity and foam stability increased with increased pressure at 5% w/v (protein samples dispersed in distilled water and adjusted to concentrations)<br>• The emulsion stability index decreased within the 40–120 MPa pressure range and slightly increased when pressure exceeded 80 MPa | Ge et al. (2021) |
| Whole cowpea flours | Z-type | P:137 MPa, with 200 µm diameter for two passes at room temperature | • Swelling capacity increased up to 107.7%<br>• Water-holding capacity increased up to 16.1%<br>• Oil-holding capacity increased up to 162.1% | Adjei-Fremah et al. (2019) |
| Hemp seed protein stabilized emulsion | Y-type | P: 0, 40, 80, 120, and 160 MPa | • The viscosity of the emulsion increased with an increase in pressure<br>• The storage modulus (G′) and loss modulus (G″) were increased | Wang et al. (2023) |
| Hempseed yogurt's | NanoGenizer | P: 80 and 160 MPa | • Water-holding capacity increased by 8%<br>• The hardness, consistency, and elasticity increased by 33.60%, 48.80%, and 0.21%, respectively<br>• Increased in the apparent viscosity<br>• Increased in the storage modulus (G′), and loss modulus (G″) | Xu et al. (2023) |

*(Continued)*

# Microfluidization of Seed Storage Proteins

**TABLE 13.1 (CONTINUED)**

**Effect of Microfluidization on the Functional, Rheological, and Thermal Properties of Seed Storage Protein**

| Product | Interaction Chamber Type | Processing Condition | Key Findings ($S_c \rightarrow S_m$) | References |
|---|---|---|---|---|
| Lotus seed protein isolate | — | P: 40, 80, 120, and 160 MPa Passed through microfluidizer thrice (three passes) | • The solubility increased with an increase in treatment pressure<br>• Emulsifying activity and stability increased with an increase in pressure | Zheng et al. (2022) |
| Perilla protein isolate from oilseed | | P: 30, 60, 90, and 150 MPa, passed through microfluidizer thrice (three passes) | • The solubility, water holding capacity, oil holding capacity, and emulsion activity index increased within the 30–120 MPa pressure range and slightly decreased when pressure exceeded 120 MPa<br>• Treatment at 90 MPa and 120 MPa increases the thermal stability | Zhao et al. (2021) |
| Walnut protein emulsion | | P: 20 MPa, passed through microfluidizer twice (two passes) | • Reduction in the viscosity from 743 to 147 mPa·s<br>• 4.5-fold increase in shear stress as the shear rate increased compared to untreated<br>• Higher denaturation temperature (154.87 °C) and enthalpy (111.76 J·g$^{-1}$)<br>• Superior homogeneity and collectivity | Ling et al. (2023) |

OBC and emulsion stability increased. However, no significant change in emulsion activity was observed compared to the control sample. The decrease in WBC could be attributed to the exposure of amino acids resulting from the high shear action of microfluidization, causing the disintegration, fragmentation, and denaturation of proteins (Shen & Tang, 2012). Emulsion activity depends on the net charge, protein conformation, and interfacial tension, while emulsion stability depends on the protein's ability to resist any physicochemical changes over time (El Nasri & El Tinay, 2007). Foaming properties, including capacity and stability, play a crucial role in assessing the foaming performance in food foams. Good foaming properties indicate that a protein possesses an optimal balance between flexibility and rigidity. During the thermal analysis of Ajmer fenugreek seeds using a differential scanning calorimeter (DSC), both the control and microfluidized samples exhibited a sharp endothermic peak in the temperature range of 40–100 °C. The microfluidized sample showed a higher mid-temperature (Tm) than the control sample, indicating that proteins are less susceptible to unfolding and denaturation at lower temperatures. The control and microfluidized samples exhibited solid-like behavior in the strain ranges of 0.1–17.18% and 0.1–44.38%, as well as stress ranges of 0–114.3 and 0–187.9 Pa, respectively. This behavior can be attributed to the higher value of the elastic modulus (G′) over the viscous modulus (G″).

Ge et al. (2021) studied the effect of microfluidization pressure (40, 80, 120, and 160 MPa) on the functional characteristics of *Eucommia ulmoides* Oliv. seed meal proteins. The results indicated that solubility increased within the 40–80 MPa pressure range but slightly decreased when the pressure exceeded 80 MPa. This slight decrease in solubility might be attributed to the exposure of more hydrophobic and nonpolar groups buried inside proteins, weakening the interaction between water and protein (Zhu et al., 2017). Water holding capacity (WHC) decreased with increased pressure in the pH range of 7–11 and increased in the pH range of 1–6. This is due to the higher pressure of microfluidization treatment under acidic conditions, resulting in more exposure to hydrophilic groups that provide additional water-binding sites.

The emulsion stability index of microfluidized *Eucommia ulmoides* Oliv. seed meal proteins decreased within the 40–120 MPa pressure range and slightly increased when the pressure exceeded 80 MPa, as compared to the control sample. Hemp seed protein-stabilized emulsion exhibited an increase in viscosity, storage modulus (G′), and loss modulus (G″) with an increase in microfluidizer operating pressure ranging from 0 to 160 MPa (Wang et al., 2023). Similarly, Xu et al. (2023) investigated the impact of microfluidization treatment (NanoGenizer) at 80 MPa on hempseed yogurt's functional and rheological properties. The results indicated an 8% increase in WHC, as well as an increase in viscosity, storage modulus (G′), and loss modulus (G″) in the microfluidized sample compared to the control (Figure 13.3).

Furthermore, in microfluidized whole cowpea flours processed through a 'Z' type interaction chamber with a 200 μm diameter for two passes at room temperature and 137 MPa, there was a notable increase in swelling capacity (up to 107.7%), water-holding capacity (up to 16.1%), and oil-holding capacity (up to 162.1%) (Adjei-Fremah et al., 2019). Microfluidized lotus seed protein isolate demonstrates an increase in solubility (36–87%), emulsifying activity (26–52%), and stability (9–26%) with rising operating pressure (40–160 MPa) (Zheng et al., 2022). The solubility of pea protein increased from 16.99% to 64.28% after microfluidization at 120 MPa (He et al., 2021). Zhao et al. (2021) studied the functional and thermal properties of perilla protein isolate from oilseed at

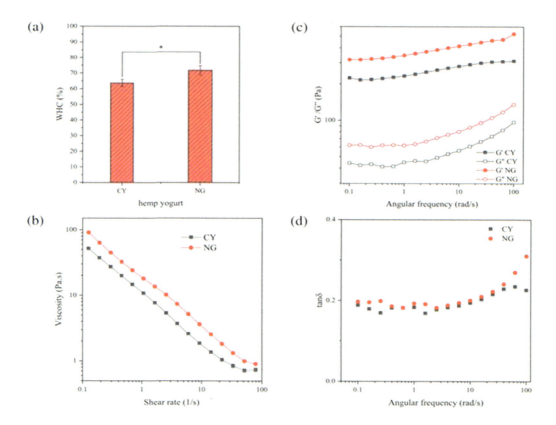

**FIGURE 13.3** Effects of microfluidization on (a) WHC, (b) apparent viscosity, (c) storage modulus G′ and loss modulus G″, and (d) loss tangent (tan δ) of control hempseed yogurt (CY) and hempseed yogurt processed with NanoGenizer (NG). *P < 0.05; **P < 0.01; ***P < 0.001. (Source: Xu et al., 2023.)

varying operating pressures (30, 60, 90, and 150 MPa) using a microfluidizer. The results indicate that solubility, WHC, oil holding capacity, and emulsion activity index increased within the 30–120 MPa pressure range but slightly decreased when the pressure exceeded 120 MPa. The maximum solubility value reached 69.59% at 120 MPa. The observed increase in solubility can be attributed to the physical forces generated by microfluidization, leading to the cleavage of hydrophobic and hydrogen bonds in proteins and enhancing the hydration of the treated sample. However, the slight decrease in solubility of the pea protein isolate (PPI) above 120 MPa may be attributed to the exposure of more hydrophobic sites and the re-aggregation of some particles into large clusters. At 120 MPa, the WHC reached its maximum value of 1.68 g/g. The increase in WHC can be attributed to the enhanced exposure of carboxyl and hydrated hydroxyl groups on the protein surface, providing more water-binding sites and facilitating improved protein–water interactions. However, beyond 120 MPa, a reduction in WHC occurred, possibly due to the exposure of hydrophobic sites induced by microfluidization treatment to water. Similarly, at 120 MPa, the oil holding capacity peaked at 5.14 g/g, followed by a slight decrease. This decrease can probably be attributed to reduced interactions between lipids and proteins, leading to increased protein re-aggregation. The foaming capacity and stability significantly increased, reaching their maximum values of 40.48% and 20.24%, respectively, at a pressure of 120 MPa. This increase could be attributed to the reduction in particle size of protein aggregates, enhancement of surface hydrophobicity, decrease in interfacial tension, and increased flexibility of proteins for adsorption at the air–water interface. Furthermore, during thermal gravimetric analysis, samples treated at 90 and 120 MPa exhibited increased thermal stability.

Ling et al. (2023) studied the effect of microfluidization on the functional, rheological, and thermal properties of walnut protein emulsion (WPE) compared to conventional homogenization. WPE was treated at 20 MPa pressure using conventional homogenization and microfluidization with two passes. The findings revealed a notable reduction in the viscosity of microfluidized WPE, decreasing from 743 to 147 mPa·s (Figure 13.4a). Simultaneously, the shear stress demonstrated a 4.5-fold increase as the shear rate escalated from 0 to 200 s$^{-1}$ (Figure 13.4b). The thermal analysis of WPE using DSC revealed that microfluidized WPE exhibited significantly higher denaturation temperature (154.87 °C) and enthalpy (111.76 J·g$^{-1}$) compared to untreated and conventional homogenized WPE. This suggests that microfluidization is a superior method for achieving enhanced thermal stability in WPE compared to conventional techniques. The physical stability of WPE was assessed using LUMiSizer. The results revealed that the transmission peak of microfluidized WPE was both lower and narrower compared to untreated and conventionally homogenized WPEs. This suggests that microfluidized WPE particles exhibit superior homogeneity and collectivity. Overall, microfluidization has been observed to enhance various functional properties of seed proteins, including

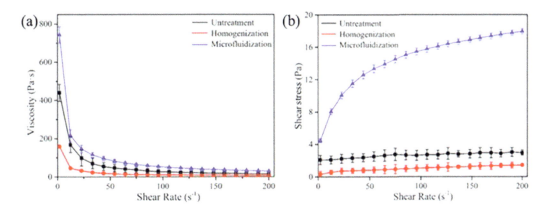

**FIGURE 13.4** Effect of homogenization and microfluidization on viscosity (a) and shear stress (b). (Source: Ling et al., 2023.)

water-holding capacity, oil-holding capacity, emulsion activity, emulsion stability, foaming capacity, thermal properties, and rheological behavior. However, it is crucial to note that the extent of these improvements is contingent upon the specific microfluidization treatment conditions and inherent characteristics of the seed proteins.

## 13.5 CONCLUSIONS AND PERSPECTIVES

Microfluidization, as a method for seed protein modification, presents a blend of advantages and limitations. Key benefits include the creation of stable products with uniform particle size distribution and consistent quality, attributed to their higher shearing rate. The technique also boasts greater repeatability, is essential for reliable processing, and has the scalability for batch and continuous operations. It can handle industrial-scale throughputs up to ~15 L/min and pressures as high as 30,000 psi, with temperature control facilitated by a cooling jacket (Microfluidics, 2024), which is crucial for preserving protein integrity.

However, the process is energy-intensive, and its microchannels are susceptible to wear, which could impact long-term efficiency and maintenance (Kumar et al., 2022). A notable limitation is the risk of over-processing at high pressures, potentially leading to undesirable structural changes in proteins. Additionally, an increase in protein particle size may occur beyond certain pressure, challenging the achievement of desired modification levels. Importantly, as a physical process, microfluidization does not introduce food safety concerns often associated with chemical and biological modification methods. This aspect makes it a safer alternative for modifying seed proteins, aligning with the growing demand for food products that are not only effective but also safe and sustainable.

In summarizing this chapter on the microfluidization of seed storage proteins, it is evident that this innovative technology plays a crucial role in addressing the growing global demand for sustainable and efficient protein sources. Microfluidization, characterized by its high-pressure treatment, significantly alters seed protein morphology and molecular size distribution, such as OPI, FPC, and RPI. These modifications, thoroughly analyzed through techniques like SDS-PAGE and SEC-HPLC, demonstrate the process's ability to enhance protein solubility, emulsification capacity, and textural properties.

The findings presented in this chapter indicate that microfluidization holds immense potential for improving the functional attributes of plant-based proteins, making them more versatile for use in various food applications. Moreover, the environmentally friendly nature of this process aligns well with the current need for sustainable food production practices. As the world progresses toward a more plant-centric dietary paradigm, microfluidization is a pivotal technology in harnessing the full potential of seed proteins, contributing significantly to innovation in food science and technology. This chapter underscores the importance of microfluidization as a transformative tool in food processing, setting the stage for future advancements in this field.

## REFERENCES

Adjei-Fremah, S., Worku, M., De Erive, M. O., He, F., Wang, T., & Chen, G. (2019). Effect of microfluidization on microstructure, protein profile and physicochemical properties of whole cowpea flours. *Innovative Food Science & Emerging Technologies, 57*, 102207. https://doi.org/10.1016/J.IFSET.2019.102207

Calicioglu, O., Flammini, A., Bracco, S., Bellù, L., & Sims, R. (2019). The future challenges of food and agriculture: An integrated analysis of trends and solutions. *Sustainability, 11*(1), 222. https://doi.org/10.3390/su11010222

Cheng, M., Li, Y., Luo, X., Chen, Z., Wang, R., Wang, T., Feng, W., Zhang, H., He, J., & Li, C. (2022). Effect of dynamic high-pressure microfluidization on physicochemical, structural, and functional properties of oat protein isolate. *Innovative Food Science & Emerging Technologies, 82*, 103204. https://doi.org/10.1016/j.ifset.2022.103204

Dhiman, A., & Prabhakar, P. K. (2021). Micronization in food processing: A comprehensive review of mechanistic approach, physicochemical, functional properties and self-stability of micronized food materials. *Journal of Food Engineering*, *292*, 110248. https://doi.org/10.1016/j.jfoodeng.2020.110248

El Nasri, N. A., & El Tinay, A. H. (2007). Functional properties of fenugreek (Trigonella foenum graecum) protein concentrate. *Food Chemistry*, *103*(2), 582–589. https://doi.org/10.1016/j.foodchem.2006.09.003

Ge, Z., Zhang, Y., Jin, X., Wang, W., Wang, X., Liu, M., Zhang, L., & Zong, W. (2021). Effects of dynamic high-pressure microfluidization on the physicochemical, structural and functional characteristics of Eucommia ulmoides Oliv. seed meal proteins. *LWT*, *138*, 110766. https://doi.org/10.1016/j.lwt.2020.110766

Ghanghas, N. (2023). *Impact of extruder die temperature and nitrogen gas injection on the physical quality of soybean protein meat analogues*. http://hdl.handle.net/1993/37851

Ghanghas, N., Mukilan, M., Sharma, S., & Prabhakar, P. K. (2020). Classification, composition, extraction, functional modification and application of rice (oryza sativa) seed protein: A comprehensive review. *Food Reviews International*, *9129*, 1–30. https://doi.org/10.1080/87559129.2020.1733596

Ghanghas, N., Mukilan, M. T., Prabhakar, P. K., & Kumar, N. (2019). Rice protein: properties, extraction, and applications in food formulation. In *Technologies for Value Addition in Food Products and Processes* (pp. 287–304). Apple Academic Press. https://doi.org/10.1201/9780429242847-12

Ghanghas, N., Nadimi, M., Paliwal, J., & Koksel, F. (2024). Gas-assisted high-moisture extrusion of soy-based meat analogues: Impacts of nitrogen pressure and cooling die temperature on density, texture and microstructure. *Innovative Food Science & Emerging Technologies*, *92*, 103557. https://doi.org/10.1016/j.ifset.2023.103557

Ghanghas, N., Prabhakar, P. K., Sharma, S., & Mukilan, M. T. (2021). Microfluidization of fenugreek (Trigonella foenum graecum) seed protein concentrate: Effects on functional, rheological, thermal and microstructural properties. *LWT*, *149*, 111830. https://doi.org/10.1016/j.lwt.2021.111830

Han, K., Feng, X., Yang, Y., Tang, X., & Gao, C. (2023). Changes in the physicochemical, structural and emulsifying properties of chicken myofibrillar protein via microfluidization. *Innovative Food Science & Emerging Technologies*, *83*, 103236. https://doi.org/10.1016/j.ifset.2022.103236

He, X., Chen, J., He, X., Feng, Z., Li, C., Liu, W., Dai, T., & Liu, C. (2021). Industry-scale microfluidization as a potential technique to improve solubility and modify structure of pea protein. *Innovative Food Science & Emerging Technologies*, *67*, 102582. https://doi.org/10.1016/j.ifset.2020.102582

Huang, X., Li, C., & Xi, J. (2023). Dynamic high pressure microfluidization-assisted extraction of plant active ingredients: A novel approach. *Critical Reviews in Food Science and Nutrition*, *63*(33), 12413–12421. https://doi.org/10.1080/10408398.2022.2101427

Jeske, S., Zannini, E., & Arendt, E. K. (2018). Past, present and future: The strength of plant-based dairy substitutes based on gluten-free raw materials. *Food Research International*, *110*, 42–51. https://doi.org/10.1016/j.foodres.2017.03.045

Jiang, Z., Sontag-Strohm, T., Salovaara, H., Sibakov, J., Kanerva, P., & Loponen, J. (2015). Oat protein solubility and emulsion properties improved by enzymatic deamidation. *Journal of Cereal Science*, *64*, 126–132. https://doi.org/10.1016/j.jcs.2015.04.010

Kavinila, S., Nimbkar, S., Moses, J. A., & Anandharamakrishnan, C. (2023). Emerging applications of microfluidization in the food industry. *Journal of Agriculture and Food Research*, *12*, 100537. https://doi.org/10.1016/j.jafr.2023.100537

Kumar, A., Dhiman, A., Suhag, R., Sehrawat, R., Upadhyay, A., & McClements, D. J. (2022). Comprehensive review on potential applications of microfluidization in food processing. *Food Science and Biotechnology*, *31*(1), 17–36. https://doi.org/10.1007/S10068-021-01010-X/METRICS

Ling, Y, Cheng, L, Bai, X, Li, Z, Dai, J, Ren, D. (2023 Jun). Effects of microfluidization on the physical and storage stability of walnut protein emulsion and beverages. *Plant Foods for Human Nutrition* 78(2):467–475. doi: 10.1007/s11130-023-01073-7

Lopes, C., Akel Ferruccio, C., de Albuquerque Sales, A. C., Tavares, G. M., & de Castro, R. J. S. (2023). Effects of processing technologies on the antioxidant properties of common bean (Phaseolus vulgaris L.) and lentil (Lens culinaris) proteins and their hydrolysates. *Food Research International*, *172*, 113190. https://doi.org/10.1016/J.FOODRES.2023.113190

Luo, S., Chan, E., Masatcioglu, M. T., Erkinbaev, C., Paliwal, J., & Koksel, F. (2020). Effects of extrusion conditions and nitrogen injection on physical, mechanical, and microstructural properties of red lentil puffed snacks. *Food and Bioproducts Processing*, *121*, 143–153. https://doi.org/10.1016/j.fbp.2020.02.002

Mandal, S., & Mandal, R. (2000). Seed storage proteins and approaches for improvement of their nutritional quality by genetic engineering. *Current Science*, *226*, 576–589.

Mert, B. (2012). Using high pressure microfluidization to improve physical properties and lycopene content of ketchup type products. *Journal of Food Engineering*, *109*(3), 579–587. https://doi.org/10.1016/j.jfoodeng.2011.10.021

Mert, I. D. (2020). The applications of microfluidization in cereals and cereal-based products: An overview. *Critical Reviews in Food Science and Nutrition*, *60*(6), 1007–1024. https://doi.org/10.1080/1040839 8.2018.1555134

Microfluidics. (2024). *M700 Series Pharma Enhanced MicrofluidizerTM Processors*. extension://efaidn-bmnnnibpcajpcglclefindmkaj/https://cdn2.hubspot.net/hubfs/2395355/IDEX%20MPT%20Group% 20Files/Microfluidics%20File%20Manager/MF-Brochures/MF-M7125_M7250-Pharma-Enhanced-BROCHURE.pdf?utm_referrer=https%3A%2F%2Fwww.microfluidics-mpt.com%2Ftyp-brochure-m700-bp-enhanced%3FsubmissionGuid%3D84c0ba00-628b-47a6-9153-7a1b0198c0bd

Nasrabadi, M. N., Sedaghat Doost, A., & Mezzenga, R. (2021). Modification approaches of plant-based proteins to improve their techno-functionality and use in food products. *Food Hydrocolloids*, *118*, 106789. https://doi.org/10.1016/j.foodhyd.2021.106789

Navaf, M., Sunooj, K. V., Aaliya, B., Sudheesh, C., Akhila, P. P., Mir, S. A., Nemtanu, M. R., George, J., Lackner, M., & Mousavi Khaneghah, A. (2023). Contemporary insights into the extraction, functional properties, and therapeutic applications of plant proteins. *Journal of Agriculture and Food Research*, *14*, 100861. https://doi.org/10.1016/j.jafr.2023.100861

Navare, S. S., Karwe, M. V., & Salvi, D. (2023). Effect of high pressure processing on selected physicochemical and functional properties of yellow lentil protein concentrate. *Food Chemistry Advances*, *3*, 100546. https://doi.org/10.1016/j.focha.2023.100546

Nivala, O., Mäkinen, O. E., Kruus, K., Nordlund, E., & Ercili-Cura, D. (2017). Structuring colloidal oat and faba bean protein particles via enzymatic modification. *Food Chemistry*, 231, 87–95. https://doi.org/10.1016/j.foodchem.2017.03.114

Ocampo-Salinas, I. O., Tellez-Medina, D. I., Jimenez-Martinez, C. and Davila-Ortiz, G. (2016). Application of high pressure homogenization to improve stability and decrease droplet size in emulsion-flavor systems. *International Journal of Environment, Agriculture and Biotechnology*, *1*(4), 646–662. https://doi.org/10.22161/ijeab/1.4.6

Osborne, T. B. (1924). *The Vegetable Proteins* (2nd ed.). Longmans, Green and Co: Paternoster Row.

Ozturk, O. K., & Turasan, H. (2022). Latest developments in the applications of microfluidization to modify the structure of macromolecules leading to improved physicochemical and functional properties. *Critical Reviews in Food Science and Nutrition*, *62*(16), 4481–4503. https://doi.org/10.1080/10408398.2021.1875981

Sharma, N., Sahil, Madhumita, Kumar, Y., & Prabhakar, P. K. (2023). Ultrasonic modulated rice bran protein concentrate: Induced effects on morphological, functional, rheological, and thermal characteristics. *Innovative Food Science & Emerging Technologies*, *85*, 103332. https://doi.org/10.1016/j.ifset.2023.103332

Shen, L., & Tang, C.-H. (2012). Microfluidization as a potential technique to modify surface properties of soy protein isolate. *Food Research International*, *48*(1), 108–118. https://doi.org/10.1016/j.foodres.2012.03.006

Singh, R., Sá, A. G. A., Sharma, S., Nadimi, M., Paliwal, J., House, J. D., & Koksel, F. (2023). Effects of feed moisture content on the physical and nutritional quality attributes of sunflower meal-based high-moisture meat analogues. *Food and Bioprocess Technology*. https://doi.org/10.1007/s11947-023-03225-8

Suhag, R., Singh, S., Kumar, Y., Prabhakar, P. K., & Meghwal, M. (2023). Microfluidization of ginger rhizome (zingiber officinale roscoe) juice: impact of pressure and cycles on physicochemical attributes, antioxidant, microbial, and enzymatic activity. *Food and Bioprocess Technology*. https://doi.org/10.1007/s11947-023-03179-x

United Nations Department of Economic and Social Affairs. (2022). *World Population Prospects 2022: Summary of Results. UN DESA/POP/2022/TR/NO. 3*. https://www.un.org/development/desa/pd/content/World-Population-Prospects-2022

Villalobos-Castillejos, F., Granillo-Guerrero, V. G., Leyva-Daniel, D. E., Alamilla-Beltrán, L., Gutiérrez-López, G. F., Monroy-Villagrana, A., & Jafari, S. M. (2018). Fabrication of Nanoemulsions by Microfluidization. In *Nanoemulsions* (pp. 207–232). Elsevier. https://doi.org/10.1016/B978-0-12-81183 8-2.00008-4

Wang, N., Wang, T., Yu, Y., Xing, K., Qin, L., & Yu, D. (2023). Dynamic high-pressure microfluidization assist in stabilizing hemp seed protein-gum Arabic bilayer emulsions: Rheological properties and oxidation kinetic model. *Industrial Crops and Products*, *203*, 117201. https://doi.org/10.1016/j.indcrop.2023.117201

Willett, W., Rockström, J., Loken, B., Springmann, M., Lang, T., Vermeulen, S., Garnett, T., Tilman, D., DeClerck, F., Wood, A., Jonell, M., Clark, M., Gordon, L. J., Fanzo, J., Hawkes, C., Zurayk, R., Rivera, J. A., De Vries, W., Majele Sibanda, L., … Murray, C. J. L. (2019). Food in the Anthropocene: The EAT–Lancet Commission on healthy diets from sustainable food systems. *The Lancet, 393*(10170), 447–492. https://doi.org/10.1016/S0140-6736(18)31788-4

Xu, J., Fan, X., Xu, X., Deng, D., Yang, L., Song, H., & Liu, H. (2023). Microfluidization improved hempseed yogurt's physicochemical and storage properties. *Journal of the Science of Food and Agriculture*. https://doi.org/10.1002/jsfa.13137

Yashwant, A. S., Kashyap, P., & Goksen, G. (2023). Recent advances in the improvement of protein-based edible films through non-thermal and thermal techniques. *Food Bioscience, 55*, 103032. https://doi.org/10.1016/j.fbio.2023.103032

Zahari, I., Ferawati, F., Helstad, A., Ahlström, C., Östbring, K., Rayner, M., & Purhagen, J. K. (2020). Development of high-moisture meat analogues with hemp and soy protein using extrusion cooking. *Foods, 9*(6), 772. https://doi.org/10.3390/foods9060772

Zhang, B., Guo, X., Zhu, K., Peng, W., & Zhou, H. (2015). Improvement of emulsifying properties of oat protein isolate–dextran conjugates by glycation. *Carbohydrate Polymers, 127*, 168–175. https://doi.org/10.1016/j.carbpol.2015.03.07

Zhang, N., Xiong, Z., Xue, W., He, R., Ju, X., & Wang, Z. (2022). Insights into the effects of dynamic high-pressure microfluidization on the structural and rheological properties of rapeseed protein isolate. *Innovative Food Science & Emerging Technologies, 80*, 103091. https://doi.org/10.1016/J.IFSET.2022.103091

Zhang, Z. H., Zhang, G. Y., Huang, J. R., Ge, A. Y., Zhou, D. Y., Tang, Y., Xu, X. B., & Song, L. (2024). Microfluidized hemp protein isolate: An effective stabilizer for high-internal-phase emulsions with improved oxidative stability. *Journal of the Science of Food and Agriculture, 104*(3), 1668–1678. https://doi.org/10.1002/JSFA.13050

Zhao, Q., Yan, W., Liu, Y., & Li, J. (2021). Modulation of the structural and functional properties of perilla protein isolate from oilseed residues by dynamic high-pressure microfluidization. *Food Chemistry, 365*, 130497. https://doi.org/10.1016/j.foodchem.2021.130497

Zheng, Y., Li, Z., Lu, Z., Wu, F., Fu, G., Zheng, B., & Tian, Y. (2022). Structural characteristics and emulsifying properties of lotus seed protein isolate-dextran glycoconjugates induced by a dynamic high pressure microfluidization Maillard reaction. *LWT, 160*, 113309. https://doi.org/10.1016/j.lwt.2022.113309

Zhu, S. M., Lin, S. L., Ramaswamy, H. S., Yu, Y., & Zhang, Q. T. (2017). Enhancement of functional properties of rice bran proteins by high pressure treatment and their correlation with surface hydrophobicity. *Food and Bioprocess Technology, 10*(2), 317–327. https://doi.org/10.1007/s11947-016-1818-7

# Index

Pages in *italics* refer to figures and pages in **bold** refer to tables.

## A

Actuator, 60, 61
Aflatoxin, 19, 71, 72
Agriculture, 69, 157, 193
Almond, 224, 225, 227–230, **231**
Antibiotic, 66, 72
Antimicrobial, 72, 169
Antioxidant, 66, 132–135, 157–158, 170, 178, 180, 181, **183**, 186, 188, 199, 212

## B

Bacillus, 86, **167**
Bioactive, 5, 128, 155–158, **167**, 169, 170, 178, 179, 181, 192, 212–220, 236
Biosensors, 11, 21, 25, 33, 35, 37, 43, 66, 86, 89–92, 99, 157

## C

Capillary action, 2, 59
Carbohydrates, 45, 155, 180, 224
Cereal, 66, 177–190, 192, 193, 216
Characterization, 8, 97, 161
Chemiluminescence, 25, 28, 64, 66, 89, 164
Chlorella, 218, 219
Colorimetric, 12, 21, 22, 60, 62, 65–73, 164, **165**
Cytometry, 10, 161

## D

Diagnostics, 7, 8, 56, 64, 73, 94, 155, 161
Digestibility, 182, 188, 189, 206, 228, **231**, 233, 234
Drug delivery, 3, 5, 7, 8, 102, 106, 112, 116, 156, 158, 160, 162, 166, **167**

## E

Electrochemical, 4, 14, 15, 29, 35, 37, 39, *40*, 43, *44*, 62–64, 66, 68, 87, 89, 90, 92, 96, 163–166
Electrokinetic, 2, 3, 13, 86
Electrophoresis, 3, 4, 13, 21, 25, 26, 27, 37, 82, 87, 89, 227, 246
ELISA, 10, 32, 33, 35, 60, 82, **168**, 206
Emulsion, 6, 115, 119, 129, 131, 142, 144, 148, 151, 155, 156, **168**, 169, 187, 201, 203–205, 226, 228, **231**, 232, 234, 235, 245–252
Encapsulation, 5, 6, 129, 132, **133**, 157, 158, 166, 169, 170, 179, 245
Enzyme, 5, 10–11, 14–16, 19, 20, 22, 24, 27, 32, 33, 35, 36, 38, 42, 45, 62, 66, 67, 69–71, 82, 85, 86, 90, 99,118, 128, 135, 136, 151, 164, 188, 212, 232, 241
Escherichia, 10, 12, 24, 66, 70, 71, 94, 137, 155, 169

Extraction, 15–17, 19, 21, 22, 42, 65, 129, 131, 132, 186, 188, 212–220, 223, 224, 227, 245

## F

Fermentation, 177, 182, 224, 225, 242
Fluorescence, 19, 21, 22, 23, 29, 35–38, 41, 42, 62–66, 82, 84, 87, 89, 96, 100, 120, **165**, 187, 228, 229
Food Safety, 7, 25, 32, 33, 35, 38, 42, 45, 55, 62, 64, 65, 68, 72, 82, 87, 88, 97, 102, 150, 155, 164, 166, **168**, 169, 208, 252
Functional properties, 144, 151, 157, 186, 187, 199, 200, 206, 223, 224, 225, 226, 228, 230, 236, 242, 246, 247, 251

## G

Ginger, 132, **133**, 135, 136, 216
Grain, 180, 181–182, **184**, 192, 216

## H

Homogenisation, 5, 6, 14, 65, 129, 131, 132, 135, 136, 137, 143, 144, 145, 146, 147, 148, 149, 150, 151, 166, 169, 178, 179, 185, 187, 192, 207, 212, 213, 215, 216, 219, 223, 224, 225, 228, 234, 235, 236, 243, 251
Hydrocolloids, **184**
Hydrolysis, 99, **133**, 188, 206, 226, 232, 236
Hydrophobicity, 4, 150, **168**, 201, 206, 225, 229, 230, **231**, 232, 233, 246, 251

## I

Iron, 115, 117, 169, 180
Irradiation, 37, 148, 149, 223, 224
Isothermal, 10, *18*, 20, 21, 23, 25, 27, 36, 37, 38, 39, 71, 82, 90

## J

Juice, 68, 98, 128, 132, **133**, 134, 135, 137, 138, 139, **165**, 215–219

## L

Lactoglobulin, 121, 144, 148, **167**, **168**
Liposomes, 112, **113**, 155, 156, 157, 158, 166, **167**, 169, 245
Lycopene, 218, 219

## M

Machine Learning, 8, 71, 102, 159, 160
Maize, 179, 180, **183**, **184**, 189, 217, 218

# Index

Metabisulphite, 137, 138
Methyl, 39, 66, 69, 94, 128, 134, 135
MicroRNA, 38, 39, 164
Microalgae, 218, **219**
Microchip, 4, 26, 27, 94, 117
Microgels, 155, 158, 166, **167**, 169
Microstructure, 149, **183**, 185, 186, 187, 204, 227, 228, 234
Microwave, 212, 217, 218
Milling, 106, 177, 181, 182, **184**, 186
Monosaccharides, 89
Morphology, 97, 118, 120, 157, 158, 160, 190, 227, 245, 246, 252
Mozzarella, 149
Muscle, 65, 200, 202
Mushroom, 135
Mycotoxins, 70

## N

Neural Network, 71, 161
Nanocrystalline, 92, 117
Nanocrystals, 68, 92, **113**
Nanomaterial, 4, 32, 35, 65, 83, 88, 91, 92, 97, 100, 101, 102, 106, 112, 113, 116, 117
Nanostructure, 42, 92
Nanotechnology, 8, 28, 68, 83, 106, 119, 219
Naringin, 134, 215
Nutraceuticals, 7, 192

## O

Oilseed, 249, 250
Orange, 67, 98, 135, 136, 137, **165**, 216, 218, 219
Organoleptic, 128, 134, 169, 215, 228
Oxidase, 33, 66, 128, 136, 137, 138

## P

Pigments, 132, **133**, 166, 213, 215, 218
Packaging, 86, 96, 150, 157, **165**, 166, **168**, 169
Parasites, 10, 38, 70
Pasta, 180, 182
Pasteurization, 137, 138, 146, 149, 225
Peptide, 70, 71, 84, 99, 227, **231**, 232, 233
Pesticide, 66, 67, 69, 99, 164
Pharmaceuticals, 8, 156, 192, 218, 219, 242
Pharmacological, 215, 217
Phosphorus, 69, 92
Phytochemicals, 128, 139, 180, 181
Plasma, 17, 58, 66, 88, 9, 95, 144, 148, 207, 223, 224
Platinum, 28
Polyacrylamide, 227, 246
Polycaprolactone, 112, 113
Polycarbonate, 164
Polymer, 32, 43, 100, 110–113, 116, 118, 131, 156, 159, **165**, **168**, 245
Polymerase, 10, 11, 20, 21, 23, 24, 32, 37, 38, 163
Polyphenols, 88, 129, 213
Polysaccharides, 129, 157, 199, 213, 217, 218, 219, 227
Polystyrene, 56, 70, 112, 117, 162
Potato, 215, 216
Potentiometric, 62, 89, 163
Powder, 4, 37, 132, 148, 149, 150, 224, 234

Preservation, 5, 6, 128, 132, 138, 146, 156, 169, 193, 199, 219, 247
Pumps, 3, 108, 113–117, 120, 163

## Q

Quartz, 3, 57, 90
Quercetin, **168**, 215

## R

Radiation, 14, 19
Recombinant, 11, 36, 37
Retrogradation, 199, 234
Rheology, 6, 148, 150, **167**, **184**, 207
Rhizome, 132, **133**, 135, 136, 216

## S

Sausage, 66, 71, 156, **165**
Seafoods, 38
Sedimentation, 187, 226
Semiconductor, 1, 67, 94
Serpentine, 14, 19, 95, 113, 117
Serum, 38, 84, 121, 144, 148, 149, 150, 188
Silicon, 3, 4, 92, 99, 110, 111, 113, 114, 117, **165**
Size reduction, 8, 64, 129, 143, 149, 178, 179, 186, 189, 193, 207, 214, 228, 235
Slurry, **184**, **231**, 234, 235
Sorghum, 179
Soybean, 72, **167**, 235
Spectroscopy, 65, 68, 69, 72, 87, 89, 120, **165**, 166, 190
Spoilage, 6, 7, 151, 178
Starch, **167**, 182, **183**, 186, 189, 199, 234
Stereolithography, 5, 111
Sucrose, 35, 84, **167**
Sugarcane, **133**, 134, 137, 138, 216, 217
Sulfhydryl, 25, 187, 201, 202, 206, 226, 229, **231**
Surface tension, 1, 59, 65, 71, 108, 178
Surfactant, 17, 23, 156, 169
Swelling, **183**, 187, 199, **248**, 250
Synergistic, 201, 219, 226, 230, 243

## T

Tertiary, 145, 202, 206, 229, 233, 247
Texture, 6, 128, **133**, 137, 177–182, 185–186, 189, 190, 192, 200, 207, 208, 226, 242
Therapeutic, 7, 106, 112, 123, 158, 227
Thermocouple, 109, 120
Thermodynamics, 85, 101
Thiamine, 145, 146
Tocopherol, 145, **168**
Tomato, **165**, 218
Toxicity, 69, 70, 97, 157
Toxins, 66, 70, 71, 100, 157
Transcription, 10, 21, 86
Transglutaminase, 230
Trypsin, 187
Tumor, 64, 156
Turbidity, 22, **133**, 134, 136, 200, 201
Turbulent, 2, 5, 6, 107, 213, 224, 226, 243, 244

## U

Ultrasound, 113, 219, 223, 228, 247
Ultraviolet, 87, 149, 228
Urease, 33
Urine, 62, 63

## V

Vaccines, 101
Vacuum, 21, 58
Valves, 3, 60, 88, 163
Vegetables, 66, 67, 128, 164
Velocity, 2, 3, 6, 129, 130, 139, 143, 179, 204, 213, 230, 243, 245
Virus, 14, 22–24, 38, 41
Viscoelastic, 150, 207
Viscosity, 1, 2, 3, 13, 59, 65, 108, 131, **133**, 138, 150, 151, 160, **183**, **184**, 186, 187, 189, 190, 200, 207, 214, 224, 226, **231**, 234, 243, **248**, 250, 251
Vitamins, 65, 128, 139, 142, 145, 146, 150, 178, 179, 180, 189
Vitro, 19, 182, 188, 206
Vivo, 118

Volatile, 71, 145, 146, 147
Volumetric, 3, 28, 59

## W

Waste, 120, 122, 131, 181, 192, 193, 220
Wettability, 200, 201
Whey, 144, 145, 148, 149, 150, **167**, **168**, 232
Wine, 68, 88

## X

Xylene, 58

## Y

Yeast, 82, 135, 136, 137
Yogurt, 169, 225, **231**, 234, **248**, 250

## Z

Zeaxanthine, 218, 219
Zein, **168**, 224
Zeolite, 68
Zinc, 99, 113–115, 117

9781032609812